MW01106778

Springer Complexity

Springer Complexity is an interdisciplinary program publishing the best research and academic-level teaching on both fundamental and applied aspects of complex systems – cutting across all traditional disciplines of the natural and life sciences, engineering, economics, medicine, neuroscience, social and computer science.

Complex Systems are systems that comprise many interacting parts with the ability to generate a new quality of macroscopic collective behavior the manifestations of which are the spontaneous formation of distinctive temporal, spatial or functional structures. Models of such systems can be successfully mapped onto quite diverse "real-life" situations like the climate, the coherent emission of light from lasers, chemical reaction-diffusion systems, biological cellular networks, the dynamics of stock markets and of the internet, earthquake statistics and prediction, freeway traffic, the human brain, or the formation of opinions in social systems, to name just some of the popular applications.

Although their scope and methodologies overlap somewhat, one can distinguish the following main concepts and tools: self-organization, nonlinear dynamics, synergetics, turbulence, dynamical systems, catastrophes, instabilities, stochastic processes, chaos, graphs and networks, cellular automata, adaptive systems, genetic algorithms and computational intelligence.

The three major book publication platforms of the Springer Complexity program are the monograph series "Understanding Complex Systems" focusing on the various applications of complexity, the "Springer Series in Synergetics", which is devoted to the quantitative theoretical and methodological foundations, and the "SpringerBriefs in Complexity" which are concise and topical working reports, case-studies, surveys, essays and lecture notes of relevance to the field. In addition to the books in these two core series, the program also incorporates individual titles ranging from textbooks to major reference works.

Editorial and Programme Advisory Board

Understanding Complex Systems

Future scientific and technological developments in many fields will necessarily depend upon coming to grips with complex systems. Such systems are complex in both their composition – typically many different kinds of components interacting simultaneously and nonlinearly with each other and their environments on multiple levels – and in the rich diversity of behavior of which they are capable.

The Springer Series in Understanding Complex Systems series (UCS) promotes new strategies and paradigms for understanding and realizing applications of complex systems research in a wide variety of fields and endeavors. UCS is explicitly transdisciplinary. It has three main goals: First, to elaborate the concepts, methods and tools of complex systems at all levels of description and in all scientific fields, especially newly emerging areas within the life, social, behavioral, economic, neuro- and cognitive sciences (and derivatives thereof); second, to encourage novel applications of these ideas in various fields of engineering and computation such as robotics, nano-technology and informatics; third, to provide a single forum within which commonalities and differences in the workings of complex systems may be discerned, hence leading to deeper insight and understanding.

UCS will publish monographs, lecture notes and selected edited contributions aimed at communicating new findings to a large multidisciplinary audience.

More information about this series at http://www.springer.com/series/5394

Janusz A. Hołyst
Editor

Cyberemotions

Collective Emotions in Cyberspace

Springer

Editor
Janusz A. Hołyst
Faculty of Physics
Warsaw University of Technology
Warsaw, Poland

ISSN 1860-0832 ISSN 1860-0840 (electronic)
Understanding Complex Systems
ISBN 978-3-319-43637-1 ISBN 978-3-319-43639-5 (eBook)
DOI 10.1007/978-3-319-43639-5

Library of Congress Control Number: 2016956713

Printed on acid-free paper

This Springer imprint is published by Springer Nature
The registered company is Springer International Publishing Switzerland

Contents

Contributors

Junghyun Ahn Media Computing Lab, Technicolor R&I, Rennes, France
Junghyun Ahn developed his expertise in 3D character animation at KAIST, VRLab in South Korea, where he received his PhD. With his doctoral background, he joined the Virtual Reality Lab at EPFL (Ecole Polytechnique Federale de Lausanne) in Switzerland and focused his research on emotion modeling and crowd animation. Later, he continued improving their work with Immersive Interaction Group (IIG) at EPFL, before joining Technicolor R&I in France, where he collaborates with VFX studios, for improving the production pipeline. His current research interests are automatic facial rig generation, real-time crowd simulation, plausible emotion expression, and immersive interaction in VR environment.

Ronan Boulic EPFL-IIG, Lausanne, Switzerland
Ronan Boulic has a PhD in computer science from the University of Rennes (1986) and a habilitation degree from the University of Grenoble (1995). He has a long experience in the real-time postural control of virtual human entities either autonomous or as avatars. Recently he has focused on scaling and retargeting strategies for controlling a wide range of virtual humans and ensuring (self-)collision avoidance. He has coauthored more than 140 publications and chaired multiple conferences: ACM Eurographics Symposium in Computer Animation (2004, 2012), Motion in Games (2010), and Joint Virtual Reality Conference (2012). He has been associate editor of *IEEE Transactions on Visualisation and Computer Graphics* (2011–2014) and is associate editor of Elsevier *Computers & Graphics* since 2015. His full profile and papers can be found on the ResearchGate portal.

Anna Chmiel Faculty of Physics, Warsaw University of Technology, Koszykowa 75, 00-662 Warsaw, Poland
She got her MSc (2006) studying the structure of companies and branches networks. After obtaining her PhD (2011) from WUT for her work on stochastic models of human behavior and their emotions in online portals, she moved to the Medical University of Vienna where she analyzed statistical properties and patterns in the comorbidity networks of two million Austrian patients. She has recently obtained a grant from the Polish National Science Centre to work as a research assistant in

the group of Katarzyna Sznajd-Weron at Wrocław University of Technology on the subject of nonstationary processes in multiplex networks.

Sabrina C. Eimler Department of Computer Science, University of Applied Sciences Ruhr West, Bottrop, Germany
Sabrina Eimler is professor for human factors and gender studies. She has an interdisciplinary background in psychology, computer science, economics, and cultural studies. Her research focuses on self-presentation and person perception in social media (e.g., Facebook, XING, LinkedIn) and computer-mediated communication, their preconditions, circumstances, and outcome. She is especially interested in the influence of gender differences and stereotypes as well as in forms of nonverbal communication in text-based computer-mediated communication (e.g., smileys). Also, she works with innovative forms of teaching in higher education, e.g., MOOCs and mobile devices in teaching. Additional interests include social aspects of human-robot interaction, positive psychology, and vicarious embarrassment as well as mobile communication and gaming (phantom phone calls, Quizduell) and film perception.

Antonios Garas ETH Zürich, Zürich, Switzerland
Dr. Garas is researcher at the Chair of Systems Design in ETH Zürich. His current research is about structural properties, stability, and efficiency of complex networks. More specifically, he studies how dynamical processes evolving on a complex system are related to fundamental properties of the system's underlying network topology. Using data-driven modeling and state-of-the-art data-mining techniques, he explores applications of these subjects to various fields ranging from physics to sociology and economy.

David García Becerra ETH Zürich, Zürich, Switzerland
David García's research focuses on computational social science, designing models, and analyzing social behavior from a quantitative perspective. His main work revolves around the topics of emotions, cultures, and political polarization, combining statistical analyses of large datasets of online interaction with agent-based modeling of individual behavior. His PhD thesis "Modeling collective emotions in online communities" provides an example of how agent-based modeling can be used to construct an integrated approach to collective emotional phenomena in cyberspace. An overview of his latest works can be found in David Garcia's website (www.dgarcia.eu).

Stephane Gobron Hes-so//ARC, Neuchatel, Switzerland
After his international studies (BSc in the USA, MSc in France, and PhD in Japan), Stephane Gobron has been doing research in France on dynamic cellular networks, GPGPU, and real-time simulation. Recently, he studied emotional models applied to computer graphics and virtual reality at EPFL with the CyberEmotions FP7 project. He is now professor at the University of Applied Sciences of Western Switzerland (HES-SO) leading the Image Processing and Computer Graphics group at HE-ARC, Neuchatel. His research focuses mainly on serious games oriented to medical

applications with several R&D projects relative to neurorehabilitation, palliative care, aging, ER, and health food. He leads the Gamification and the Health at Work consortiums, and the Gamification & Serious Game Symposium first held in Neuchatel in July 2016. He is a conference and journal editorial committee member and the author of "WebGL par la practice," PPUR http://webgl3d.info/.

Janusz A. Hołyst Faculty of Physics, Warsaw University of Technology, Warsaw, Poland
Janusz Hołyst is full professor at the Faculty of Physics, Warsaw University of Technology, where he leads the *Center of Excellence for Complex Systems Research* and the lab *Physics in Economy and Social Sciences*. His current research includes simulations of evolving networks, information processing in hierarchical systems, models of collective opinion and emotion formation, econophysics, and equilibrium and nonequilibrium statistical physics. He is one of the pioneers in applications of physical methods to economic and social systems and maintains a close collaboration with many institutes in Germany, Italy, UK, Switzerland, Japan, and USA where he spent over 6 years as visiting professor, fellow of Humboldt Foundation, or guest scientist. He is editor of the *European Physical Journal B* and of the *Journal of Computational Science*. He is a co-founder and acting chairman of KRAB (National Council for Research Project Coordinators) as well as a co-founder and the former chairman of FENS (Physics in Economy and Social Sciences, Division of the Polish Physical Society). Janusz Hołyst was coordinator of the CYBEREMOTIONS Project.

Robert Hillmann Department of Systems Analysis, Berlin, Germany
Dr. Robert Hillmann has studied industrial engineering with a focus on information and communication technologies. During his work as a scientific assistant at TU Berlin, he was involved in different projects covering research topics like business process analysis, knowledge management, dynamic social network analysis and digital collaboration networks. In his PhD thesis he deals with the dissemination of link attributes in digital social networks providing an analysis framework and corresponding software to explore multiple facets of link attribute dissemination in electronic social networks. After a recent job change he is now working on design and implementation of business oriented IT-applications and their process alignment.

Arvid Kappas Jacobs University Bremen, Bremen, Germany
Arvid Kappas is professor of psychology and dean at Jacobs University. He obtained his PhD (social psychology) in 1989 at Dartmouth College and has held positions at the University of Geneva, Laval University, and Hull University, before joining Jacobs in 2003. His research focuses on human emotions, nonverbal communication, and affective computing, and he has published extensively in these areas. He is currently president of the International Society for Research on Emotion, and fellow of the Association for Psychological Science and is active in various societies, such

as in the executive board of the Association for the Advancement of Affective Computing and in editorial functions of journals, such as the steering committee of *IEEE Transactions on Affective Computing*.

Nicole C. Krämer Department of Computer Science and Applied Cognitive Science, Social Psychology: Media and Communication, University of Duisburg-Essen, Duisburg, Germany
Nicole Krämer is professor for Social Psychology: Media and Communication at the University Duisburg-Essen. She has a background in social and media psychology. Her research interests include human-computer interaction and computer-mediated communication, especially social media. More specifically her research focuses on forms and effects of social media usage, related to impression management, self-disclosure, or social comparison. Additionally, she analyses social effects of virtual agents and robots.

Dennis Küster Focus Area Diversity, Jacobs University Bremen, Bremen, Germany
Dennis Küster received his PhD in psychology from Jacobs University Bremen, Germany, where he is presently a postdoctoral fellow. His research interests focus on the psychophysiology of online communication, and the role of implicit social contexts for psychophysiological and facial responding, as well as the role of online social contexts for self-presentation. He has previously published on psychophysiological methods and has more recently expanded his research in the context of large-scale interdisciplinary approaches studying the role of empathy and engagement in learning with robotic tutors.

German Neubaum Department of Computer Science and Applied Cognitive Science, Social Psychology: Media and Communication, University of Duisburg-Essen, Duisburg, Germany
German Neubaum is researcher in the team of Social Psychology: Media and Communication at the University Duisburg-Essen. He has an interdisciplinary background in psychology, computer science, and economics. His research focuses on psychological determinants and effects of the use of social media technologies. More specifically, he is interested in how people monitor, form, and express opinions on societal issues online as well as how people can experience the use of social media platforms as emotionally beneficial. His works also include applied research interests in the fields of health and crisis communication.

Georgios Paltoglou Faculty of Science and Engineering, School of Mathematics and Computer Science, University of Wolverhampton, Wolverhampton, UK
Dr. Georgios Paltoglou is a senior lecturer at the School of Technology of the University of Wolverhampton and member of the Statistical Cybermetrics Research Group. His research interests include sentiment analysis and opinion mining, machine learning, and information retrieval. He has over 18 journal publications, 3 book chapters, and 18 publications in peer-reviewed conferences and workshops. His publications have been cited over 1000 times, and he has an h-index of 14 (source: Google Scholar).

Stefan Rank Department of Digital Media, Antoinette Westphal College of Media, Arts and Design, Drexel University, Philadelphia, PA, USA
Stefan Rank is an assistant professor in Drexel University's Digital Media Department. His research focuses on the development and application of computational models of personality and affective behavior, at different levels of granularity, for users and synthetic characters, intelligent software agents, game design, interactive narrative, and human-computer interaction. He received his doctoral degree from the Vienna University of Technology in 2009 at the Austrian Research Institute for Artificial Intelligence and was involved in several European multi-disciplinary research projects. He is also a member of the executive committee and serves as portal editor of the Association for the Advancement for Affective Computing.

Bernard Rimé Psychological Sciences Research Institute, University of Louvain, Louvain-la-Neuve, Belgium
Bernard Rimé is an emeritus and invited professor of psychology at the University of Louvain. His research addresses the social psychology of emotion with a particular interest for the strong links existing between emotional experience and social communication. His studies documented the fact that emotions powerfully stimulate communication, that emotional information propagates across social networks, and that the social sharing of emotions impacts upon social ties. His current research examines individual and collective effects of collective emotional expression in mass gatherings such as civil or religious ceremonies, commemorations, collective festivities, and sporting, musical, folk, or sociopolitical events.

Frank Schweitzer ETH Zürich, Zürich, Switzerland
The research of Frank Schweitzer can be best described as data-driven modeling of complex systems, with particular emphasis on social, socio-technical, and socio-economic systems. He is interested in phenomena as diverse as user interaction in online social networks, collective decisions in animal groups, failure cascades and systemic risk in economic networks, and the rise and fall of scientific collaborations. In his methodological approach, he combines the insights from big data analysis with the power of agent-based computer simulations and the strength of rigorous mathematical models. Frank Schweitzer is a founding member of the ETH Risk Center and editor in chief of *ACS (Advances in Complex Systems)* and *EPJ Data Science*.

Julian Sienkiewicz Faculty of Physics, Warsaw University of Technology, Warsaw, Poland
Dr. Julian Sienkiewicz is assistant professor at the Faculty of Physics, Warsaw University of Technology. He got his MSc (2004) in physics from WUT for studying statistical properties of Polish public transport networks. After obtaining his PhD (WUT, 2010) for examining scaling relations of path lengths in complex networks, he has used statistical physics approaches (such as Ising model or maximal entropy principle) to explore social phenomena of isolation and emotional communication. In February 2016, he has completed a 17-month post doctoral study in Eduardo

Altmann's group at MPI-PKS Dresden where he worked on the impacts of lexical and sentiment factors on the popularity of scientific papers.

Quentin Silvestre Genview, EPFL Innovation Park, Lausanne, Switzerland
Quentin Silvestre got his university degree at the University of Nancy (France) in computer science with a specialty in digital images and sounds. After few years of creating websites, he went back to his specialty in 3D rendering at the EPFL VRLab and IIG. His main research themes were crowd simulation and facial emotion in real-time rendering. Then he has created a 3D game for kids to discover classical music in a nice hide-and-seek game, representing forest' animals hidden in a clearing. Now he is developing a 3D application for business key performance indicators visualization.

Marcin Skowron Austrian Research Institute for Artificial Intelligence, Vienna, Austria
Marcin Skowron is a research scientist at the Austrian Research Institute for Artificial Intelligence and in Johannes Kepler University's Computational Perception Department. He received an MSc degree from Gdansk University, Poland, and a PhD degree from Hokkaido University, Japan, in 2000 and 2005, respectively. His research interests are in cognitive and affective sciences, human-computer interactions, and artificial intelligence. In the EU FP7 project CYBEREMOTIONS, he led the research on the development and evaluation of affective human-computer interaction systems and the analysis of data on human-computer and human-human online interactions.

Paweł Sobkowicz Faculty of Physics, Warsaw University of Technology, Warsaw, Poland
Paweł Sobkowicz has a PhD in solid state theory obtained at the Institute of Physics, Polish Academy of Sciences. He has left the academic world nineteen years ago, but continues to work on topics related to physics-based description of various social phenomena. The interest points cover agent based models, opinion and emotion dynamics, effectiveness of information transfer, and models of social processes in research.

Milovan Šuvakov Institute of Physics, University of Belgrade, Zemun-Belgrade, Serbia
Milovan Šuvakov is research scientist at the Institute of Physics Belgrade. His research interests are numerical analysis and modeling in several fields including complex systems, social networks, kinetic theory, celestial mechanics, and nanoscience. He received his BSc in physics from the University of Belgrade (Serbia) in 2004, MSc in physics from the University of Belgrade (Serbia) in 2006, and PhD in physics from Jožef Stefan Postgraduate School (Slovenia) in 2009. He has obtained his PhD under the supervision of Prof. Dr. Bosiljka Tadić, as part of Marie Curie Training and Research Network, MTRN-CT-2004-005728 "Unifying Principles in Non-Equilibrium Pattern Formation." He worked in Petnica Science Center as head of the Mathematical Department from 2010 to 2012.

Bosiljka Tadić Department of Theoretical Physics, Jožef Stefan Institute, Ljubljana, Slovenia
Bosiljka Tadić is research scientist at full professor rank at the Jozef Stefan Institute. Her research focuses on theoretical and numerical study of complex dynamical systems and networks. Substantively, her studies include the stochastic dynamics on complex networks, processes of self-assembly and self-organization in physics and social dynamics, and emergent collective phenomena in complex systems. Methodologically, her research involves concepts of nonequilibrium statistical physics, theory of complex graphs and their applications across different scales from nanoscience to the web, analysis of large empirical datasets, and agent-based modeling of socio-technological interactions. She is a member of international editorial boards for several journals in interdisciplinary physics and computational mathematics and advisory board member for two leading conferences in statistical physics.

Daniel Thalmann Institute for Media Innovation, Nanyang Technological University, Singapore, Singapore
Daniel Thalmann is a pioneer in research on virtual humans. His current research interests include real-time virtual humans in virtual reality, crowd simulation, and 3D interaction. Daniel Thalmann has been the founder of the Virtual Reality Lab (VRLab) at EPFL, Switzerland. He is coeditor in chief of the *Journal of Computer Animation and Virtual Worlds* and member of the editorial board of 6 other journals. Daniel Thalmann was program chair and cochair of several conferences including IEEE VR, ACM VRST, ACM VRCAI, CGI, and CASA. Daniel Thalmann has published more than 600 papers in graphics, animation, and virtual reality. He is coeditor of 30 books and coauthor of several books including *Crowd Simulation* (second edition 2012) and *Stepping Into Virtual Reality* (2007, Springer). He received his PhD in computer science from the University of Geneva (1977) and an honorary doctorate (honoris causa) from Paul Sabatier University in Toulouse (2003). He also received the Eurographics Distinguished Career Award (2010), the 2012 Canadian Human Computer Communications Society Achievement Award, and the CGI Career Award 2015.

Mike Thelwall School of Mathematics and Computer Science, University of Wolverhampton, West Midlands, UK
Mike Thelwall's research interests are in sentiment analysis, quantitative methods and software for social web analysis from a social science perspective, and research evaluation. His sentiment analysis software can be found at http://sentistrength.wlv.ac.uk. He is the founder and head of the Statistical Cybermetrics Research Group and is a professor of information science at the University of Wolverhampton. He is an associate editor of the *Journal of the Association for Information Science and Technology* and sits on three other editorial boards.

Matthias Trier Department of IT Management, Copenhagen Business School, Frederiksberg, Denmark
Dr. Matthias Trier is associate professor at Copenhagen Business School. He studies phenomena related to electronic communication and social influence effects in

online media within and outside the organization with a mixed-methods approach that blends quantitative, qualitative, and social network analytical methods. Example topics include the implementation/appropriation of social media, online participation, framing of electronic discourses (e.g., from a management perspective), information transfer, dissemination processes, or bottom-up community emergence as a part of knowledge management initiatives. One special methodological focus is on developing an event-driven method for dynamic network analysis. The corresponding software Commetrix.net enables research into emerging structures and dynamic processes of networking among people.

Nan Wang EPFL-IIG, Lausanne, Switzerland
Nan Wang got his university degree at the Communication University of China (Beijing) in 2006 and his master's research degree in France, at the Department of Informatique of INSA de Lyon in 2007. After one further year in EURECOM as a research engineer working in computer vision and image processing, he joined the Centre de Robotic of Ecole des Mines de PARIS and started his PhD research with Prof. Philippe Fuchs on the topic of immersive 3D interaction and 3D perception. In early 2012, he got his PhD and moved to the Immersive Interaction Group of EPFL as a postdoctoral research associate. Nan Wang's research interests are human-computer interaction, particularly in 3D visualization, immersive interaction, motion capture, and analysis.

Chapter 1
Introduction to Cyberemotions

Janusz A. Hołyst

1.1 Introduction

Even a beginner in psychology or sociology agrees that emotions can crucially influence our life and they can do it even without our awareness. "No need to get emotional about it" is a response often heard when trying to calm someone down. But the fact is that we do need to get emotional. Emotions are the backbone of family life; they enable us to react in dangerous situations; they create the group experience of a football crowd, a pop concert or a historical event. Emotions are part of what makes us human. We cannot escape from them since they belong to the core of our human nature. Indeed psychologists define emotions as bodily and social processes, shaped by cultural influences and biological constraints, that are best understood in an evolutionary context (Leventhal and Scherer 1987, see also Kappas, Chap. 3 of this book). A physicist or a mathematician may however be surprised to learn that despite a long history of research on emotional phenomena psychologists can agree neither on the number of basic emotions nor on the number of variables that are needed to describe emotional states. It is after all largely accepted that emotions should refer not only to the feeling component, but also to an expressive behaviour in the face or the voice as well as to bodily changes, such as differences in sweating, the frequency with which the heart beats or shifts in the activation of certain brain parts. In brief, emotions exist both in our mind and body (if one has the right to separate these two).

With information technology occupying such a central part in all our lives, it's important to ask whether there are emotions in cyberspace too? Since cyberspace is just another human space, it's bound to have an emotional context. However, it possesses special features that make social interactions between people different

J.A. Hołyst (✉)
Faculty of Physics, Warsaw University of Technology, Koszykowa 75, 00-662 Warsaw, Poland
e-mail: jholyst@if.pw.edu.pl

J.A. Hołyst (ed.), *Cyberemotions*, Understanding Complex Systems,
DOI 10.1007/978-3-319-43639-5_1

from those taking place in the offline world. One difference is the often much shorter lifetime of e-communities compared to their offline counterparts. Since participants of internet forums or discussion groups are also less bounded by local social norms, they may interact more quickly and express their feelings more often. The internet, for example, is well-known as a site for the expression of strong emotions, for example in "flaming". It also hosts many environments in which multiple participants engage with others.

What's to be gained by studying such phenomena? Well, firstly understanding how emotions and intuition interact with information technology could help us to build better ICT systems. Humans might also benefit from ICT systems that were able to react emotionally and, ultimately, that were sensitive to emotions. More fundamentally, modelling emotions in artificial systems might also add to our understanding of humans at a psychological level. Emotions are complex processes. Behaviour, expression, physiological changes in the brain and in the body at large, motivational processes, and subjective experience are just some of the factors involved. These components are only loosely related (what researchers call 'exhibiting low coherence'). They are constrained by our biology but also constantly shaped and modulated by social and cultural contexts. And there is constant mutual interaction with processes such as attention, perception, and memory. All of this serves to make the study of emotions a challenging field of science.

Being partly social processes emotions influence our group relations and vice versa—they are dependent on social interactions. Emotional experiences are frequently shared with members of social networks which strengthens social ties. This echoes the dynamics of epidemic propagation since reproduction of the sharing process enables propagation of emotions and results in emotion multiplication as well as the emergence of collective emotional patterns. In this way emotions very generally elicit a process of social sharing of emotional experience (Rimé, Chap. 4). When an external emotional impact simultaneously influences members of a social group it provokes a number of discussions where emotions are reinforced and people sharing the same emotions get a feeling of belonging to the same community.

In the present digitalised world the spreading of emotions becomes easier since new kinds of social links are possible (Carr et al. 2012; O'Reilly 2005) and they can be much effortlessly created as compared to face to face contacts (Walther 2011, Krämer et al., Chap. 2). It follows that social sharing can be reduced to simple post forwarding or re-tweeting which may lead to emergence of collective emotions in larger groups. Such states can be also elicited by an external event that has attracted attention a large group of internet users, e.g., the death of a media star or a special sporting event (Chołoniewski et al. 2015).

Members of e-communities frequently make use of their apparent anonymity to express their opinions and emotions in a more open way (Suler 2004). Unfortunately, this advantage of on-line communication possesses also a darker side: receivers of negative emotional messages can suffer from a serious emotional discomfort when the posts target their fragile points—an action commonly known as cyber-bullying (Berson et al. 2004). The effect is enhanced during synchronized attacks of so-called haters and can result in depression or even suicide for victims (Sourander et al. 2010).

Studies of emotional processes in human brain and body are frequently performed in special labs using psychophysiological methods such as facial electromyography and electrodermal activity (Küster and Kappas, Chap. 5). This kind of methodology is impossible to apply when one is interested in the emotional behaviour of social groups with millions members. In the case of e-communities such large scale studies are feasible due to *sentiment analysis that deals with the computational treatment of expressions of private states in written text* (Quirk et al. 1985, Paltoglou and Thelwall, Chap. 6). Two main sentiment analysis approaches are *machine learning algorithms* and *lexicon-based solutions*.

Machine learning methods of sentiment analysis exploit pre-annotated data (e.g., a set of sentences or longer text segments) that serve as examples of specific emotional contents. During the initial training phase special machine learning algorithms such as Naive Bayes, Maximum Entropy or Support Vector Machines try to discover hidden rules describing the presence of a given sentiment in the text. Once such rules have been found they can be used for automatic detection of sentiments in a text that has not been annotated by humans. Because of the necessary training phase these kind of algorithms are called *supervised methods*. Their main weakness is a dependence on a pre-annotated text set that can be non-representative for other domains, e.g., algorithms trained on political discussions can have a limited efficiency for detecting sentiments in product reviews (Paltoglou and Thelwall, Chap. 6)

Lexicon-based methods do not need a training phase with pre-annotated texts. Instead they use *lexicons*, i.e., emotional dictionaries that include information on the emotional contents of some words. The methods also apply lexical rules that take into account how emotions are expressed in a given language. The final estimation of sentiments is dependent not only of detected emotional words but also on the presence (and position) of special *prose signals* (Paltoglou and Thelwall, Chap. 6) such as negation, capitalization, exclamation marks, emoticons, intensifiers and diminishers.

Although lexicon-based sentiment detection algorithms can operate without a training phase and are called unsupervised methods one can enhance their efficiency by adding additional training. In such a case the algorithm optimizes its lexicon term weights for a specific set of human-coded texts. The *SentiStrength* algorithm uses idiom and emoticon lists as well several other rules such as increasing sentiment strength when CAPITAL letters are found. SentiStrength can score each can text with dual positive and negative scales simultaneously which corresponds to psychological observations of humans. In fact one can experience positive and negative emotions at the same time (Norman et al. 2011). Properties of the SentiStrength algorithm are discussed in detail in this book (Thelwall, Chap. 7).

The availability of large datasets containing sentiment scores has encouraged physicists to study emotional patterns using tools from statistical physics. One of the fundamental issues is the detection of emotional sharing in e-communities. It is well known that physical interactions between atoms or molecules can lead to the emergence of collective states with an appropriate order parameter (Stanley 1987). It is however a priori not clear if the concept of order parameter can be introduced for social systems with emotional interactions. One of the special features of on-line discussions is the fact that different participants rarely present their emotions (or opinions) at the same time. It follows that a quantitative description of emotion sharing should take into account *time series* of consecutive values corresponding to emotional valencies, arousal and other emotional components rather than from a single number describing the mean value of an emotion observable.

The *order* of appearance of different emotional comments contains significant information about the social sharing of emotions and the strength of emotional interactions. When a few negative comments increases the chance of the next comment being negative then the time series is persistent and one can anticipate the presence of emotional interactions related to sharing negative emotions. The level of this persistence can be measured in many ways, e.g., by appropriate conditional probabilities (Hołyst et al., Chap. 8) Hurst exponents (García et al., Chap. 10) or spectra of emotional avalanches observed in social networks (Tadić et al., Chap. 11). Data-driven agent-based models of virtual emotional human can describe these features provided that an appropriate way of *emotional communication* (García et al., Chap. 10) and/or parameters of hidden topology of underlying social community (Tadić et al., Chap. 11) are assumed. It is interesting that the level of emotional interaction does not only influence a strength of individual social links but emotions can be also a kind of fuel for the existence of the on-line group and in many cases the decay of this fuel leads to group death (Sienkiewicz et al., Chap. 9).

Studies of communication in specific on-line groups take into account the time-dependent topology of the emerging social network and resulting patterns of bi–directional emotional relations. Such observations give evidence (Trier and Hillmann, Chap. 12) of *an emotional balance* for individual agents, i.e., a given agent expresses both negative and positive emotions. There is however a strong positive or negative emotional polarization of specific *links* between different agents. The polarization possesses usually a reciprocal (symmetrical) character and when triangles of interacting agents are studied then a Heider balance (Heider 1946) known from groups communicating in a direct way is observed (friend of my friend is my friend, enemy of friend is my enemy etc.).

It has been known since the times of Darwin (1872) that human beings and animals (unconsciously) communicate their current state of mind through facial expressions and body gestures. However, currently a majority of on-line communities rely on text messages as a medium of communication and it is quite possible that it will remain so even if supported by images and video. The popularity of this communication form has its roots in degree effectiveness of the written form of language for conveying meaning with a small number of signs. Nonetheless, as raised by Morris (1978) text exchanges filter out the large range of non-verbal communication channels and it turn can lead to ambiguity in case the text content lacks redundancy about the expressed meaning. Therefore it is often just essential to use of a complementary *visual* communication channel in the form of an *animated 3D avatar* to convey the potential emotions carried by the text messages. For the above stated reasons scientists struggle to provide an automated mapping of an emotion model to real-time facial expressions and body movements (Boulic et al., Chap. 13).

Another impact of research on emotions in on-line communities is the development of Interactive artificial systems (IAS) that would take into account sentiments during human-computer communication (Skowron and Rank 2014). In principle, interactive affective systems can communicate with users directly, provide new content to a group of users, and provide reports on group activities identifying interaction patterns. IAS aim to model the affective dimensions of multi-party interactions, to simulate potential future changes in group dynamics, and, in part based on that information, to suitably respond to utterances both on the content and the affect level (Skowron et al., Chap. 14)

1.2 Book Description

The book is divided into four distinct parts corresponding to different layers: (I) Foundations, (II) Sentiment analysis (III) Modeling and (IV) Applications.

The first part is a vivid discourse between psychologists, social psychologists and psychophysiologists regarding the role and function of (cyber)emotions. As most of us are frequent users of such applications as YouTube or Facebook it is essential to find out what is the status of the current scientific knowledge about social and emotional aspects of those applications. This subject is tackled by Nicole Krämer et al. in the first chapter. Yet, a probably even more fundamental question is next asked by Arvid Kappas: *how we define emotions?* Moreover, the author comes also with a working definition of "cyberemotions"—a common denominator of all the works that constitute this very book. As already mentioned, another important aspect of emotions is them being the trigger of social sharing process. The general framework of this phenomenon is widely discussed by Bernard Rimé in Chap. 4 along with some perspectives on cyberemotions. The final chapter of "Foundations" touches the problem of measurement of emotions in individuals—Dennis Küster and Arvid Kappas give the psychophysiologist's point of view, focusing on multi-modal assessment of emotions in the lab. Here we can find out about paradigms particularly tailored for research of cyberemotions which are illustrated with concrete examples of recorded data.

Although the next part, "Sentiment analysis" may look like quite disjoint from its predecessor, it is a key input for the following "Modelling" and "Applications" sections. From Georgios Paltoglou and Mike Thelwall, authors of Chap. 6, we learn about the methods that concern computational detection and extraction of opinions, beliefs and emotions in written text. Those solutions come from of lexicon-based algorithms (i.e., previously created dictionary) or machine learning approaches that automatically or semi-automatically learn to detect the affective content of text. In Chap. 7 Mike Thelwall recapitulates those considerations by presenting the successful fruit of their lab—the SentiStrength program—along with the ways it can be refined different topics, contexts or languages. The chapter also briefly describes some studies that have applied SentiStrength to analyse trends in Twitter and You Tube comments.

The authors of four chapters that form the "Modelling" part had a particularly difficult task: being physicists, specializing in complex systems, to analyse the on-line sentiment-annotated data taking into account the theories of emotions and specific phenomena connected to them. The group from Warsaw University of Technology resolved this issue by first showing signs of collective emotions and their homophily in clusters of messages (Chap. 8) and then addressing the problem of non-stationarity in on-line discussions, underlying the necessity of an emotional imbalance for the thread to last (Chap. 9). On the other hand, David García et al. present in Chap. 10 a carefully designed, elegant and complete agent-based model that combines two main dimensions of emotions—valence and arousal. An application of this modelling framework to different online communities is shown as well as the way it reproduces the properties of collective emotional states in different media (Amazon, IRC channels). The framework is then adapted by Bosiljka Tadić et al. (Chap. 11) to successfully reproduce the stylized facts observed in the empirical data of online social networks, indicating group behaviour arising from individual emotional actions of agents.

The fascinating "Application" part gives the reader a glimpse of the possibilities that come when we introduce emotional components to computer applications. Chapter 12 by Matthias Trier and Robert Hillmann is a good example of the intersection between social and computer sciences by setting the scope of interest on network motif analysis. In particular, the authors show that users develop polarized sentiments towards individual peers but keep sentiment in balance on their ego-network level. Graphically most advanced chapter in this book—Chap. 13 by Ronan Boulic et al.—brings forward the issue of non-verbal facial and body communication, focusing on the instantaneous display of a detected emotion with a human avatar, reconsidering the expressive power of asymmetric facial expression for better conveying a whole range of complex and ambivalent emotions. In the final Chap. 14 Marcin Skowron et al. follow the line of futuristic examples of emotion application by presenting how interactive affective systems can be used to study the role of emotion in online communication at the micro-scale, i.e. between individual users or between users and artificial communication partners.

1.3 CYBEREMOTIONS Project

The book presents in a large part research results of CYBEREMOTIONS project (Collective Emotions in Cyberspace, 2009–2013) that was an EU Large Scale Integrating Project within the 7th Framework Programme in FET ICT domain Theme 3: Science of complex systems for socially intelligent ICT. The project associated nearly 40 scientists from Austria (Österreichische Studiengesellschaft für Kybernetik), Germany (Jacobs University and Technische Universität Berlin), Great

Britain (University of Wolverhampton), Poland (Warsaw University of Technology and Gemius SA), Slovenia (Jožef Stefan Institute), and Switzerland (ETH Zürich and École Polytechnique Fédérale de Lausanne).

The main objectives of CYBEREMOTIONS were to understand the role of collective emotions in creating, forming and breaking up ICT mediated communities and to prepare the background for next generation of emotionally-intelligent ICT services. Project Partners collected data on emotions in e-communities and psycho-physiological data on emotions evoked by on-line discussions, developed data-driven models of cyberemotions, and created emotion-related software.

CYBEREMOTIONS met successes in all domains of its activity. Data collected in the Project from blogs, forums, portals, IRC channels, `Twitter` and `MySpace` are ranked among "the 70 Online Databases that Define Our Planet" (TechnologyReview 2010). Also several experimental setups provided a microscopic view on affective processes associated with reading and writing contents on the Internet. Our `SentiStrength` (SentiStrength 2016) program is considered as a one of the most advanced tools in sentiment detection. Theoretical models based on active agents approach, complex networks and stochastic processes are able to describe several stylized facts for emotional dynamics in social groups communicating by Internet.

Outputs of the Project were used for creating new affective dialog systems as interactive tools as well as semi-automated simulation of facial expression through a user's 3D avatar that can facilitate the process of the online affective communication. The Project delivered network visualization tools for effective demonstration of emotional contagion processes in social networks.

Acknowledgements I would like to express my gratitude to all the Partners of CYBEREMOTIONS Consortium for a very fruitful collaboration during the whole Project period and for their contributions to this book. I am also indebted to external experts that prepared additional materials presented in Chaps. 2 and 4. Last but not least I am very thankful to Dr. Julian Sienkiewicz who did a splendid job by compiling and synchronizing individual files in the single volume. This work was supported by a European Union grant by the 7th Framework Programme, Theme 3: Science of complex systems for socially intelligent ICT. It is part of the CyberEmotions project (contract 231323). The work was also supported by Polish Ministry of Science Grant 1029/7.PR 631 UE/2009/7. I acknowledge in addition a Grant from The Netherlands Institute for Advanced Study in the Humanities and Social Sciences (NIAS) and support from European Union COST TD1210 KNOWeSCAPE action.

References

Berson, I.R., Berson, M.J., Ferron, J.M.: Emerging risks of violence in the digital age: lessons for educators from an online study of adolescent girls in the United States. J. Sch. Violence **1**(2), 51–71 (2002). doi:10.1300/j202v01n02_04

Carr, C.T., Schrock, D.B., Dauterman, P.: Speech acts within Facebook status messages. J. Lang. Soc. Psychol. **31**(2), 176–196 (2012). doi:10.1177/0261927X12438535

Chołoniewski, J., Sienkiewicz, J., Hołyst, J.A., Thelwall, M.: The role of emotional variables in the classification and prediction of collective social dynamics. Acta Phys. Pol. A **127**(3-A), A21–A28 (2015). doi:10.12693/APhysPolA.127.A-21

Darwin, C.: The Expression of the Emotions in Man and Animals. Murray, London (1872)
Heider, F.: Attitudes and cognitive organization. J. Psychol. **21**, 107–112 (1946). doi:10.1080/00223980.1946.9917275
Leventhal, H., Scherer, K.: The relationship of emotion to cognition: a functional approach to a semantic controversy. Cognit. Emot. **1**(1), 3–28 (1987). doi:10.1080/02699938708408361
Morris, D.: Manwatching: A Field Guide To Human Behaviour. Triad Panther, London (1978)
Norman, G.J., Norris, C.J., Gollan, J., Ito, T.A., Hawkley, L.C., Larsen, J.T., Cacioppo, J.T., Berntson, G.G.: The neurobiology of evaluative bivalence. Emot. Rev. **3**(3), 349–359 (2011). doi:10.1177/1754073911402403
O'Reilly, T.: What Is Web 2.0 - Design Patterns and Business Models for the Next Generation of Software. Retrieved from http://oreilly.com/pub/a/web2/archive/what-is-web-20.html?page=1 (2005)
Quirk, R., Greenbaum, S., Leech, G., Svartvik, J.: A Comprehensive Grammar of the English Language. Longman, New York (1985)
SentiStrength: http://sentistrength.wlv.ac.uk/ (2016)
Skowron, M., Rank, S.: Interacting with collective emotions in e-communities. In: Scheve, C.V., Salmela, M. (eds.) Collective Emotions. Oxford University Press, Oxford (2014). doi:10.1093/acprof:oso/9780199659180.003.0027
Sourander, A., Klomek, A.B., Ikonen, M., Lindroos, J., Luntamo, T., Koskelainen, M., Ristkari, T., Helenius, H.: Psychosocial risk factors associated with cyberbullying among adolescents. Arch. Gen. Psychiatry **67**(7), 720–728 (2010). doi:10.1001/archgenpsychiatry.2010.79
Stanley, H.E.: Introduction to Phase Transitions and Critical Phenomena. Oxford University Press, Oxford (1987)
Suler, J.: The online disinhibition effect. Cyberpsychol. Behav. **7**(3), 321–326 (2004). doi:10.1089/1094931041291295
The 70 Online Databases that Define Our Planet: Retrieved from https://www.technologyreview.com/s/421886/the-70-online-databases-that-define-our-planet/ (2010)
Walther, J.B.: Theories of computer-mediated communication and interpersonal relations. In: Knapp, M.L., Daly, J.A. (eds.) The Handbook of Interpersonal Communication, 4th edn., pp. 443–479. Sage, Thousand Oaks (2011)

Part I
Foundations

Chapter 2
A Brief History of (Social) Cyberspace

Nicole C. Krämer, German Neubaum, and Sabrina C. Eimler

2.1 Introduction

Within the last 20 years, the world has witnessed fundamental developments in the ways in which humans interact with each other: Besides talking to our family at the dinner table, exchanging ideas concerning work at the office, and sending letters to our friends living abroad, nowadays, we have the possibility to initiate a video conference with our loved ones via `Skype`, collaborate with colleagues in digital rooms such as `Google Docs`, and reveal our latest thoughts to our friends on `Facebook`.

The aim of this chapter is to describe how the current status quo of cyber communication evolved and how its analysis has developed both theoretically and empirically. We focus on the development of interpersonal communication via cyberspace and the accompanying socio-emotional processes from a psychological perspective. However, we will not consider evidence concerning task-oriented or consumer-oriented Internet use.

First, we will describe how Internet applications have developed, and comment on specific features and corresponding affordances for the users. Furthermore, we will illustrate how theories of computer-mediated communication have been extended to enable new developments to be taken into account. In a second step, cyberspace usage is systematized from the users' perspective, with a differentiation

---

N.C. Krämer (✉) • G. Neubaum
Department of Social Psychology: Media and Communication, University of Duisburg-Essen, Forsthausweg 2, 47057 Duisburg, Germany
e-mail: nicole.kraemer@uni-due.de; german.neubaum@uni-due.de

S.C. Eimler
Department of Computer Science, Human Factors and Gender Studies, University of Applied Sciences Ruhr West, Lützowstrasse 5, 46236 Bottrop, Germany
e-mail: sabrina.eimler@hs-ruhrwest.de

J.A. Hołyst (ed.), *Cyberemotions*, Understanding Complex Systems,
DOI 10.1007/978-3-319-43639-5_2

being drawn between users' needs and motives, usage patterns, produced content, affective mechanisms of usage and effects of interacting via social media. Finally, we describe implications as well as conclusions and suggest directions for future research.

2.2 Characterization of Social Cyberspace From Then Until Now

Although originally meant for purposes of military application and research, the Internet quickly became a global phenomenon and has rapidly been rededicated from a mere task-oriented instrument into a medium of interpersonal communicative, relational and emotional exchange. In the early stages of the World Wide Web, communication was primarily one-sided, with users playing a rather passive role, consuming editorial or professional, but mostly static, text-based content.

During the 1990s, the term Web 2.0, which is attributed to O'Reilly (2005), was established, referring to an opening and flexibilization of the Internet into a more participatory medium. In line with this, one-to-many and many-to-many communication gained more importance, since users were no longer passive, but rather actively generated content on their own. With the rise of Web 2.0 applications, technical barriers were broken down and platforms made available, the technical operation of which also allowed inexperienced users to create, organize and publish content (cf. user-created content, OECD 2007; Lai and Turban 2008). In this way, Internet users were no longer merely consumers, but producers of information—which inspired the term *Prosumer* (which combines the words producer and consumer; Ritzer and Jurgenson 2010).

One of the central characteristics of the emerging platforms that are associated with the Web 2.0 paradigm is that they easily facilitate social behavior (O'Reilly 2005). The social orientation of these media is emphasized by the terms social media or social web, which can nowadays be understood as established synonyms for Web 2.0. Social media can facilitate social interaction and make interpersonal communication more efficient by overcoming limitations in space and time and providing issue-specific spaces for communication and action (see Walther 2011).

Today, users can communicate via various online applications and services: Newsboards, rating platforms (`Qype`, `Tripadvisor`), wikis, blogs and micro-blogging services, social networking sites (SNS), chats (including specific chats with doctors or other experts), newsletters, newsfeeds and video platforms offer a variety of opportunities for these purposes—and all (often) without knowing the interlocutor (from face-to-face contexts). Many applications which were developed in recent years and are now taken for granted by today's Internet user community, like Facebook, knowledge-sharing or video-sharing platforms, go far beyond the original text-based chat and email communication or personal web pages of the early days of the Internet. In the following, the currently most prominent platforms which are used to communicate and connect with other people are presented.

In particular, social networking sites like Facebook for private usage and LinkedIn for professional usage have seen rising membership numbers worldwide, and have become an integral part of life for many people. Users can create a profile on Facebook containing a variety of personal information; they can connect with friends, "like", comment on and share (multimedia) content ranging from plain text status updates to videos documenting life events. Founded in 2004, it has since grown to be the largest online social network. While in 2008, Facebook had about 100 million users, this number has rapidly grown ever since—in March 2016, Facebook counted 1.65 billion monthly active users (Facebook 2016). Besides the rapid growth in memberships, Facebook has managed to introduce a variety of features to its users, leading to a number of changes in functionality which have not always remained without protests from the users (for a historical perspective on Facebook and other social networking sites, see Ellison et al. 2007; Boyd and Ellison 2007).

LinkedIn was founded in 2003 and has since become the most widely used business networking site. Like Facebook, LinkedIn allows members to create a personal profile and connect with other members of the network. With 433 million members in 2016, the platform has experienced an extremely strong growth in memberships since 2009, when it had about 37 million members.

WordPress, established between 2001 and 2005, helps people to administer their personal blog or website and has become one of the most widely used blogging systems. The organization provides live views of commenting, posts, and likes around the globe.[1] In July 2016, more than 409 million users visited more than 22.2 billion WordPress sites across the globe. While in October 2006, 595,412 posts were made, the number of monthly posts reached 69,795,604 posts in July 2016. Meanwhile, 25 % of all webpages in the Internet are WordPress sites.

The micro-blogging service Twitter was established in 2006 for internal communication purposes. Users are able to post and read messages of a maximum of 140 characters (so-called "tweets"). In September 2015, Twitter had more than 190 million unique visitors every month and more than 135,000 new users signing up each day, with over 58 million tweets created on a daily basis (Statistic Brain 2015).

The video-sharing platform YouTube, which is used to upload, watch, rate and comment on videos, was launched in 2005 and acquired by Google in 2006. It now has over one billion unique visits each month, with visitors from 56 countries. According to YouTube statistics, the service is used by almost one-third of all Internet users (more than a billion users worldwide) and is available in 76 different languages. The number of daily users has increased by 40 % since March 2014 leading to hundreds of millions of hours of video material being watched on YouTube every day (YouTube 2016). YouTube's enormous success and popularity is

[1] See http://en.wordpress.com/stats/.

partly due to the growth of platforms like Twitter and Facebook, from where people post links with references to videos on YouTube.

With the ubiquity of mobile communication in individuals' everyday lives, the online application `Instagram` has enjoyed growing popularity since its launch in October 2010. More than 400 million people across the world use this service (Instagram 2016), take pictures and videos of everyday experiences and immediately share them (using their smartphone) with their friends on Facebook or Twitter, since Instagram can be integrated into these social networking platforms.

In parallel to these developments, the research on cyberpsychology has also evolved (Barak and Suler 2008). Early theorists argued that an emotional and satisfying relational communication is impossible due to a lack of cues referring to person and context characteristics in mere text-based computer-mediated communication. As one of the first theories, *social presence theory*, suggested by Short et al. (1976), originally concentrated on the absence of nonverbal cues in audio and video conferencing, but was later applied to text-based computer-mediated communication (Hiltz et al. 1978; Rice 1984, 1983, see Walther 2011, for a review). In contrast to face-to-face communication, this theory suggests that the individually perceived social presence decreases from video conferencing, to audio conferencing, to mere text-based communication, due to a stepwise reduction of cues upon which participants can rely. In (traditional) computer-mediated communication settings, like email or chat, the communication partner is only textually present and is not represented by a variety of cues perceivable in face-to-face interaction. As a consequence, the social category to which an interaction partner belongs, e.g., gender, social class or ethnic background, and the norms arising from this, are less salient compared to immediate interpersonal nonverbal feedback. This, in turn, may have negative and positive consequences. On the one hand, deindividuation and reduced sociality may lead to norm violation and anti-social behavior (cf. toxic disinhibition effect: Lapidot-Lefler and Barak 2012, cyberbullying: Li et al. 2012; for an overview of cyberbullying and cyberthreats, see Willard 2007). On the other hand, this fostered hope for a reduction in social inhibitions, leading to more openness, the inclusion of shy or disabled people, and the promotion of equality of stigmatized groups (Carstensen 2009; Ebo 1998).

Similarly, the *lack of social context cues hypothesis* (Kiesler 1987; Kiesler et al. 1984; Siegel et al. 1986; Sproull and Kiesler 1986) argues that in face-to-face communication, certain verbal, nonverbal, and situational cues are important in defining the social context of an interaction. Through this awareness regarding the social context, participants in a communicative exchange infer behavioral and self-presentational norms (and behave accordingly). Nonverbal and visual social and socio-demographic information like a person's age, appearance, education, and status, which influence how we perceive and evaluate a person, are reduced in computer-mediated communication due to the lack of co-presence (in the sense of the interlocutor not being physically present in the same place). The resulting limited availability of facial expression, gesture, visual and further behavioral information, along with the increased (possible) anonymity, hinders accurate and reliable judgments (Dubrovsky et al. 1991; Siegel et al. 1986; Sproull and Kiesler 1986).

Meanwhile, there is a large body of research illustrating that it is indeed possible to shape interpersonal relations in computer-mediated communication that are comparable to face-to-face interactions. This assertion is supported by numerous examples of friendships and partnerships that have evolved via the Internet, calling into question the suggested lack of social and emotional quality.

Social information processing theory, posited by Walther (1992), suggests that people readily provide social background information and also infer relevant social cues in text-based computer mediated communications (CMC) from verbal information like linguistic, stylistic, and time-related cues (Walther 2007), as people are motivated to reduce uncertainty and are as interested in establishing positive relationships in computer-mediated communication as they are in face-to-face communication. Provided there is enough time, relationships in computer-mediated communication can reach the same levels of intimacy as in face-to-face communication.

Based on the *social identity model of deindividuation* by Reicher et al. (1995), and extending the ideas suggested by *social information processing theory*, Walther's (1996) *hyperpersonal communication* perspective argues that impressions can even go beyond those formed in typical face-to-face encounters. The lack of physical cues can provoke a higher sensitivity towards the (minimal amount of) given information, which is over-evaluated by the partners. Against the background of a cognitive construction process, people may develop hyper-positive and also hyper-negative views of their communication partner. This can, however, generally lead to higher degrees of interpersonal intimacy and connection (Jiang et al. 2011).

Today, both the negative and the positive aspects of communication in cyberspace are targeted, and the insight that the psychology of cyberspace does not differ from the psychology of real life (Barak and Suler 2008) is becoming increasingly accepted. In line with this, theories of interpersonal communication are applied to explain behavior and experiences in cyberspace. Moreover, the Internet is used to observe and re-examine all kinds of psychological models and theories (Amichai-Hamburger 2002; Barak 2007; Barak and Suler 2008; McKenna and Bargh 2000).

2.3 Media Psychological Perspectives on Social Interaction in Today's Cyberspace

Considering that cyberspace has become increasingly social and has extended the ways in which individuals can interact on the Internet, the question arises of how individuals might initiate, maintain, and experience social interaction in today's cyber communities. In the last decade, an overwhelming body of empirical research has addressed these questions and has provided illuminating insights into the socio-emotional mechanisms underlying the use of social media online. From

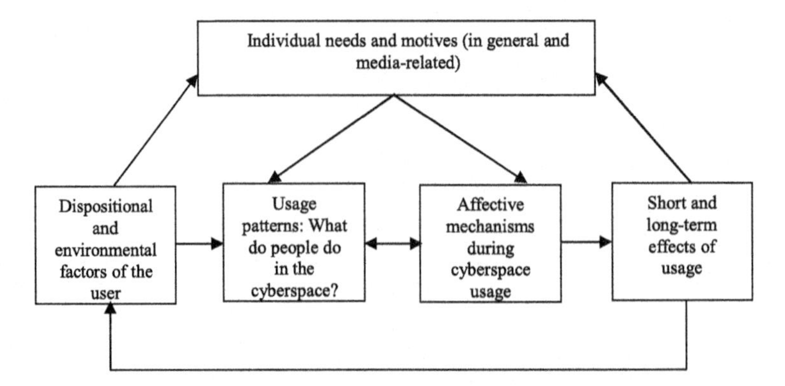

Fig. 2.1 Model regarding cyberspace usage from a media psychological perspective (based on Katz et al. 1974; Slater 2007; Valkenburg and Peter 2013)

a media psychological point of view, theoretical frameworks were posited in order to analyze communication processes targeting what individuals do with media and, in turn, how they might be affected by media (see Katz et al. 1974; Slater 2007; Valkenburg and Peter 2013), which can be also adapted to the context of online communication. According to these approaches, we first have to take into account that dispositional (e.g., personality traits) and environmental (e.g., subjective norms) variables influence the individual's motivational system, and thus shape people's needs. These needs might drive the individual's selection of specific cyberspace platforms and how these are used. Focusing on the concrete media usage situation, we propose that people's usage patterns constantly interact with their subjective experiences during exposure to cyberspace (in this chapter with a special focus on affect-based experiences). However, this interaction might be considerably affected by users' needs and expectations toward media use. The relationship between usage patterns and the accompanying affective mechanisms might lead to short- and long-term effects, which, in turn, shape the dispositional, environmental and motivational factors of the individual (see Fig. 2.1 for an overview of the proposed model regarding usage of cyberspace). Taking the model's constituents as a structure, the following sections summarize recent empirical evidence on the micro-perspective on cyberspace usage. Since social networking sites such as Facebook represent a highly popular form of social cyberspace, findings related to this kind of today's cyber-communities are a key focus of the subsequent sections.

2.3.1 *Individual Needs and Motives*

As mentioned above, the Internet has become a social space. However, the question remains of which needs and motives influence the use of and attitudes towards the above-mentioned Internet applications. Which socio-emotional needs are met by social media applications and what motives drive their use? Why do people create profiles on Facebook and willingly provide personal information and make their networks visible to others? What benefits do they gain from sharing knowledge in wikis and discussing on Twitter, and what kind of gratification do they receive from uploading or sharing (emotional) content such as videos on YouTube?

Based on a three-dimensional theory of interpersonal behavior suggested by Schutz (1966) and Ho and Dempsey (2010) introduced a conceptual framework for why people share online content. This framework also fits nicely on a more general level to explain needs and motives that drive Internet behavior.

The framework comprises several different needs that are, in a sense, mutually dependent:

The first, the need for inclusion, which is related to the need to be part of a group, refers to the fundamental social nature of humans (Baumeister and Leary 1995; Cacioppo and Patrick 2008; Fiske 2009), which may have promoted the enormous and enduring success of social media applications to date. Known under the terms *need to belong*, relatedness or need for affiliation, the "need to form and maintain enduring interpersonal attachments" (Baumeister and Leary 1995, p. 522) is pervasive (but varies inter-individually; Leary, M.R., Kelly, K.M., Cottrell, C.A., Schreindorfer, L.S., 2001, Individual differences in the need to belong, Unpublished manuscript, Wake Forest University, Winston-Salem). Meanwhile, it has also been found to be influential with regard to the antecedents and style of social networking site usage (Gangadharbatla 2008; Krämer et al. 2013; Reich and Vorderer 2013).

Krämer et al. (2013) argue that the need to belong has explanatory value for the frequency and motives of social networking site usage. In an online survey with 1162 participants, they investigated to what extent Facebook can be considered a source for satisfying the need to belong. Results show that a high perceived importance of social contacts (need to belong subscale 1) is indeed associated with a high number of friends—in real life and on Facebook—on the one hand, and is also associated with more frequent Facebook use on the other hand. The fear of losing relationships (need to belong subscale 2) is positively correlated with the motivation to use Facebook to exchange information on the one hand and as a function of peer pressure on the other hand. Both the fear of losing contacts and the perceived importance of social contacts predict habitual Facebook use and the motivation to stay in contact with other users. Similarly, in a survey with 286 participants, Reich and Vorderer (2013) found that motivations to use social networking sites and usage patterns are significantly influenced by the individual need to belong (also see Chap. 4).

A second need is the need to be individualistic and different from others. This can be related to the notion of self-concept and impression management, i.e. "Impression management (also called self-presentation) refers to the process by which individuals attempt to control the impressions others form of them" (Leary and Kowalski 1990, p. 34). In line with the need to belong, individuals are interested in conveying a positive image to others, which, in turn, positively affects how we see ourselves. In this way, the maintenance of personal homepages (Dominick 1999), blog entries, Twitter and Instant Messaging (Thiel 2005) or profiles on social networking sites like Facebook can be considered as platforms for meeting this need for self-presentation (Haferkamp and Krämer 2009; Krämer and Winter 2008; Marder et al. 2016).

Furthermore, Ho and Dempsey posited the need to be altruistic which, with reference to Schutz (1966), they describe as "the need to maintain a satisfactory relationship, leading individuals to engage in behaviors related to intimacy, warmth and emotional involvement" (Ho and Dempsey 2010, p. 1001). This need was found to be related to the amount of online content individuals forward to others: The higher the need to be altruistic, the more they share information with others online (Ho and Dempsey 2010). Finally, the need for control (mapped on the need for personal growth), which "relates to feelings of competence, achievement, influence, and accomplishment" (Ho and Dempsey 2010, p. 1002) is a relevant factor in interpersonal relationships. On the one hand, this can be associated with Internet self-efficacy (Gangadharbatla 2008) and other skill-related aspects, while on the other hand, together with the other needs, it might be acted out on platforms like wikis or blogs, by illustrating one's knowledge or sharing experience, or just by deciding whom to include in one's list of contacts.

2.3.2 Usage Patterns

Another way to adopt a psychological perspective on how users socio-emotionally deal with social media is to observe how they use these applications. Here, it has been asked what users actually do when exposing themselves to social media applications. Numerous studies also target the psychological determinants (in the sense of dispositions or personality attributes) which influence these specific usage patterns:

With regard to general usage patterns, most recent studies still show an increase in social media usage: As an example, a survey among North American adult Internet users showed that 72 % use Facebook, 25 % LinkedIn, 23 % Twitter, 31 % Pinterest, 28 % Instagram, 10 % Tumblr (Pew Research Center 2015).

Since there is a lack of statistics mapping common usage patterns worldwide, we refer to national studies to outline the activities which users perform on social media by way of example: A survey conducted in Germany with 1366 Internet users (Busemann and Gscheidle 2012) showed that individuals use cyber-communities such as Facebook on a daily basis to inform themselves about their network (36 %), to chat with other users (34 %), to send private messages (32 %), to write comments on entries of other users (25 %) or to browse through others' picture galleries (19 %). Thus, consuming information about others as well as private communication seem to be the dominant activities on social networking sites. This study also showed that despite the participatory nature of today's cyber-communities, users are more likely to consume than to generate content on these platforms. Studies in North America yield similar patterns: After logging the data of 269 Facebook users, it was revealed that in the 1-month period of observation, users received more than they gave throughout all Facebook functions. Thus, a typical Facebook user receives more private messages than she/he sends, is more likely to have her/his content liked than she/he likes others' content, receives more friend requests than she/he sends and so forth. This pattern emerges due to the special group among Facebook users whom the authors name "power users." Power users (approximately 20–30 % of all Facebook users) are much more productive (e.g., by sending friend requests, liking content, posting pictures) than other Facebook users (Hampton et al. 2012). These results give rise to the assumption that usage patterns of cyber-communities might vary between users and might be traceable to individual factors. Here, especially gender and personality aspects have been targeted.

2.3.2.1 Sex Differences in Usage

With regard to sex differences in usage, females have been found to be more likely to use Facebook than males (Duggan and Brenner 2013). Moreover, females are more prone to generate status updates and comment on entries and pictures of other users than males (Hampton et al. 2011). While males were found to use social networking sites for pragmatic purposes such as searching for friends and informing themselves, females more intensively pursue interpersonal communication as well as self-presentation strategies, for instance by creating a thorough profile (Haferkamp et al. 2012; McAndrew and Jeong 2012).

2.3.2.2 Personality Traits as Determinants of Specific Usage Patterns

Additionally, a number of studies analyzed whether specific personality aspects predict usage of social networking sites. In this regard, extraversion and openness to experience seem to be especially likely to serve as predictors of SNS use (Correa et al. 2010; Krämer and Winter 2008; Ross et al. 2009; Zywica and Danowski 2008; Winter et al. 2014). More interesting with regard to the topic of the present chapter is the finding that emotional stability also influences social networking site usage. Emotional stability was identified as a negative predictor, with people with greater levels of neuroticism and negative affectivity being shown to be more likely to use social networking sites, particularly among men (Correa et al. 2010). Similarly, it was demonstrated that higher levels of neuroticism were related to the usage of Web social services such as chat rooms (Hamburger and Ben-Artzi 2000) and instant messaging (Ehrenberg et al. 2008). However, not only quantity of usage but also quality has been observed to be related to emotional stability: Buechel and Berger (2012) found emotionally unstable social networking site users to be more likely to share self-related and emotional information in their status updates than emotionally stable users. Recently, however, Utz et al. (2012) have criticized the fact that research largely concentrates on the five-factor model of personality and introduced the concept of the need for popularity in terms of the personal "motivation to do certain things in order to appear popular" (p. 38) which is related to narcissism. Indeed, the authors' data show that the need for popularity predicts self-centered social networking site behaviors such as profile enhancement and the disclosure of feelings. By combining survey data and a content analysis of Facebook users' status updates, Winter et al. (2014) showed that higher levels of narcissism and need to belong go hand in hand with more intimate content in one's status update.

Summing up, diverse social behaviors online are traceable to specific user characteristics such as emotional stability, the level of narcissism, the need to belong, or the need for popularity. Since these psychological factors are related to the patterns of people revealing their feelings or thoughts and presenting themselves in public and private online spaces, a consideration of how users' online self-disclosure can be characterized seems to be of pivotal importance (see Kim and Dindia 2011).

2.3.2.3 Specifics of Online Self-Disclosure

Self-disclosure in the sense of revealing private information about oneself, varying in amount (e.g., quantity of information), breadth (e.g., how many different topics or facets of the self are disclosed?), and depth (how intimate is the information?), is a prerequisite for building and maintaining relationships with other humans (Altman and Taylor 1973; Cozby 1973; Greene et al. 2006).

In earlier computer-mediated communication research, the breadth and depth of users' self-disclosure online were found to be greater than in face-to-face communication (Joinson 2001; Tidwell and Walther 2002). While these results were interpreted in the context of anonymity and the compensation of reduced social cues of computer-mediated communication, the question arises of how the enhancement of self-disclosure identified in earlier studies may apply to present-day online settings, in which users might be more identifiable and confronted with more visual stimuli on platforms such as Facebook.

Self-Disclosing and Presentation Patterns on Social Networking Site Profiles
Focusing on social networking sites and their affordances for self-disclosure, there are two key features which might be of particular interest: the users' profiles and their entries on their digital wall (the so-called status updates).

Numerous studies on social networking sites have explored how individuals use their personal online profiles in terms of self-disclosure. Based on Goffman's (1959) work on selective self-presentation and its application to computer-mediated communication (Walther 2007), scholars quickly recognized that social networking profiles serve as appropriate instruments for users to selectively disclose key information on their personality and thus "manage" the impressions the audience form of them (Boyd and Ellison 2007; Donath and Boyd 2004; Jung et al. 2007; Krämer and Winter 2008). On their Facebook profiles, for instance, users reveal information about their favorite music, movies, their romantic status, sexual orientation, favorite books as well as pictures of friends and themselves (Nosko et al. 2010; Tufekci 2008). The way in which individuals present themselves was found to be related to users' personality traits: While a higher amount of profile details and a more experimental style of profile pictures (e.g., making a face or striking a pose) are associated with users' higher self-efficacy concerning their self-presentation (individuals who feel confident and competent about their offline self-presentation; Krämer and Winter 2008), the self-promoting nature of profile pictures, profile description and status updates is positively related to users' self-reported narcissism scores (Buffardi and Campbell 2008; Mehdizadeh 2010; Winter et al. 2014).

Another important realm in which to study self-disclosure patterns consists of users' status updates (i.e., short notes on current activities, feelings and thoughts that are presented on the user's profile/timeline) as a popular feature on social networking sites: By means of a content analysis of 254 status updates, Carr et al. (2012) demonstrated that 60 % of status messages are *expressive* speech acts, in which users reveal emotional experiences relating to situations or to other people. 39 % of status updates were still coded as *assertive* in the sense of providing general information about themselves. The analysis also showed that 21 % of the status messages included humorous content. Including time cues of the status updates in their analysis, the authors ascertained that users update their status on Facebook once every 5 days. Thus, status updates seem to offer an appropriate means to express emotions and reveal diverse self-related information on a regular basis. Carr et al. concluded "that even in a mass medium, users seem to use status messages as interpersonal communicative tools, constructing messages with rich socioemotional information" (p. 188).

Focusing on the concrete emotions which users express in status updates, an analysis by Kramer and Chung (2011) showed that especially positive emotions were a critical part of users' status messages across the countries of Australia, Canada, the US and the UK. This finding was underlined by a further study which revealed that status updates contain more positive than negative emotions (Lin and Qiu 2012). However, the expression of emotions was related to the individual number of friends: Users with a lower number of friends tend to express more negative emotions than users with a higher number of friends. The assumption that users might tailor their emotional expression pattern to the specific audiences on social networking sites was explored by Bazarova et al. (2013). Comparing the language style within status updates (non-directed communication), private messages and wall posts (directed communication), the authors were able to establish that while positive emotions were present in an equal amount in all communication channels on social networking sites, negative emotions were expressed significantly less in status updates than in wall posts or private messages. However, the expression of positive emotions in public status updates was related to self-presentational motives of users. Given these findings, Bazarova and colleagues reasoned that expressing emotions on Facebook might imply the strategic nature of publicly persuading others of a "happy self" rather than to get connected with others.

Other studies focused on what is revealed in comments on social networking sites: Thelwall et al. (2010) analyzed to what extent positive and negative emotions are disclosed in user profile comments within the social networking site MySpace. To this aim, they performed a sentiment text analysis on 819 user comments, classifying the valence (positive vs. negative) and strength (weak vs. strong) of emotional words or phrases. Results showed that the emotions are omnipresent in users' comments: Two thirds of comments contained positive emotions, while 20 % included negatively valenced phrases. The strength of these emotional comments was predominantly classified between 1 and 3 on a Likert scale from 1 (weak) to 5 (strong). Furthermore, gender differences were identified: Females tend to produce and receive positively valenced comments more than males do. However, no gender differences were found with regard to negative emotions. These results suggest social networking sites, in this case MySpace, to be a platform for disclosure of positive affect, which is mostly expressed by women. However, the findings do not reveal the actual feelings of the users—they only encompassed the emotions expressed (see also Chap. 3). Here, the authors suggest that the decision to display specific emotions might be determined predominantly by the user' s social role rather than her/his actual feelings and emotions.

The reported empirical evidence on usage patterns in social cyberspaces might provide first clues as to the psychological mechanisms behind specific activities. Users pursue relational and self-presentational concerns while using social media on the Internet. Moreover, the fact that users are prone to share their emotions and experiences with the cyber-community can also be seen as an indication that social media might also serve purposes of personal emotion regulation. In the following, we will take a closer look at affective mechanisms underlying individuals' social media usage. In this case, we assume that these internal affective processes interact with users' observable usage patterns (cf. Fig. 2.1).

2.3.3 *Affective Mechanisms During Usage*

Psychological functions of self-disclosure of emotions have been widely studied, especially in Western culture. Researchers have found that "the disclosure of traumatic and emotional experiences can promote physical and psychological health" (Pennebaker 1995, p. 8; Pennebaker and Chung 2011, see also Chap. 4). In this line, Tamir and Mitchell (2012) demonstrate that disclosing information about the self is intrinsically rewarding. They suggest that self-disclosure represents an event with intrinsic value which is comparable to the primary rewards offered by food or sex. Additionally, in terms of a theoretical framework that may address the question of why people aspire to disclose themselves by publishing their thoughts and feelings on social networking sites, the concept of social sharing of emotions (e.g., Rimé et al. 1991) can make a contribution. According to this approach, human beings need to express emotional states in order to make sense of them, for example "after an emotional event, people are expected repetitively to seek social opportunities to verbalise their experience" (Rimé et al. 1991, p. 437). However, this approach inherently includes a more social view, in that it is seen as essential that there is a (potentially reacting) receiver of the verbalization. In an interesting recent study, Deters and Mehl (2013) suggest a potential reason for the beneficial aspects of self-disclosure especially on social networking sites: They followed an experimental approach to explore the socio-emotional effects of online self-disclosure and asked participants, for 1 week, to post status messages on Facebook more often than they commonly did. Here, the authors showed that posting status updates diminishes the sense of being lonely. This effect, however, was not explained by the proportion of direct feedback (in terms of comments) which users receive from others, but rather by their perceived connectedness with others. Thus, "the act of writing itself—in the absence of any direct effects status updates may have on one's social network—might create a feeling of connectedness" (p. 5). Further evidence corroborates the prevalence and patterns of sharing emotions through social media and the conditions under which the sharing process can be psychologically beneficial: Facebook users were generally found to prefer sharing negative emotions and intimate content via private messages than via commonly more public channels such as status updates (Bazarova et al. 2015). The same study also showed that users perceived greater overall satisfaction after disclosing emotions, the more satisfied they were with how their network has replied to their sharing message. Burke and Develin (2016) characterized the content of people's replies to emotionally disclosing messages on Facebook and found that the emotional disclosures (especially negative ones) are met with more and longer responses.

Moreover, not only the act of verbalizing the emotions per se, but also the opportunity to reflect on the self and to provide self-promotional content might lead to positive emotional outcomes: Gonzales and Hancock (2011) demonstrated that after individuals had browsed their profiles, they reported a higher self-esteem than individuals who were asked to look at themselves in a mirror (which serves as experimental manipulation in social psychology, yielding objective self-awareness; Fejfar and Hoyle 2000). Participants' self-esteem was even higher when they had edited or updated their profile during the experiment. The authors explained the effects of social networking sites profiles on self-esteem, building on the concept of selective self-presentation online (cf. Walther 2007): Since social networking site profile affordances give users the chance to select the facets of their self they want to show, the displayed information may reflect the positive aspects of their personality. Therefore, the exposure to this self-generated content on one's own profile might enhance one's self-evaluations in terms of a higher self-esteem.

However, not all mechanisms that have been discussed lead to beneficial outcomes. Especially when focusing on the reception of profiles or status updates of other people, negative emotions have also been described. Here, considering that individuals are prone to evaluate themselves by monitoring the social environment and comparing themselves with the prevailing standards (Festinger 1954), it has been assumed that cyber-communities might also be used for social comparisons. In line with this, Haferkamp and Krämer (2011) focused on the consequences of these processes. Using an experimental approach, the authors showed that looking at profiles of rather attractive strangers is more strongly related to less positive affect and less satisfaction with one's own body image afterwards than looking at profiles of less attractive people. Moreover, males were found to be less satisfied with their career (in terms of real-ideal discrepancy) after looking at a profile of a successful person than males who browsed a profile of a less successful user. While this experimental study suffers from the shortcomings of an artificial setting (e.g., because users would usually look at acquaintances rather than strangers), there are also online studies which confirm the results: A survey among 425 undergraduate students revealed that those participants who had been using Facebook for a longer time believed that others' lives were happier, and agreed less with items such as "life is fair" (Chou and Edge 2012). A similar pattern was also found for users who spent more time on Facebook every week—these users also tended to perceive other users as being happier with better lives. Such perceptions, however, can even lead to feelings of envy: A recent survey demonstrated that the sense of envy mediates the relationship between consuming content about others on Facebook and one's own life satisfaction (Krasnova et al. 2013). However, users still seem to be able to regulate their emotions by carefully selecting the profiles they spend time on: Johnson and Knobloch-Westerwick (2014) showed that on social networking sites, people with a negative mood spend more time on downward than on upward comparisons.

Altogether, the mechanisms already imply that outcomes of interactions within social media might have positive or detrimental effects, which are specified in the next chapter (Chap. 5).

2.3.4 Short-Term and Longitudinal Effects

With regard to short-term effects, but especially concerning longitudinal effects, it has been debated whether social media or social networking site usage increases life satisfaction or can have negative effects. A decisive component of this research line is the question of whether social networking site usage can help to build so-called social capital, which helps the user to receive emotional support.

Indeed, several studies reveal a relation between intensity of Facebook use and students' life satisfaction (Valenzuela et al. 2009). Moreover, in a much-cited article, Ellison et al. (2007) reported findings on self-esteem and well-being. Using the Rosenberg Self-Esteem Scale (Rosenberg 1965) and an adjusted form of the Satisfaction with Life Scale (Diener et al. 1997; Pavot and Diener 1993), they assessed 286 undergraduate students' level of well-being and self-esteem. Their results show that Facebook use is positively related to well-being—especially for those people with low self-esteem and low life satisfaction. The authors suspect that this can be explained by the fact that those who use Facebook intensely reported having higher bridging social capital (i.e., superficial social contacts) than those who used Facebook less. Thus, Facebook use may be helping to provide increased information and opportunities for social contact, meaning that particularly students who have low satisfaction and low self-esteem can benefit from Facebook usage. Furthermore, data from the Pew Research Center (Hampton et al. 2011) demonstrate that Facebook users perceive higher levels of social and emotional support in their lives than other Internet users and non-Internet users. This might be due to the fact that Facebook users report having significantly more close relationships than other Internet users and non-Internet users. Recently, a laboratory study assessing people's psychological experiences and affective states during Facebook use showed that the more people interact with other users on this platform (e.g., by commenting on others' messages), the closer they feel to these interaction partners and the more positive are their affective states (Neubaum and Krämer 2015). In line with this, a considerable body of empirical work has dealt with the concept of social capital in the sense of benefits people draw from interpersonal relationships in the context of social networking sites (e.g., Ellison et al. 2007, 2011). The concept was based on Putnam's (2000) differentiation between bridging social capital, in terms of exchanging information with loose contacts ("weak ties") within one's networks, and bonding social capital, referring to benefits from communication with close friends and family ("strong ties"). Social and emotional support represents a form of benefit that is especially derived from bonding social capital (see also Wellman and Wortley 1990, who show that tie strength is the strongest predictor of feelings of emotional support). The fact that loose contacts might indeed not be sufficient to increase social and emotional support is shown by Kalpidou et al. (2011): These authors demonstrated that at least first-year students do not benefit from intense Facebook usage and its opportunity to connect socially: For them, the number of Facebook friends was negatively associated with emotional and academic adjustment. Only for students in later college life was the number of Facebook friends positively related to social adjustment and attachment to one's institution. In line with this, recent research has demonstrated that people perceive their strong ties as more valuable—in terms of both emotional as well as informational support given—while the weak ties are experienced as dispensable (Krämer et al. 2014). At the same time, however, they are not willing to delete considerable parts of the contacts deemed as expendable (Krämer et al. 2015).

But what mechanisms are responsible for the finding that a larger network eventually leads to more positive emotions and overall well-being? Here, three aspects can be listed that have been addressed in different studies: (a) positive feedback by others, (b) social support after the disclosure of (emotionally) negative experience, and (c) the users' active engagement in social networking sites. Ad (a) As early as 2006, during the infancy of research into social networking sites, Valkenburg et al. (2006) published a survey with the aim of finding out about the general effects of social networking site usage on adolescents' self-esteem and well-being. Their structural equation model showed that the tone of feedback to one's profile was the only factor influencing self-esteem, in that positive feedback was favorable for people's self-esteem and well-being and negative feedback was harmful and lowered self-esteem. Against the background of 78 % of the respondents reporting to receive predominantly positive feedback, Valkenburg et al. (2006) conclude that for adolescents, "the use of friend networking sites may be an effective vehicle for enhancing their self-esteem" (p. 589). Trepte and Reinecke (2013) additionally showed that the more rewards users receive in terms of social capital, the more willingly they show self-disclosure. Ad (b) With regard to social support after the disclosure of (emotionally) negative experience, Buechel and Berger (2012) demonstrated that users' subjective well-being increased especially when they perceived social support with regard to the disclosure of negative experience. Similarly, Lee et al. (2013) showed that, for lonely people, self-disclosure positively influences social support, and social support positively influences well-being.

In another line of research, studies on disaster communication show that various forms of social media are used to cope with emotionally negative experiences. Blogs, online support groups, but also social networking sites, are used to share emotions in times of crisis (e.g., natural disasters such as Hurricane Katrina; Macias et al. 2009) and to support others (Barak 2010). A recent study by Neubaum et al. (2014) showed that condolence and grief were communicated most frequently in the aftermath of the stampede at the Love Parade 2010 in Germany and that the motives for using social networking sites were predominantly emotional processing and offering (emotional) support. Ad (c) Hinting at the importance of users' active engagement in social networking sites, Burke et al. (2010) demonstrated, by means of a survey combined with an analysis of user activities based on server logs, that greater use of social networking sites is especially associated with increased social well-being, social capital, and reduced loneliness when engagement with Facebook is active rather than passive. Therefore, in sum, several moderators have been identified which affect whether the usage of social networking sites leads to beneficial emotional outcomes.

However, the tendency to experience greater well-being and positive feelings does not seem to apply to every user group: Nowadays, older adults are also increasingly using social networking sites (Madden 2010), and these online communities have therefore been suggested as potential venues for older people and retirees to establish high-quality social interaction in order to prevent them from aging alone (Sundar et al. 2011). However, this assumption has not been validated by empirical data. By means of a survey, Sundar et al. (2011) tested whether the intensity of social networking site usage is related to the perceived quality of life. Results did not reveal any correlation between perceived quality of life and the usage of social networking sites.

Additionally, several studies have focused on emotional outcomes of using social media. In a laboratory study, Wise et al. (2010) observed how individuals use social networking sites: 29 participants' social searching, i.e., "an extractive social information-seeking strategy" (p. 560), and social browsing, i.e., "passive social information-seeking strategy" (p. 560), activity was recorded by screen capturing software and subsequently coded. Additionally, physiological measures were recorded as well as facial EMG. Participants were instructed to spend 5 min on Facebook.com, without leaving the page. According to the findings, users spent more time on active extracting of social information (social searching) than on those pages categorized as social browsing pages. Independently of the way in which Facebook was used, sympathetic arousal dropped over time. While there was no notable difference between social searching and browsing activity with regard to evidence of unpleasantness (in the facial EMG data), physiological measures indicated that pleasantness was greater during social searching activities (active extracting).

Although social searching in particular is emotionally positive, clearly the reception of other profiles can also have detrimental effects. Lewis and West (2009) reported that, for example, looking at the profile of an ex-boyfriend may cause negative feelings. As summarized above, social comparison processes can also lead to more negative feelings after the reception of other people' s profiles. Therefore, effects of social networking sites do not appear to be always positive. In line with this, Tokunaga (2011) presented a typology of 10 negative events that social networking site users may encounter. The three event types that people experienced most commonly are deleting messages that can be accessed publicly, being ignored or denial of friend requests, or the identification of disparities in the top friends' lists' ranking. Additionally, even more serious events can lead to negative emotional outcomes. In Europe, 6 % of young people aged from 9 to 16 reported having experienced a hurtful or nasty incident against them on the Internet during the 12 months before the survey (Livingstone et al. 2011). Unquestionably, harnessing actions on the Internet can have dramatic effects on the victim's affective system and health. Thus, empirical research has revealed that cyberbullying can lead to states of anger, embarrassment, defenselessness, and depression (Ortega et al. 2009) and can longitudinally reduce life satisfaction (Sumter et al. 2012). Similarly, Internet addiction in terms of the "phenomenon of the inability to control Internet usage followed by problems in daily life and psychological domains" (Pawlikowski et al. 2013, p. 1213) can lead to serious negative consequences. And indeed, pathological online behavior has not only been observed for online gaming or cyber pornography but also concerning social networking site usage: With regard to online communication addiction, a 6-month longitudinal survey among adolescents revealed that users' self-reported frequency of chatting and instant messaging positively predicted compulsive Internet use at the second measurement time point (van den Eijnden et al. 2008). Moreover, a significant relationship was identified between frequency of instant messaging and feelings of depression 6 months later. Thus, excessive online communication can interfere with teenagers' psychological well-being from a longitudinal perspective.

Altogether, while numerous positive consequences and opportunities have been identified, several risks for well-being and life satisfaction have also been described. Thus, just like in "real life", interpersonal interaction via the Internet brings with it opportunities as well as risks.

2.4 Concluding Thoughts

This review of research conducted over the past years illustrating how people use and are affected by social cyberspace points at a number of facets of social cyberspace: Socio-emotional processes reaching from self-reflection (while comparing with others), regulating self-esteem, generating social capital (maintaining strong ties and deepening weak ties) to well-being and life satisfaction. In sum, the review shows that within the context of social networking site usage, social psychological aspects as well as emotions are relevant in several ways. With regard to motives, social connection, communication and self-presentation have been identified as crucial. Concerning the social sharing of emotions, content analyses demonstrate that, in general, more positive than negative emotions are shared. As determinants for sharing emotions, social role (e.g., women display more positive emotions than men) and the tendency to tailor the expression of emotion to the specifics of the network (i.e., size and density of the network), have been identified. Most importantly, with regard to the effects of social networking site usage, findings show that especially people with low self-esteem and low life satisfaction benefit from usage of these platforms and experience more positive emotions. In particular, bonding social capital seems to be responsible here—especially when these "strong ties" provide positive feedback and emotional support via social networking sites. However, especially when users are confronted with profiles of happy, successful or attractive people, usage of social networking sites may also have detrimental emotional effects, by means of social comparison.

Although, as the findings show, the concept of emotions is intertwined in various ways with usage and effects of social networking sites, research conducted so far has rarely focused on emotions. This is also reflected in the assessment of emotions in the studies summarized above: Mostly, emotions are not assessed directly, but instead, concepts like life satisfaction or well-being are considered. Furthermore, if emotions are targeted directly, a differentiation is merely made between positive and negative emotions (see also Chap. 3). Therefore, future studies should focus more explicitly on emotions and differentiate between different emotional facets. Emotional processes not only have to be considered for the individual user but also need to be embedded within a complex social context. They should not only be targeted at an individual level but also be studied by taking the perspective of an interpersonal or group level. So far, research has benefited from interdisciplinary collaborations, bringing together perspectives from communication science, media and developmental psychology, sociology and computer science, to name but a few. With a view to these complex tasks ahead, scholars should be encouraged to intensify these collaborations in order to gain insights into why humans behave according to certain patterns online. The present volume testifies to how different disciplines might contribute to tackling the complexity of social processes in cyberspace on a larger scale.

It might, however, be a methodological challenge to assess emotions in the context of social networking site usage. One possibility that has already been used is to assess feelings by self-report measures (either by means of questionnaires or interviews). What is largely missing is the use of physiological measures and experimental settings (see also Chap. 5). Longitudinal research is desirable and could be conducted using web-based diaries in order to assess the variability of experiences while using social media. Given the results summarized here, it can be expected that considering emotions as a crucial aspect of SNS usage might help to explain the success of social cyber-communities. Therefore, future research needs to accept at least two new challenges: (1) To explore the (longitudinal) emotional effects not only on an individual but also on a collective level, and (2) to maximize the environmental validity of evidence by adapting research questions to the continuously changing socio-technical features and corresponding affordances in cyberspace.

References

Altman, I., Taylor, D.: Social Penetration: The Development of Interpersonal Relationships. Holt/Rinehart and Winston, New York (1973)

Amichai-Hamburger, Y.: Internet and personality. Comput. Hum. Behav. **18**(1), 1–10 (2002). doi:10.1016/S0747-5632(01)00034-6

Barak, A.: Phantom emotions: psychological determinants of emotional experiences on the Internet. In: Joinson, A., McKenna, K., Postmes, T., Reips, U. (eds.) The Oxford Handbook of Internet Psychology, pp. 303–330. Oxford University Press, Oxford (2007). doi:10.1093/oxfordhb/9780199561803.013.0020

Barak, A.: The psychological role of the internet in mass disasters: past evidence and future planning. In: Brunet, A., Ashbaugh, A.R., Herbert, C.F. (eds.) NATO Science for Peace and Security Series - E: Human and Societal Dynamics vol 72: Internet Use in the Aftermath of Trauma, pp. 23–43. IOS Press, Amsterdam (2010). doi:10.3233/978-1-60750-626-3-23

Barak, A., Suler, J.: Reflections on the psychology and social science of cyberspace. In: Barak, A. (ed.) Psychological Aspects of Cyberspace: Theory, Research, Applications, pp. 1–12. Cambridge University Press, Cambridge (2008)

Baumeister, R.F., Leary, M.R.: The need to belong: desire for interpersonal attachments as a fundamental human motivation. Psychol. Bull. **117**(3), 497–529 (1995). doi:10.1037//0033-2909.117.3.497

Bazarova, N.N., Taft, J., Choi, Y., Cosley, D.: Managing impressions and relationships on Facebook: self-presentational and relational concerns revealed through the analysis of language style. J. Lang. Soc. Psychol. **32**(2), 121–141 (2013). doi:10.1177/0261927X12456384

Bazarova, N.N., Choi, Y.H., Sosik, V.S., Cosley, D., Whitlock, J.: Social sharing of emotions on Facebook: channel differences, satisfaction, and replies. In: Proceedings of the 18th ACM Conference on Computer Supported Cooperative Work and Social Computing (CSCW '15) (2015)

Boyd, D.M., Ellison, N.: Social network sites: definition, history, and scholarship. J. Comput.-Mediat. Commun. **13**(1), 210–230 (2007). doi:10.1111/j.1083-6101.2007.00393.x. Retrieved from http://jcmc.indiana.edu/vol13/issue1/boyd.ellison.html

Buechel, E., Berger, J.A.: Facebook therapy? Why people share self-relevant content online. In: Society for Consumer Psychology Winter Conference, Las Vegas (2012). doi:10.2139/ssrn.2013148

Buffardi, L.E., Campbell, W.E.: Narcissism and social networking web sites. Personal. Soc. Psychol. Bull. **34**(10), 1303–1314 (2008). doi:10.1177/0146167208320061

Burke, M., Develin, M.: Once more with feeling: supportive responses to social sharing on Facebook. In: Proceedings of the 19th ACM Conference on Computer Supported Cooperative Work and Social Computing (CSCW '16) (2016)

Burke, M., Marlow, C., Lento, T.: Social network activity and social well-being. Paper presented at the CHI, Atlanta, GA (2010)

Busemann, K., Gscheidle, C.: Web 2.0: Habitualisierung der Social Communitys. Ergebnisse der ARD/ZDF-Onlinestudie 2012. [Web 2.0: Habitualization of social communities. Results from the ARD/ZDF online study 2012]. Media Perspektiven **7–8**, 380–390 (2012)

Cacioppo, J.T., Patrick, B.: Loneliness: Human Nature and the Need for Social Connection. WW Norton and Company, New York (2008)

Carr, C.T., Schrock, D.B., Dauterman, P.: Speech acts within Facebook status messages. J. Lang. Soc. Psychol. **31**(2), 176–196 (2012). doi:10.1177/0261927X12438535

Carstensen, T.: Gender in trouble in Web 2.0: gender relations in social networks sites, wikis and weblogs. Int. J. Gen. Sci. Technol. **1**(1), 105–127 (2009)

Chou, H.T., Edge, N.: "They are happier and having better lives than I am": the impact of using Facebook on perceptions of others' lives. Cyberpsychol. Behav. Soc. Netw. **15**(2), 117–121 (2012). doi:10.1089/cyber.2011.0324

Correa, T., Hinsley, A.W., Zúñiga, H.G.: Who interacts on the web?: the intersection of users' personality and social media use. Comput. Hum. Behav. **26**(2), 247–253 (2010). doi:10.1016/j.chb.2009.09.003

Cozby, P.C.: Self-disclosure: a literature review. Psychol. Bull. **79**, 73–91 (1973)

Deters, F.G., Mehl, M.R.: Does posting Facebook status updates increase or decrease loneliness? An online social networking experiment. Soc. Psychol. Personal. Sci. **4**(5), 579–586 (2013). doi:10.1037/h0033950

Diener, E., Suh, E., Oishi, S.: Recent findings on subjective well-being. Indian J. Clin. Psychol. **24**(1), 25–41 (1997)

Dominick, J.R.: Who do you think you are? Personal home pages and self-presentation on the World Wide Web. J. Mass Commun. **76**(4), 646–658 (1999). doi:10.1177/107769909907600403

Donath, J., Boyd, D.: Public displays of connection. BT Technol. J. **22**(4), 71–82 (2004). doi:10.1023/B:BTTJ.0000047585.06264.cc

Dubrovsky, V.J., Kiesler, S., Sethna, B.N.: The equalization phenomenon: status effects in computer-mediated and face-to-face decision making groups. Hum.-Comput. Interact. **6**(2), 119–146 (1991). doi:10.1207/s15327051hci0602_2

Duggan, M., Brenner, J.: The demographics of social media users - 2012. Pew Internet and American Life Project. Retrieved from http://www.pewinternet.org/Reports/2013/Social-media-users.aspx (2013)

Ebo, B. (ed.): Cyberghetto or Cybertopia? Race, Class, and Gender on the Internet. Praeger, London (1998)

Ehrenberg, A., Juckes, S., White, K.M., Walsh, S.P.: Personality and self-esteem as predictors of young people's technology use. Cyberpsychol. Behav. **11**(6), 739–741 (2008). doi:10.1089/cpb.2008.0030

Ellison, N.B., Steinfield, C., Lampe, C.: The benefits of Facebook "friends:" social capital and college students' use of online social network sites. J. Comput.-Mediat. Commun. **12**(4), 1143–1168 (2007). doi:10.1111/j.1083-6101.2007.00367.x

Ellison, N.B., Steinfield, C., Lampe, C.: Connection strategies: social capital implications of Facebook-enabled communication practices. New Media Soc. **13**(6), 873–892 (2011). doi:10.1177/1461444810385389

Facebook Key Facts: Retrieved from http://newsroom.fb.com/Key-Facts (2016)

Fejfar, M.C., Hoyle, R.H.: Effect of private self-awareness on negative affect and self-referent attribution: a quantitative review. Personal. Soc. Psychol. Rev. **4**(2), 132–142 (2000). doi:10.1207/S15327957PSPR0402_02

Festinger, L.: A theory of social comparison processes. Hum. Relat. **7**(2), 117–140 (1954). doi:10.1177/001872675400700202

Fiske, S.T.: Social Beings: Core Motives in Social Psychology. Wiley, New York (2009)

Gangadharbatla, H.: Facebook me: collective self-esteem, need to belong, and Internet self-efficacy as predictors of the iGeneration's attitudes toward social networking sites. J. Interact. Advert. **8**(2), 5–15 (2008)

Goffman, E.: The Presentation of Self in Everyday Life. Anchor Books, New York (1959)

Gonzales, A.L., Hancock, J.T.: Mirror, mirror on my Facebook wall: effects of exposure to Facebook on self-esteem. Cyberpsychol. Behav. Soc. Netw. **14**(1–2), 79–83 (2011). doi:10.1089/cyber.2009.0411

Greene, K., Derlega, V.J., Mathews, A.: Self-disclosure in personal relationships. In: Vangelisti, A., Perlman, D. (eds.) The Cambridge Handbook of Personal Relationships, pp. 409–427. Cambridge University Press, New York (2006). doi:10.1017/CBO9780511606632.023
Haferkamp, N., Krämer, N.C.: "When I was your age, Pluto was a planet": impression management and need to belong as motives for joining groups on social networking sites. Paper presented at the annual meeting of ICA 2009 (International Communication Association), Chicago, IL, May 2009
Haferkamp, N., Krämer, N.C.: Social comparison 2.0: examining the effects of online profiles on social-networking sites. Cyberpsychol. Behav. Soc. Netw. **14**(5), 309–314 (2011). doi:10.1089/cyber.2010.0120
Haferkamp, N., Eimler, S.C., Papadakis, A.-M., Kruck, J.V.: Men are from Mars, women are from Venus? Examining gender differences in self-presentation on social networking sites. Cyberpsychol. Behav. Soc. Netw. **15**(2), 91–98 (2012). doi:0.1089/cyber.2011.0151
Hamburger, Y.A., Ben-Artzi, E.: The relationship between extraversion and neuroticism and the different uses of the Internet. Comput. Hum. Behav. **16**(4), 441–449 (2000). doi:10.1016/S0747-5632(00)00017-0
Hampton, K.N., Goulet, L.S., Rainie, L., Purcell, K.: Social networking sites and our lives. Pew Internet and American Life Project. Retrieved from http://pewinternet.org/~/media//Files/Reports/2011/PIP%20-%20Social%20networking%20sites%20and%20our%20lives.pdf (2011)
Hampton, K.N., Goulet, L.S., Marlow, C., Rainie, L.: Why most Facebook users get more than they give. Pew Internet and American Life Project. Retrieved from http://www.pewinternet.org/~/media//Files/Reports/2012/PIP_Facebook%20users_2.3.12.pdf (2012)
Hiltz, S.R., Johnson, K., Agle, G.: Replicating Bales' problem solving experiments on a computerized conference: a pilot study. Research Report No 8, New Jersey Institute of Technology, Computerized Conferencing and Communications Center, Newark (1978)
Ho, J.Y.C., Dempsey, M.: Viral marketing: motivations to forward online content. J. Bus. Res. **63**(9–10), 1000–1006 (2010). doi:10.1016/j.jbusres.2008.08.010
Instagram: Introducing Instagram for Windows Phone. Retrieved from http://blog.instagram.com/tagged/instagram_news (2016)
Jiang, L., Bazarova, N.N., Hancock, J.T.: The disclosure-intimacy link in computer-mediated communication: an attributional extension of the hyperpersonal model. Hum. Commun. Res. **37**(1), 58–77 (2011). doi:10.1111/j.1468-2958.2010.01393.x
Johnson, B.K., Knobloch-Westerwick, S.: Glancing up or down: mood Management and selective social comparisons on social networking sites. Comput. Soc. Behav. **41**, 33–39 (2014). doi:10.1016/j.chb.2014.09.009
Joinson, A.N.: Self-disclosure in computer-mediated communication: the role of self-awareness and visual anonymity. Eur. J. Soc. Psychol. **31**(2), 177–192 (2001). doi:10.1002/ejsp.36
Jung, T., Youn, H., McClung, S.: Motivations and self-presentation strategies on Korean-based "Cyworld" weblog format personal homepages. Cyberpsychol. Behav. **10**(1), 24–31 (2007). doi:10.1089/cpb.2006.9996
Kalpidou, M., Costin, D., Morris, J.: The relationship between Facebook and the well-being of undergraduate college students. Cyberpsychol. Behav. Soc. Netw. **14**(4), 183–189 (2011). doi:10.1089/cyber.2010.0061
Katz, E., Blumler, J.G., Gurevitch, M.: Utilization of mass communication by the individual. In: Blumler, J.G., Katz, E. (eds.) The Uses of Mass Communications. Current Perspectives on Gratifications Research, pp. 19–32. Sage, Beverly Hills (1974)
Kiesler, S.: The hidden messages in computer networks. McKinsey Q. **3**, 13–26 (1987)
Kiesler, S., Siegel, J., McGuire, T.W.: Social psychological aspects of computer-mediated communication. Am. Psychol. **39**(10), 1123–1134 (1984). doi:10.1037//0003-066X.39.10.1123
Kim, J., Dindia, K.: Online self-disclosure: a review of research. In: Wright, K.B., Webb, L.M. (eds.) Computer-Mediated Communication in Personal Relationships, pp. 156–180. Peter Lang, New York (2011)

Kramer, A.D.I., Chung, C.K.: Dimensions of self-expression in Facebook status updates. In: Proceedings of the 2011 International Conference on Weblogs and Social Media (ICWSM), pp. 169–176 (2011)
Krämer, N.C., Winter, S.: Impression management 2.0. self-presentation on social networking sites and its relationship to personality. J. Med. Psychol. **20**(3), 106–116 (2008). doi:10.1027/1864-1105.20.3.106
Krämer, N.C., Hoffmann, L., Fuchslocher, A., Eimler, S.C., Szczuka, J.M., Brand, M.: Do I need to belong? Development of a scale for measuring the need to belong and its predictive value for media usage. Paper presented at the annual meeting of ICA 2012 (International Communication Association), Phoenix (2013)
Krämer, N.C., Rösner, L., Eimler, S.C., Winter, S., Neubaum, G.: Let the weakest link go! Empirical explorations on the relative importance of weak and strong ties on social networking sites. Societies **4**, 785–809 (2014). doi:10.3390/soc4040785
Krämer, N.C., Hoffmann, L., Eimler, S.C.: Not breaking bonds on Facebook–Mixed–Methods research on the influence of individuals' need to belong on 'Unfriending' behavior on Facebook. Int. J. Dev. Sci. **9**(2), 61–74 (2015). doi:10.3233/DEV-150161
Krasnova, H., Wenninger, H., Widjaja, T., Buxmann, P.: Envy on Facebook: a hidden threat to users' life satisfaction? In: Proceedings of 11th International Conference on Wirtschaftsinformatik (WI), Leipzig (2013)
Lai, L.S., Turban, E.: Groups formation and operations in the Web 2.0 environment and social networks. Group Decis. Negot. **17**(5), 387–402 (2008). doi:10.1007/s10726-008-9113-2
Lapidot-Lefler, N., Barak, A.: Effects of anonymity, invisibility, and lack of eye-contact on toxic online disinhibition. Comput. Hum. Behav. **28**(2), 434–443 (2012). doi:10.1016/j.chb.2011.10.014
Leary, M.R., Kowalski, R.M.: Impression management: a literature review and two-component model. Psychol. Bull. **107**(1), 34–47 (1990). doi:10.1037//0033-2909.107.1.34
Lee, K.T., Noh, M.J., Koo, D.M.: Lonely people are no longer lonely on social networking sites: the mediating role of self-disclosure and social support. Cyberpsychol. Behav. Soc. Netw. **6**(16), 413–418 (2013). doi:10.1089/cyber.2012.0553
Lewis, J., West, A.: 'Friending': London-based undergraduates' experience of Facebook. New Med. Soc. **11**(7), 1209–1229 (2009). doi:10.1177/1461444809342058
Li, Q., Smith, P.K., Cross, D.: Research into cyberbullying: context. In: Li, Q., Cross, D., Smith, P. (eds.) Cyberbullying in the Global Playground: Research from International Perspectives, pp. 3–12. Wiley-Blackwell, Chichester (2012)
Lin, H., Qiu, L.: Sharing emotion on Facebook: network size, density, and individual motivation. In: Proceedings of the 2012 Annual Conference Extended Abstracts on Human Factors in Computing Systems (CHI 2012) (2012)
Livingstone, S., Haddon, L., Görzig, A., Ólafsson, K.: Risks and Safety on the Internet: The Perspective of European Children. Full Findings. EU Kids Online, LSE, London (2011)
Macias, W., Hilyard, K., Freimuth, V.: Blog functions as risk and crisis communication during Hurricane Katrina. J. Comput.-Mediat. Commun. **15**(1), 1–31 (2009). doi:10.1111/j.1083-6101.2009.01490.x
Madden, M.: Older adults and social media. Social networking use among those ages 50 and older nearly doubled over the past year. Pew Internet and American Life Project. Retrieved from http://pewinternet.org/~/media//Files/Reports/2010/Pew%20Internet%20-%20Older%20Adults%20and%20Social%20Media.pdf (2010)
Marder, B., Joinson, A., Shankar, A., Thirlaway, K.: Strength matters: self-presentation to the strongest audience rather than lowest common denominator when faced with multiple audiences in social network sites. Comput. Hum. Behav. **61**, 56–62 (2016). doi:10.1016/j.chb.2016.03.005
McAndrew, F.T., Jeong, H.S.: Who does what on Facebook? Age, sex, and relationship status as predictors of Facebook use. Comput. Hum. Behav. **28**(6), 2359–2365 (2012). doi:10.1016/j.chb.2012.07.007

Mehdizadeh, S.: Self-presentation 2.0: Narcissism and self-esteem on Facebook. Cyberpsychol. Behav. Soc. Netw. **13**(4), 357–364 (2010). doi:10.1089/cyber.2009.0257

McKenna, K.Y.A., Bargh, J.A.: Plan 9 from cyberspace: the implications of the internet for personality and social psychology. Personal. Soc. Psychol. Rev. **4**(1), 57–75 (2000). doi:10.1207/S15327957PSPR0401_6

Neubaum, G., Krämer, N.C.: My friends right next to me: a laboratory investigation on predictors and consequences of experiencing social closeness on social networking sites. Cyberpsychol. Behav. **18**, 443–449 (2015). doi:10.1089/cyber.2014.0613

Neubaum, G., Rösner, L., Rosenthal-von der Pütten, A.M., Krämer, N.C.: Psychosocial functions of social media usage in a disaster situation: a multi-methodological approach. Comput. Hum. Behav. **34**, 28–38 (2012). doi:10.1016/j.chb.2014.01.021

Nosko, A., Wood, E., Molema, S.: All about me: disclosure in online social networking profiles: the case of FACEBOOK^{TM}. Comput. Hum. Behav. **26**(3), 406–418 (2010). doi:10.1016/j.chb.2009.11.012

Organisation for Economic Co-operation and Development: Participative web and user-created content: Web 2.0, wikis and social networking. OECD, London. Retrieved from: http://www.biac.org/members/iccp/mtg/2008-06-seoul-min/9307031E.pdf (2007)

O'Reilly, T.: What Is Web 2.0 - Design Patterns and Business Models for the Next Generation of Software. Retrieved from http://oreilly.com/pub/a/web2/archive/what-is-web-20.html?page=1 (2005)

Ortega, R., Elipe, P., Mora-Merchán, J.A., Calmaestra, J., Vega, E.: The emotional impact on victims of traditional bullying and cyberbullying. A study of Spanish adolescents. Z. Psychol./J. Psychol. **217**(4), 197–204 (2009). doi:10.1027/0044-3409.217.4.197

Pavot, W., Diener, E.: Review of the satisfaction with life scale. Psychol. Assess. **5**(2), 164–172 (1993). doi:10.1037/1040-3590.5.2.164

Pawlikowski, M., Altstötter-Gleich, C., Brand, M.: Validation and psychometric properties of a short version of Young's Internet Addiction Test. Comput. Hum. Behav. **29**(3), 1212–1223 (2013). doi:10.1016/j.chb.2012.10.014

Pennebaker, J.W. (ed.): Emotion, Disclosure, and Health. American Psychological Association, Washington, DC (1995)

Pennebaker, J.W., Chung, C.K.: Expressive writing and its links to mental and physical health. In: Friedman, H.S. (ed.) Oxford Handbook of Health Psychology, pp. 417–437. Oxford University Press, New York (2011)

Pew Research Center: Social Networking Use. Retrieved from http://www.pewresearch.org/data-trend/media-and-technology/social-networking-use/ (2015)

Putnam, R.: Bowling Alone: The Collapse and Revival of American Community. Simon and Schuster, New York (2000)

Reich, S., Vorderer, P.: Individual differences in need to belong in users of social networking sites. In: Moy, P. (ed.) Communication and Community, pp. 129–148. Hampton Press, New York (2013)

Reicher, S.D., Spears, R., Postmes, T.: A social identity model of deindividuation phenomena. Eur. Rev. Soc. Psychol. **6**(1), 161–198 (1995). doi:10.1080/14792779443000049

Rice, R.E.: Mediated group communication. In: Rice, R.E. (ed.) The New Media: Communication, Research, and Technology, pp. 129–156. Sage, Beverly Hills (1984)

Rice, R.E., Case, D.: Electronic message systems in the University: a description of use and utility. J. Commun. **33**(1), 131–152 (1983). doi:10.1111/j.1460-2466.1983.tb02380.x

Rimé, B., Mesquita, B., Philippot, P., Boca, S.: Beyond the emotional event: six studies on the social sharing of emotion. Cognit. Emot. **5**(5–6), 435–466 (1991). doi:10.1080/02699939108411052

Ritzer, B., Jurgenson, N.: Production, consumption, prosumption. The nature of capitalism in the age of the digital "prosumer". J. Consum. Cult. **10**(1), 13–36 (2010). doi:10.1177/1469540509354673

Rosenberg, M.: Society and the Adolescent Self-Image. Princeton University Press, Princeton (1965)

Ross, C., Orr, E.S., Sisic, M., Arseneault, J.M., Simmering, M.G., Orr, R.R.: Personality and motivations associated with Facebook use. Comput. Hum. Behav. **25**(2), 578–586 (2009). doi:10.1016/j.chb.2008.12.024

Schutz, W.C.: FIRO: A Three Dimensional Theory of Interpersonal Behavior. Holt, Rinehart and Winston, New York (1966)

Short, J., Williams, E., Christie, B.: The Social Psychology of Telecommunications. Wiley, London (1976)

Siegel, J., Dubrovsky, V., Kiesler, S., Mcguire, T.W.: Group processes in computer-mediated communication. Organ. Behav. Hum. Decis. **37**(2), 157–187 (1986). doi:10.1016/0749-5978(86)90050-6

Slater, M.D.: Reinforcing spirals: the mutual influence of media selectivity and media effects and their impact on individual behavior and social identity. Commun. Theor. **17**(3), 281–303 (2007). doi:10.1111/j.1468-2885.2007.00296.x

Sproull, L., Kiesler, S.: Reducing social context cues: electronic mail in organizational communication. Manag. Sci. **32**(11), 1492–1512 (1986). doi:10.1287/mnsc.32.11.1492

Statistic Brain: Twitter Statistics. Retrieved from http://www.statisticbrain.com/twitter-statistics/ (2015)

Sumter, S.R., Baumgartner, S.E., Valkenburg, P.M., Peter, J.: Developmental trajectories of peer victimization: off-line and online experiences during adolescence. J. Adolesc. Health **50**(6), 607–613 (2012). doi:10.1016/j.jadohealth.2011.10.251

Sundar, S.S., Oeldorf-Hirsch, A., Nussbaum, J.F., Behr, R.A.: Retirees on Facebook: Can online social networking enhance their health and wellness? In: Proceedings of the 2011 Annual Conference Extended Abstracts on Human Factors in Computing Systems (CHI EA'11), pp. 2287–2292 (2011)

Tamir, D.I., Mitchell, J.P.: Disclosing information about the self is intrinsically rewarding. Proc. Natl. Acad. Sci. USA **109**(21), 8038–8043 (2012). doi:10.1073/pnas.1202129109

Thelwall, M., Wilkinson, D., Uppal, S.: Data mining emotion in social network communication: gender differences in MySpace. J. Am. Soc. Inf. Sci. Technol. **61**(1), 190–199 (2010). doi:10.1002/asi.21180

Thiel, S.M.: "IM Me" Identity construction and gender negotiation in the world of adolescent girls and instant messaging. In: Mazzarella, S.R. (ed.) Girl Wide Web. Girls, the Internet, and the Negotiation of Identity, pp. 179–201. Peter Lang Publishing, New York (2005)

Tidwell, L.C., Walther, J.B.: Computer-mediated communication effects on disclosure, impressions, and interpersonal evaluations: getting to know one another a bit at a time. Hum. Commun. Res. **28**(3), 317–348 (2002). doi:10.1111/j.1468-2958.2002.tb00811.x

Tokunaga, R.S.: Friend me or you'll strain us: understanding negative events that occur over social networking sites. Cyberpsychol. Behav. Soc. Netw. **14**(7–8), 425–432 (2011). doi:10.1089/cyber.2010.0140

Trepte, S., Reinecke, L.: The reciprocal effects of social network site use and the disposition for self-disclosure: a longitudinal study. Comput. Hum. Behav. **29**(3), 1102–1112 (2013). doi:10.1016/j.chb.2012.10.002

Tufekci, Z.: Can you see me now? Audience and disclosure regulation in online social network sites. Bull. Sci. Technol. Soc. **28**(1), 20–36 (2008). doi:10.1177/0270467607311484

Utz, S., Tanis, M., Vermeulen, I.: It is all about being popular: the effects of need for popularity on social network site use. Cyberpsychol. Behav. Soc. Netw. **15**(1), 37–42 (2012). doi:10.1089/cyber.2010.0651

Valenzuela, S., Park, N., Kee, K.F.: Is there social capital in a social network site?: Facebook use and college students' life satisfaction, trust, and participation. J. Comput.-Mediat. Commun. **14**(4), 875–901 (2009). doi:10.1111/j.1083-6101.2009.01474.x

Valkenburg, P.M., Peter, J.: The differential susceptibility to media effects model. J. Commun. **63**(2), 221–243 (2013). doi:10.1111/jcom.12024

Valkenburg, P.M., Peter, J., Schouten, A.P.: Friend networking sites and their relationship to adolescents' well-being and social self-esteem. Cyberpsychol. Behav. **9**(5), 584–590 (2006). doi:10.1089/cpb.2006.9.584

van den Eijnden R.J.J.M., Meerkerk, G.-J., Vermulst, A.A., Spijkerman, R., Engels, R.C.M.E.: Online communication, compulsive Internet use, and psychosocial well-being among adolescents: a longitudinal study. Dev. Psychol. **44**(3), 655–665 (2008). doi:10.1037/0012-1649.44.3.655

Walther, J.B.: Interpersonal effects in computer-mediated interaction: a relational perspective. Commun. Res. **19**(1), 52–90 (1992). doi:10.1177/009365092019001003

Walther, J.B.: Computer-mediated communication: impersonal, interpersonal, and hyperpersonal interaction. Commun. Res. **23**(1), 3–43 (1996). doi:10.1177/009365096023001001

Walther, J.B.: Selective self-presentation in computer-mediated communication: hyperpersonal dimensions of technology, language, and cognition. Comput. Hum. Behav. **23**(5), 2538–2557 (2007). doi:10.1016/j.chb.2006.05.002

Walther, J.B.: Theories of computer-mediated communication and interpersonal relations. In: Knapp, M.L., Daly, J.A. (eds.) The Handbook of Interpersonal Communication, 4th edn., pp. 443–479. Sage, Thousand Oaks (2011)

Wellman, B., Wortley, S.: Different strokes from different folks: community ties and social support. Am. J. Sociol. **96**(3), 558–588 (1990). doi:10.1086/229572

Willard, N.: Cyberbullying and Cyberthreats: Responding to the Challenge of Online Social Aggression, Threats, and Distress. Research Press, Champaign (2007)

Winter, S., Neubaum, G., Eimler, S.C., Gordon, V., Theil, J., Herrmann, J., Meinert, J., Krämer, N.C.: Another brick in the Facebook wall – How personality traits relate to the content of status updates. Comput. Hum. Behav. **34**, 194–202 (2014). doi:10.1016/j.chb.2014.01.048

Wise, K., Alhabash, S., Park, H.: Emotional responses during social information seeking on Facebook. Cyberpsychol. Behav. Soc. Netw. **13**(5), 555–562 (2010). doi:10.1089/cyber.2009.0365

YouTube Statistics: Retrieved from http://www.youtube.com/yt/press/statistics.html (2016)

Zywica, J., Danowski, J.: The faces of facebookers: investigating social enhancement and social compensation hypotheses; Predicting Facebook and offline popularity from sociability and self-esteem, and mapping the meanings of popularity with semantic networks. J. Comput.-Mediat. Commun. **14**(1), 1–34 (2008). doi:10.1111/j.1083-6101.2008.01429.x

Chapter 3
The Psychology of (Cyber)Emotions

Arvid Kappas

3.1 Introduction

It is somewhat of a cliché that researchers on emotions cannot agree on a common definition of the *emotion* concept (see Cornelius 1996). However, there is nothing special or unusual about *emotion* as a psychological construct, with regard to the difficulties of clearly separating it from other related processes or concepts. Non-scientific folk-representations of many key psychological concepts, such as *behavior*, *desire*, or *attitudes*, also do not easily map to clear-cut scientific definitions. And the scientific definitions of these other concepts are either quite divergent from each other, or, if they are only implicit, betray vast differences. The term *cognition,* for example, is used by some authors to mean *thinking*, and for others is basically synonymous with *any* type of *information processing* in the brain—a considerable difference indeed (Kappas 2006).

Such divergences between scientific and pre-scientific emotion concepts, as well as some exceptions from the current modal understanding of emotions in the scientific community, lead occasionally to confusion in understanding how emotion scientists conceive of affective processes, how they measure them, and how they separate them from related concepts. As this cannot be discussed in each of the following chapters in the present volume, these issues shall be briefly discussed here. Unfortunately, it is not possible to circumvent the conceptual quagmire by creating *ex-novo* arbitrary definitions that only deal with objective observables, because all emotion researchers somehow have to deal with the fact that one of the important aspects of emotions is the subjective component—how something *feels*. At the same time, it is now largely accepted that emotion should refer not only to the feeling component, but also to behaviors, such as expressive behavior in the face or the

A. Kappas (✉)
Jacobs University Bremen, Campus Ring 1, 28759 Bremen, Germany
e-mail: a.kappas@jacobs-university.de

J.A. Hołyst (ed.), *Cyberemotions*, Understanding Complex Systems,
DOI 10.1007/978-3-319-43639-5_3

voice. Furthermore, bodily changes, such as changes in sweating or the frequency with which the heart beats, as well as changes in brain activation are also associated with emotions. At the outset of this book on CYBEREMOTIONS it appears relevant to highlight some of the difficulties in defining and measuring emotions. I will then present a proposal to define the term CYBEREMOTIONS.

While most treatises on the history of the scientific study of emotions typically anchor their narrative at the publication of Charles Darwin's *The Expression of the Emotions in Man and Animals* (Darwin 1872), there have been many discussions in philosophy, but also in physiology and related disciplines predating Darwin's work. In fact, already at the end of the nineteenth century there were so many texts trying to capture the essence of emotions to lead William James in 1890 to proclaim exasperatedly

> But as far as 'scientific psychology' of the emotions goes, I may have been surfeited by too much reading of classic works on the subject, but I should as lief read verbal descriptions of the shapes of the rocks on a New Hampshire farm as toil through them again. They give one nowhere a central point of view, or a deductive or generative principle. They distinguish and refine and specify in infinitum without ever getting on to another logical level. (p. 448)

Since the seminal publications of William James that proposed a new theory of emotions, there has been a waxing and waning of interest in emotions in psychology and neighboring disciplines, such as sociology (see von Scheve and von Luede 2005). A renaissance could be seen in the 1960s, starting with the founding works on appraisal theory that finally dealt in detail with the causation of emotion (Arnold 1960; e.g., Lazarus et al. 1970; see Scherer et al. 2001), the interest in the role of arousal for the sensation of emotions (Schachter and Singer 1962), and the role of facial activity (e.g., Ekman 1972). This new wave of emotion research triggered several parallel lines of inquiry which led to a further proliferation of emotion concepts. In 1981, Kleinginna and Kleinginna published an attempt to take stock of over 100 definitions of emotions (including so-called 'critical statements'). Based on primary and secondary emphases of the definitions, they identified 11 different categories of definitions and finally attempted to provide a 'consensual definition'

> Emotion is a complex set of interactions among subjective and objective factors, mediated by neural/hormonal systems, which can (a) give rise to affective experiences such as feelings of arousal, pleasure/displeasure; (b) generate cognitive processes such as emotionally relevant perceptual effects, appraisals, labeling processes; (c) activate widespread physiological adjustments to the arousing conditions; and (d) lead to behavior that is often, but not always, expressive, goal-directed, and adaptive. (p. 355)

As is obvious, this definition is rather wide, and perhaps not fully satisfying, because it does not really become clear what is or what is not an emotion. In other words, the definition did not permit to identify whether changes in any of the components involved in emotion would constitute an instance of an emotion or not. In the following years, a sense of frustration could be felt among emotion researchers and various attempts were made at increasing the cohesion of emotion concepts in research and theory. For example, the founding of the *International Society for Research on Emotion* in 1984, was arguably an attempt to increase communication between theorists and to facilitate the development of a greater consensus. Despite continuing discussions highlighting differences in definitions, I have argued in the past that the core concepts of emotion would be shared by a majority of researchers in the field when focusing on a process view (Kappas 2002). However, more recently a qualitative study interviewing key researchers in the field demonstrated that there are still considerable differences regarding how they understand what emotions are, how they come about, what functions they serve, etc. (Izard 2010; see also Kappas 2013; Ekman 2016).

When talking about emotions in individuals, I will refer to *responses to external or internal stimuli in the central nervous system (brain) or peripheral nervous system, that can manifest in changes in physiological activity, behavior, including expression, as well as psychological changes, such as changes in the motivation and readiness to do certain things, such as approaching or avoiding something, and associated changes in feeling state.* The story is more complicated because these changes are associated with regulatory processes that might terminate itself (Kappas 2013), and are typically explicitly or implicitly social, that is they are interindividual processes. However, for the initial discussion I will focus on the first element of the definition—emotions as responses. As this is of central importance for all CYBEREMOTIONS studies and models, the question of how these responses come about is particularly important. Thus, I will discuss the elicitation of emotions.

3.2 The Elicitation of Emotions: The Role of Cognition

The classical discussions of how emotions are elicited are muddled by the aforementioned definition issue. If the term *emotions* is used synonymously with feeling state, then someone might propose that emotion is elicited by the perception of arousal states and context information, as in the classical model by Schachter and Singer (1962). However, for most current emotion theorists the arousal is already an important element in the emotion cascade and the important question becomes, why was there a change in the arousal state in the first place? Why did the brain modulate activation levels? The Schachter and Singer theory cannot answer this (see Reisenzein 1983). It is here that the concept of *appraisal* comes into play. According to *appraisal theory*, external or internal events are appraised in the light of their implications for current needs and goals. This process can be very fast and automatic, but it may also happen slowly as a function of conscious reflection and thought. In most cases, these different processes will happen at the same time, as the brain has the capacity for parallel processing of emotion relevant stimuli in circuits that differ in complexity due to their phylogenetic origin (see Leventhal and Scherer 1987; Kappas 2001; Scherer 2001).

While some stimuli might be biologically hardwired elicitors, or at least prepared, such as certain facial expressions, rapidly approaching objects, intense stimulation in any sensation channel, text on a computer screen certainly is not. As the study of CYBEREMOTIONS oftentimes involves people responding by typing text to what other people wrote. it is important to reflect briefly on how the word on a page or a screen might change the timing of heart beats or a subjective feeling state. This question is not trivial and it goes to the heart of appraisal theory, as to how emotions spread and organized themselves over the Internet. Reading a rejection letter is likely to provoke an emotion in *me*, provided it is a letter that relates to *me* being rejected—less so if it is about *you* being rejected. Particularly, if I do not know you. However, if you are my spouse, or my enemy, I might have a reaction! How can this be explained? The early key theorists of appraisal, Magda Arnold and Richard Lazarus emphasized that *personal relevance* is a key component of emotional appraisal. The assumption is that an apple might be emotionally relevant if you are hungry, but not necessarily if you are not. The cake might be emotionally relevant if you like it, but it makes a difference if you are on a diet or not, even if you would like it. Again, this is not trivial, because this proposed mechanism is the explanation, why there is no fixed relationship between a stimulus **X** and a response **Y**. The relationship is dependent on the appraisal which might differ from person **a** to person **b**, or within person **a** at different times as a function of a bodily or psychological state.

To go one step further—assuming that a particular emotional cascade has been triggered by the appraisal of stimulus **X**, this does not imply that a particular behavior *must* follow. A fixed relationship between a stimulus and a behavior is, according to the definitions of psychologists, either an unconditional reflex, or a conditioned response. Emotions are different. Emotions *bias* the likelihood of the occurrence of a behavior. Subjectively, they inform us and change the way we might look at the situation using basic modulation processes in the brain, but just because I want to run away does not mean that I have to. Scherer argued that this decoupling of stimulus and response is one of the key evolutionary advances that emotions provide to other means of adapting to changes in the environment (see Leventhal and Scherer 1987). In other words, flexibility in behavior, while providing a guiding function, allows for the adaptation to a variety of situations. We can feel the tension when seeing a scary animal in the movies, but we are not forced to run out of the theatre. But because the potential danger posed by the animal has been registered at some level of the brain, certain adaptive changes have likely been triggered in the organism and so we might experience symptoms, such as a change in heart rate. It has been argued that this is the reason that we enjoy scary movies, or roller coasters. Part of our brain knows that we are safe, while another part triggers automatic responses to danger and we can enjoy the rush the body provides, while being safe at the same time.

However, what is arguably a great selection success for evolution, also can be a headache for the researcher. Flexibility and decoupling mean that one cannot conclude that the presence or absence of a behavior is necessarily indicative of the presence of an emotion. For example, I can write at this very moment "I am sad" when actually I am not sad. This means you cannot take the phrase "I am sad" in the previous sentence to take that the author of the chapter was sad. In this special case I provided you with an explanation of my actions, so you know the speech act was not really reflecting my feeling state. What if I wrote "I am angry right now"—am

I or am I not? You cannot tell. Without the proper context of my situation at this very moment you cannot tell. But would there have been a way to know whether I feel angry, had you been here while I wrote the sentence—or could you, even after the fact? This relates to the issue that emotions are characterized by responses in multiple response systems, such as changes in the activation of the autonomic nervous system (ANS). There might be changes in facial behavior, or in the force with which I hit the keys of the keyboard. All of these differ with regard to the degree to which I can voluntarily influence them. Clearly, writing a particular text is largely volitional. The other end of the spectrum consists of changes in pupillary diameter, or salivation in my mouth. These are difficult to control voluntarily. Thus, one could argue that if one wants to know how someone feels, one should not ask the person, but instead measure pupillary diameter. Unfortunately, the relationship between these different levels of response is not fixed either. The next section will discuss this in more detail. Furthermore, what if I cannot have access to these "objective measures" because I am interested in how emotions spread on the Internet and all I have is what people write? I want to raise three different possibilities how to deal with the uncertainty whether written text really is a reliable indicator of affective processes:

1. *EXPERT JUDGEMENTS OF TEXT:* If I make a conceptual distinction between what people write and what they feel, I can at least measure whether emotional content is embedded in the text regardless of whether this reflects necessarily at that moment how a person felt. This is the approach of *sentiment mining* that explicitly focuses on the text. The criterion here is whether experts or large numbers of judges agree that "I am sad" is negative (they do). Thus, I can state that "I am sad" is negative (see Chaps. 6 and 7). Furthermore, if I can show that messages that contain negativity sent by one person lead to the production of other messages that contain negativity sent by other persons, I can objectively describe how these spreading, propagation, or damping phenomena develop as a function of structural and temporal features—this is what many of the chapters in this book demonstrate. However, it is important that the propagation of text features alone is no proof that the relay nodes of the network, the human readers and writers, are actually experiencing different affective states.
2. *LANGUAGE STYLE:* There are aspects of written texts that are hardly under volitional control and thus are reliable indicators of person characteristics, including, potentially, affective state. One of the most interesting developments in recent years, in this context, was the development of the Linguistic Inquiry and Word Count software (LIWC; e.g., Tausczik and Pennebaker 2010; see also Pennebaker 2011). What is interesting about the LIWC is that it does not only look for the usage of nouns that are generally thought to convey meaning, such as in *sentiment mining*, but also in the usage of function words, such as pronouns or articles, or more generally, not only language content, but *language style*. It is assumed that language style is largely unreflected and has, in the meanwhile, been shown to correlate to certain emotion related states, such as depression (Pennebaker 2011). LIWC has been used in a variety of studies and certain

patterns were found to correlate with person characteristics. Often these make sense, but they are difficult to predict. In other words, large studies validating LIWC for non-obvious associations between affective states and language style are still required—in these cases there needs to be an assessment of ground truth and matching language style (see also point 3).

3. People can be studied in the laboratory and different objective measures can be taken while they read and write. Then the text can be related to the other indicators of affective state in the writer (sender) or the reader (receiver). This has been done within the CYBEREMOTIONS project and relevant data are presented in the Chap. 5 in the present volume. This experimental approach is critical for the establishment of ground truth, whether for statistical validation, or for training algorithms. Once this has been done, one can estimate the reliability of emotion estimates in the wild—for example in actual Internet conversations.

3.3 The Relationship of Emotions, Attitudes, and Opinions in the Appraisal Context

Several recent theorists have proposed related appraisal models that vary regarding how many dimensions of appraisal are required to account for the vast differentiation of emotional states that we know. An overview can be found in Scherer et al. (2001). Appraisal theory also clarifies how other psychological constructs, such as opinions or attitudes, relate to the elicitation of emotions. These more static representations about how a person relates to certain objects or states are important elements in the appraisal process. For example, I might *like* a particular author and so I respond positive to a gift of a book by that author, if I did not own it before. If I do not like the author, then the emotional response is different. Thus, the elicitation of the emotion relates to a situation where a particular person gifts an object to which I have a particular relationship. It may sound somewhat complicated to the non-psychologist, but it is not. In short, the idea is that we are very good and efficient in interpreting the meaning of events in terms of appraisals, in the light of attitudes and opinions we have developed regarding many concrete and abstract objects, events, or notions. If, in general, I tend to like vanilla ice cream—I have a positive attitude towards vanilla ice cream. In the actual situation when I am eating vanilla ice cream and I am asked how I feel, chances are that I feel good because I am eating something that I like. If I have a negative opinion of a particular politician, I am inclined not to agree with statements s/he makes. Listening to the politician then might create a negative emotion, such as anger, because I have a negative opinion. In both of these examples, the more static relationship between objects and how we tend to evaluate them is described as "attitude" or "opinion"—the appraisal of an ongoing situation and the following emotional state is typically in the light of the attitude or opinion we have. Of course, the attitude does not predict all responses, for example, we might feel positively, if something to which we have a negative relationship is not evaluated as negative as we had assumed. The appraisal in all of these cases relates to the on-the-spot process of evaluation in real time.

In fact, the argument has been made that we are so good at making these interpretations that they are in fact not causal to emotion, but rather explanations for us to make sense of the world (Parkinson 1997, 2012; but see Moors 2013). However, this relates also to a relatively specific interpretation of what cognitive processes are required to appraise a particular event or situation. While it may sound tautological—if different stimuli lead to different responses, then there must have been a distinction at the brain level, whether we would be aware of it or not. It is also an interesting corollary that, if the assumptions of appraisal theory are correct, people can (and apparently will) conclude from an emotional reaction what attitudes or opinions someone has of a particular event or situation (Hareli and Hess 2010).

In summary, it is assumed that emotional responses to Internet content are elicited via information processing that contemporary psychologists would refer to as *appraisal*. This process is likely to involve aspects to which we have conscious access, but it can also be automatic and fast, because of schemata that we have acquired (see Leventhal and Scherer 1987). In all cases, it is the *meaning* of Internet contents that elicits the cascade of emotional responses. The cascade involves a variety of levels, such as subjective experience, physiological responses in the brain and the body at large, changes in ongoing behavior, including expressive behavior, as well as changes in action readiness. One of the big challenges to emotion researchers is that the cohesion between all of these emotional components is low. Because of this, it is very difficult to make an inverse inference from the presence of one of the components to the presence of a particular emotion. This is discussed in the next section.

3.4 Cohesion of Emotional Components is Low

Arguably the most influential emotion theorist in the twentieth century is Paul Ekman. His research has had an immense impact on methods and concepts and is still today very important in the context of the new applications of emotion theory in the context of affective computing. It is well known that Ekman posits the existence of a small number of basic emotions that are characterized by several criteria, including a typical and universal facial expression. The criteria are, according to Ekman and Cordaro (2011, p. 365):

1. Distinctive universal signals.
2. Distinctive physiology.
3. Automatic appraisal.
4. Distinctive universals in antecedent events.
5. Presence in other primates.
6. Capable of quick onset.
7. Can be of brief duration.

8. Unbidden occurrence.
9. Distinctive thoughts, memories, and images.
10. Distinctive subjective experience.
11. Refractory period filters information available to what supports the emotion.
12. Target of emotion unconstrained.
13. The emotion can be enacted in either a constructive or destructive fashion.

According to Ekman, these criteria are fulfilled for seven emotions: Anger, Contempt, Disgust, Fear, Happiness, Sadness, and Surprise. Distinctiveness and stereotypy of these seven emotions, as proposed by Ekman, are derived from theoretical notions put forward by his mentor Silvan Tomkins in the 1960s. Tomkins had argued that there are eight primary affects, each of which is controlled by a unique sub-cortical affect program (e.g., Tomkins and McCarter 1964). One can imagine these programs like packages that involve all components. Once the program is triggered a cascade will start that is relatively similar between people and also within individuals. This is the theoretical backdrop to the assumption that measurement of one or two of the components would allow predictions regarding all other components because they are part of the affect program. Fifty years later, the same idea is prevalent. In applied contexts, such as in affective computing, the idea that measuring one component, for example facial behavior, is necessarily a reliable indicator for another component, for example how someone feels, is well accepted. Of course, this view is also consistent with naive or folk-theories that would hold that someone who smiles is happy, or that someone who cries is sad. Unfortunately, this is, apparently, not how things are in reality!

With regard to psychophysiological responses, there had been a waxing and waning of beliefs, whether specific emotions would have clear and discrete patterns of activation. While William James (1890) held that there should be such patterning, because he believed that the subjective experience, the feeling, is based on the perception of different patterns of bodily changes, a problem to identify such patterns led researchers in the twentieth century eventually to the notion that physiological responses are not patterned and that many emotions (and possibly also non-emotional states) share a non-specific autonomic arousal (see also Reisenzein 1983). In recent years, the notion of general and unspecific arousal has again been challenged (e.g., Ekman et al. 1983), but the problem is that empirically no clear patterns seemed to emerge either across paradigms and researchers [see Rimé et al. (1990), Philippot (Peripheral differentiation of emotion in bodily sensations, physiological changes, and social schemata. Unpublished Ph.D. Thesis. Université catholique de Louvain, 1992)]. More recently, systematic investigations showed that cohesion between different components was in fact only low (e.g., Mauss and Robinson 2009; Reisenzein et al. 2013). In other words, the presence of a particular facial expression is not a reliable predictor for how that person feels, or, how a person reports to feel is not a reliable predictor for changes in physiological activity.

Why cohesion between affective components could be low, is a difficult question and cannot be answered today. Ekman, had already early on (e.g., Ekman 1972) proposed how cultural display rules could interfere with expressive displays and thus decouple expression and feeling (and potentially physiological responses). However, this "neuro-cultural theory" predicted that in social isolation, i.e., when people were alone in a room, perfect cohesion between the components was to be expected. Fridlund (1991) demonstrated that this was not the case and Hess et al. (1995) further demonstrated that neither the neuro-cultural theory, nor Fridlund's counter position "behavioral ecology" could account for the empirical findings (see Kappas 2003). There is presently no satisfying theory that understands the relationship of feelings and expression in a way that could predict *who* shows *what, when,* and to *whom.* There is an increasing number of authors who argue against a program or package concept of how emotional components relate to each other, but rather a dynamical systems view that sees periods of increased correlation or coactivity across components an emergent phenomenon (see Kappas 2013; but also Scherer 2001).

The consequence is that today, the measurement of affective responses is typically done in a multimodal fashion, where self-reported subjective experience, changes in physiological activity, and expression are all measured and related to each other. A gold standard does not exist (see also Kappas 2003; Chap. 5).

A systematic analysis of previous research led Mauss and Robinson (2009) to the conclusion that the issue of low cohesion might be exacerbated by the desire of researchers to look for discrete emotional states. If one would consider a low-dimensional affective space as a theoretical background, cohesion could be considered much less of a problem. Thus, the following section compares the dimensional approach with the discrete emotion approach.

3.5 Discrete Emotions vs. Dimensional Approaches to Emotional States

There are numerous emotion theories with a specific number of basic emotions or primary affects. Some propose 6, 7, 8, 10, or even higher numbers of such discrete states. But which is the right number? How many emotions are there (see Scherer 2001; Ellsworth 2014)? A different way to think of emotions is to see whether there are underlying dimensions that distinguish different affective states. From Wilhelm Wundt in the nineteenth century, to modern theories, there have been attempts to characterize emotions as points in a two- or three-dimensional space.

The most important dimension in affective space is the hedonic valence. For some authors this is the essence of emotions. All emotional states are either positive or negative. While some authors, such as Ekman (see Ekman and Cordaro 2011) count surprise as an emotional state and surprise is, at least, initially not clearly positive or negative, there are really few exceptions to the notion that all emotions may fall somewhere along a dimension of pleasantness.

The second dimension is arousal or activation. Any affective state can be characterized as more or less aroused. Together, these two dimensions have been

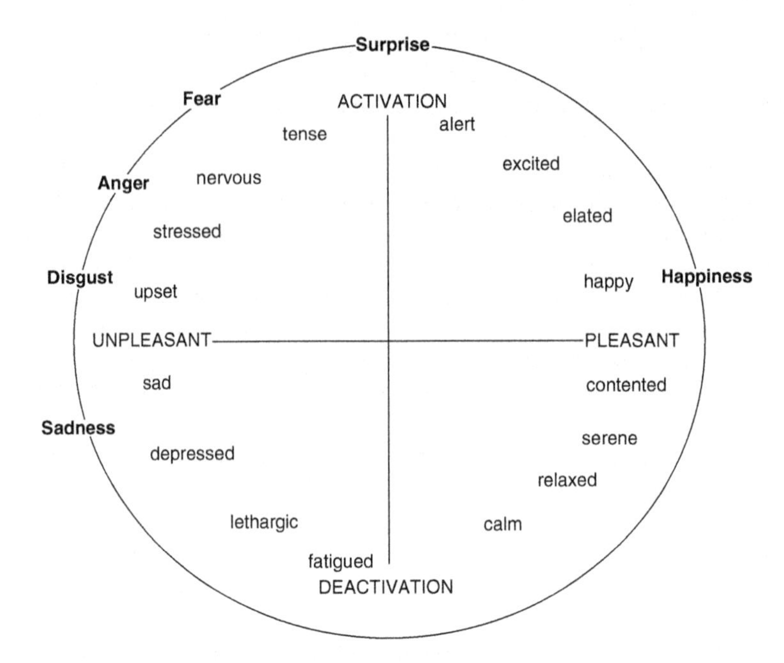

Fig. 3.1 Core Affect with dimensions of pleasure-displeasure (*horizontal axis*) and degree of arousal (*vertical axis*) from Russell and Barrett (1999, p. 808). Reproduced with permission from the American Psychological Association (APA)

shown to distinguish between emotion words but also facial expressions (see Fig. 3.1). In fact, "basic emotions", as well as terms that are typically seen as being emotional but not part of any of the "big" emotion theories, can be well placed in such a two dimensional space. *Core Affect Theory* (e.g., Russell 2003) assumes that these two dimensions help to define the primitives of all affective states.

As mentioned in the previous section, studies looking at the coherence of emotional components favor a dimensional approach. Certain behavioral measures, such as the activation of the facial muscle *Corrugator Supercilii*, that pulls the brows together and down, seem to correlate quite well with the subjectively rated valence of various objects, such as images or texts (Cacioppo et al. 1986; Larsen et al. 2003), or, as we have shown in my laboratory, posts from Internet forums (e.g., Kappas et al. 2010, see Chap. 5). According to theorists, such as Russell (2003), affective experience is constructed out of primitives in a particular context. This is a very elegant way to account for the subjective existence of emotional categories, which are shaped by culture and social context, but an underlying mechanism that requires only few dimensions to distinguish an infinite number of affective states.

While *Core Affect Theory* discusses mainly two dimensions, in line with classical approaches of psychological construction, some theories postulate a third dimension of power or dominance (see Bradley and Lang 1994). To clarify: One might be afraid or angry, with both states relating to negative states with high arousal. A third dimension of power would clearly help to differentiate both states—fear would be a low power and anger a high power state. In fact, appraisal theory (e.g., Scherer 2001) requires only a few more dimensions to differentiate all imaginable affective states.

Furthermore, while most dimensional theories consider hedonic valence to be a unidimensional construct, there are also some theorists who hold that there is reason to consider positive and negative affect independently from one another. Here the assumption would be that the object of an emotion might be positive and negative at the same time. For example eating a wonderful chocolate cake while being on a diet would involve positive *and* negative feelings at the same time. In fact, there is some evidence from current neuroscience that suggests that there is not a single structure of the brain that computes how positive or negative something is, but instead, that how positive something is might be processed in a different brain locations than something negative and that, more general, there are distinctive processes for positive and negative throughout the nervous system (Norman et al. 2011). Because of this, some of the approaches discussed in other chapters of these volume, use bivariate measurements of hedonic valence/pleasantness. For example, the SentiStrength program (Thelwall et al. 2010) will provide a value each for positive and negative affective content. However, at this point, there is no convergence in the research community on whether in general it is more useful to consider valence as a univariate or bivariate construct.

To be clear, Cyberemotions are neither necessarily discrete nor do they necessarily require a dimensional model. These are choices the researcher makes when designing studies, defining variables, and developing models. In a way, there are times when both discrete emotions as well as dimensional measurements are both useful at the same time (see also Kappas 2003). However, particularly with view to the dependent measures used in the CYBEREMOTIONS project, which often relates to short, informal, samples of text, it was decided early on that dimensional models would be used. Continuous measurement of facial activity, or of autonomic changes in the laboratory further lends itself to a dimensional approach, as there are no criteria where one emotion would start and another would stop (see Chap. 5). Similarly, when describing in text (see Chaps. 6 and 7) it made much more sense to consider degrees of how positive or negative a post might be rather than using discrete categories. Consider a response in an online interaction: "That sucks!"—this would be clearly negative, but what discrete emotion would fit here? Thus, there are clear and pragmatic reasons to consider a dimensional framework rather than a discrete emotions framework in the context of studying online affect.

3.6 Emotion and Regulation

When it comes to emotions, cause and effect are not always easy to disentangle. In fact, any discussion on emotions that is based on the assumption that an organism tends to be typically in a neutral state, may be affected by an external or internal

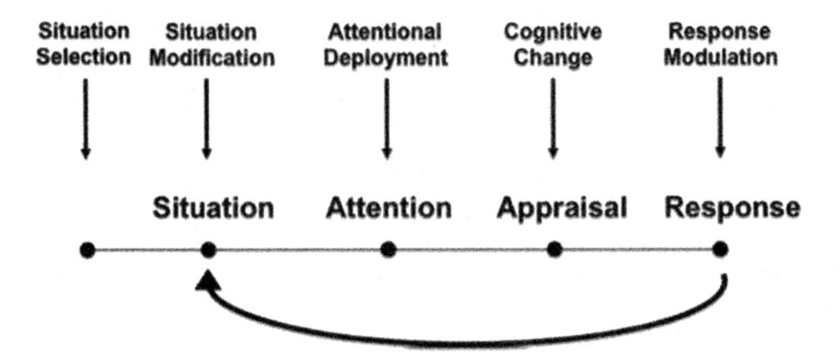

Fig. 3.2 Gross' process model of emotion regulation indicating different points in time at which emotion regulation might interact with ongoing emotions (Gross 2013, p. 360). Reproduced with permission of the American Psychological Association (APA)

stimulus and then recovers back to baseline, does not take into account that emotions are typically self-regulating (Kappas 2011, 2013). Regulation can take place even before any emotional episode—someone may choose to avoid an emotion-eliciting situation, or, contrariwise, go out of their way to put themselves into such a situation. From the moment that an emotion-eliciting stimulus is present, or an expected stimulus is absent, there are a variety of processes possible. Typically, negative states are avoided or down-regulated, while positive states are sought out or maintained. In interpersonal contexts, there is much regulation where, depending on whether empathic processes are involved; similarly, negative and positive states in others are regulated. One might want, for example, to reduce the intensity of a negative state of someone who one cares for—in this context one would try to comfort the other. In turn, one might actually want to increase the negative state of a person one does not like, either because of something the person said or did (in which case there would be a clear link to previous speech acts), or because one does not like the person or something the person represents, such as an outgroup. In this latter scenario the dynamics might be less clear based on an analysis of what was said, because it is a context that moderates the whole interaction (Fig. 3.2).

Schweitzer and García (2010; see also Garcia et al. 2016) present an agent based model of collective emotions in online communities. In this model, arousal serves as a driving function for the communication of emotion. This process can be interpreted in a larger framework of intra- and interpersonal emotion regulation, where the motivation to respond to a situation online, such as one or more posts from one or more others, can be framed as an attempt to regulate the own emotion. The difference to a model that favors a contagion metaphor is that (a) it is oftentimes difficult to understand what the function of emotion contagion might be and (b) a contagion metaphor works very well if the "same" emotion spreads. But this might not necessarily be the case. Instead, depending on the goal of the intervention, a different emotion might be present in the original sender and the responder. A gloating post, might be perceived as negative and arousing and cause anger. Here, it does not make sense to think of a specific emotion spreading over a network, instead, the interpersonal relationship and specific situational contexts are required to interpret why a particular post elicits a particular response. A multilevel emotion regulation framework is here particularly promising as an explanatory context (Kappas 2013).

3.7 A Working Definition of Cyberemotions

I have discussed various theoretical issues relevant to the study of Cyberemotions, such as how emotions in general can or should be defined or measured. However, the concept of Cyberemotions has not yet been formally been introduced. I will conclude this chapter with a proposal for the definition of Cyberemotions.

Over the course of the CYBEREMOTIONS project there were many internal discussions among consortium partners regarding how to define CYBEREMOTIONS. It became evident that CYBEREMOTIONS could refer to the research project itself, a research area, and a psychological/behavioral phenomenon. Here are definitions for the three different usages of the term.

1. (Project) *CYBEREMOTIONS* was a large scale research project, supported by a 7th Framework European Union grant, Theme 3: Science of complex systems for socially intelligent ICT, involving more than 40 researchers. The exact name of the project was *CYBEREMOTIONS: Collective Emotions in Cyberspace*. It was active from 2009 to the final review in the summer of 2013. The present volume involves contributions by most of the partners of the CYBEREMOTIONS consortium. The website of the project cyberemotions.eu will continue to exist for several years (see also the introductory Chap. 1 of this volume).
2. (Research domain) *CYBEREMOTIONS* is a research domain that studies observable and analyzable phenomena related to any means of communication provided by the Internet—such as text, sound, visual, or any combination of these—that are related to emotional processes in individuals or groups. In this sense, many research activities linked to affective processes in the context of the Internet can be seen as belonging to this research domain, independent of the CYBEREMOTIONS project or consortium.
3. (Psychological process) "Cyberemotions" refers to affective processes in social networks, involving mediated communications that are influenced by emotional states of individuals and that, in turn, may lead to the elicitation or modulation of emotional states of individuals in networks and in consequence to state changes in e-communities as a whole. Cyberemotions do not imply homogeneous distributions of a specific affective state in networks. Network characteristics may affect the dynamic properties of Cyberemotions. They are typically, but not exclusively linked to external events and ongoing exchanges also involve off-line communications.

This working definition answers the question whether Cyberemotions differ from other emotions. Clearly, many intrapersonal aspects of any emotional process, whether the affective responses are triggered by a dangerous animal, a book, a movie, or a post in a social network are the same. They relate to psychological processes that are shaped by social and cultural context within biological constraints of the individual and the species. However, the idea is that there are phenomena unique to Cyberemotions that relate to the propagation of emotions in networks, where it makes sense to consider that the resulting processes are not just the sum of individual affective processes, but the product of the interplay of intra-personal and inter-personal processes in networks. Because network properties affect the dynamics, both at the micro-level within the individual and the level of e-communities, it is useful to consider these as processes that are overlapping, but not redundant with offline interpersonal emotions as they have been studied in the past.

Because of this added layer of complexity and the trickle down effects of network structures, phenomena can emerge that relate, for example, to sudden collective actions, both with regard to negative affect ("shit-storms"; Cyberbullying) or positive affect (for example sudden massive outpouring of support in case of disasters), Cyberemotions merit to be considered as distinguishable affective phenomena bridging different levels of organization (see Kappas 2002, 2013).

The present volume will report a plethora of empirical research approaches and different types of models that will highlight the importance of considering Cyberemotions with view to the interaction of large numbers of individuals in Cyberspace. Despite of all the constant changes of technology and platforms, these processes are likely here to stay for a while. Just as psychological and behavioral consequences emerged from the reorganization of humans in larger cities, thousands of years ago, there are consequences of creating high-speed social networks that are truly global in scope and that blend off- and online interactions using a variety of media in a rich tapestry that is blending aspects of interpersonal communication and interaction with mass communication. In this context it is important to understand that the tendency to share emotional experiences in communication is a common phenomenon that predates such technological developments (see Chap. 4).

Understanding these processes requires truly multidisciplinary approaches to emotion research (Kappas 2002), that bridge psychology, social sciences, biology, and physics on the one hand and interact with engineers and computer scientists on the other. There are issues at stake of understanding and predicting emergent emotional responses of individuals and systems, as well as possibly finding ways to intervene to prevent destructive phenomena that not only might interfere with effective communication, but that may even challenge the existence of e-communities and have manifest impact on health and well-being of individuals.

Acknowledgements This work was supported by a European Union grant by the 7th Framework Programme, Theme 3: Science of complex systems for socially intelligent ICT. It is part of the CyberEmotions project (contract 231323).

References

Arnold, M.B.: Emotion and Personality, 2 vols. Columbia University Press, New York (1960)

Bradley, M.M., Lang, P.J.: Measuring emotion: the self-assessment-manikin and the semantic differential. J. Behav. Ther. Exp. Psychiatry **25**(1), 49–59 (1994). doi:10.1016/0005-7916(94)90063-9

Cacioppo, J.T., Petty, R.E., Losch, M.E., Kim, H.S.: Electromyographic activity over facial muscle regions can differentiate the valence and intensity of affective reactions. J. Pers. Soc. Psychol. **50**(2), 260–268 (1986). doi:10.1037//0022-3514.50.2.260

Cornelius, R.R.: The Science of Emotion. Prentice-Hall, Upper Saddle River, NJ (1996)

Darwin, C.: The Expression of the Emotions in Man and Animals. Murray, London (1872)

Ekman, P.: Universals and cultural differences in facial expressions of emotion. In: Cole, J. (ed.) Nebraska Symposium on Motivation, vol. 19, pp. 207–283. University of Nebraska Press, Lincoln, NE (1972)

Ekman, P.: What scientists who study emotions agree about. Perspect. Psychol. Sci. **11**(1), 31–34 (2016). doi:10.1177/1745691615596992

Ekman, P., Cordaro, D.: What is meant by calling emotions basic. Emot. Rev. **3**(4), 364–370 (2011). doi:10.1177/1754073911410740

Ekman, P., Levenson, R.W., Friesen, W.V.: Autonomic nervous system activity distinguishes among emotions. Science **221**(4616), 1208–1210 (1983). doi: 10.1126/science.6612338

Ellsworth, P.C.: Basic emotions and the rocks of New Hampshire. Emot. Rev. **6**(1), 21–26 (2014). doi:10.1177/1754073913494897

Fridlund, A.J.: The sociality of solitary smiles: effects of an implicit audience. J. Pers. Soc. Psychol. **60**(2), 229–240 (1991). doi:10.1037/0022-3514.60.2.229

Garcia, D., Kappas, A., Küster, D., Schweitzer, F.: The dynamics of emotions in online interactions. R. Soc. Open Sci. 3, 160059 (2016). doi:10.1098/rsos.160059

Gross, J.J.: Emotion regulation: taking stock and moving forward. Emotion **13**(3), 359–365 (2013). doi:10.1037/a0032135

Hareli, S., Hess, U.: What emotional reactions can tell us about the nature of others: an appraisal perspective on person perception. Cognit. Emot. **24**(1), 128–140 (2010). doi:10.1080/02699930802613828

Hess, U., Banse, R., Kappas, A.: The intensity of facial expression is determined by underlying affective state and social situation. J. Pers. Soc. Psychol. **69**(2), 280–288 (1995). doi:10.1037/0022-3514.69.2.280

Izard, C.E.: The many meanings/aspects of emotion: definitions, functions, activation, and regulation. Emot. Rev. **2**(4), 363–370 (2010). doi:10.1177/1754073910374661

James, W.: Principles of Psychology, vol. 2. Henry Holt, New York (1890)

Kappas, A.: A metaphor is a metaphor is a metaphor: exorcising the homunculus from appraisal theory. In: Scherer, K.R., Schorr, A., Johnstone, T. (eds.) Appraisal Processes in Emotion: Theory, Methods, Research, pp. 157–172. Oxford University Press, New York (2001)

Kappas, A.: The science of emotion as a multidisciplinary research paradigm. Behav. Process. **60**(2), 85–98 (2002). doi:10.1016/S0376-6357(02)00084-0

Kappas, A.: What facial expressions can and cannot tell us about emotions. In: Katsikitis, M. (ed.) The Human Face: Measurement and Meaning, pp. 215–234. Kluwer, Dordrecht (2003)

Kappas, A.: Appraisals are direct, immediate, intuitive, and unwitting ... and some are reflective ... Cognit. Emot. **20**(7), 952–975 (2006). doi:10.1080/02699930600616080

Kappas, A.: Emotion and regulation are one! Emot. Rev. **3**(1), 17–25 (2011). doi:10.1177/1754073910380971

Kappas, A.: Social regulation of emotion: messy layers. Front. Psychol. **4**, 51 (2013). doi:10.3389/fpsyg.2013.00051

Kappas, A., Küster, D., Theunis, M., Tsankova, E.: Cyberemotions: subjective and physiological responses to reading online discussion forums. In: Poster presented at the 50th Annual Meeting of the Society for Psychophysiological Research, Portland, OR (2010)

Kleinginna, P.R., Kleinginna, A.M.: A categorized list of emotion definitions with suggestions for a consensual definition. Motiv. Emot. **5**(4), 345–379 (1981). doi:10.1007/BF00992553

Larsen, J.T., Norris, C.J., Cacioppo, J.T.: Effects of positive and negative affect on electromyographic activity over the zygomaticus major and corrugator supercilii. Psychophysiology **40**(5), 776–785 (2003). doi:10.1111/1469-8986.00078

Lazarus, R.S., Averill, J.R., Opton, E.M. Jr.: Towards a cognitive theory of emotion. In: Arnold, M.B. (ed.) Feelings and Emotions: The Loyola Symposium, pp. 207–232. Academic, New York (1970)

Leventhal, H., Scherer, K.: The relationship of emotion to cognition: a functional approach to a semantic controversy. Cognit. Emot. **1**(1), 3–28 (1987). doi:10.1080/02699938708408361

Mauss, I.B., Robinson, M.D.: Measures of emotion: a review. Cognit. Emot. **23**(2), 209–237 (2009). doi:10.1080/02699930802204677

Moors, A.: On the causal role of appraisal in emotion. Emot. Rev. **5**(2), 132–140 (2013). doi:10.1177/1754073912463601

Norman, G.J., Norris, C.J., Gollan, J., Ito, T.A., Hawkley, L.C., Larsen, J.T., Cacioppo, J.T., Berntson, G.G.: The neurobiology of evaluative bivalence. Emot. Rev. **3**(3), 349–359 (2011). doi:10.1177/1754073911402403

Parkinson, B.: Untangling the appraisal-emotion connection. Personal. Soc. Psychol. Rev. **1**, 62–79 (1997)

Parkinson, B.: Piecing together emotion: sites and time-scales for social construction. Emot. Rev. **4**(3), 291–298 (2012). doi:10.1177/1754073912439764

Pennebaker, J.W.: The Secret Life of Pronouns: What Our Words Say About Us. Bloomsbury Press, New York (2011)

Reisenzein, R.: The Schachter theory of emotion: two decades later. Psychol. Bull. **94**(2), 239–264 (1983). doi:10.1037/0033-2909.94.2.239

Reisenzein, R., Studtmann, M., Horstmann, G.: Coherence between emotion and facial expression: evidence from laboratory experiments. Emot. Rev. **5**(1), 16–23 (2013). doi:10.1177/1754073912457228

Rimé, B., Philippot, P., Cisamolo, D.: Social schemata of peripheral changes in emotion. J. Pers. Soc. Psychol. **59**(1), 38–49 (1990). doi:10.1037/0022-3514.59.1.38

Russell, J.A.: Core affect and the psychological construction of emotion. Psychol. Rev. **110**(1), 145–172 (2003). doi:10.1037/0033-295X.110.1.145

Russell, J.A., Barrett, L.F.: Core affect, prototypical emotional episodes, and other things called emotion: dissecting the elephant. J. Pers. Soc. Psychol. **76**(5), 805–819 (1999). doi:10.1037/0022-3514.76.5.805

Schachter, S., Singer, J.E.: Cognitive, social, and psychological determinants of emotional state. Psychol. Rev. **69**(5), 379–399 (1962). doi:10.1037/h0046234

Scherer, K.R.: Appraisal considered as a process of multi-level sequential checking. In: Scherer, K.R., Schorr, A., Johnstone, T. (eds.) Appraisal Processes in Emotion: Theory, Methods, Research, pp. 92–120. Oxford University Press, New York and Oxford (2001)

Scherer, K.R., Schorr, A., Johnstone, T. (eds.): Appraisal Processes in Emotion: Theory, Methods, Research. Oxford University Press, New York and Oxford (2001)

Schweitzer, F., García, D.: An agent-based model of collective emotions in online communities. Eur. Phys. J. B **77**(4), 533–545 (2010). doi:10.1140/epjb/e2010-00292-1

Tausczik, Y.R., Pennebaker, J.W.: The psychological meaning of words: LIWC and computerized text analysis methods. J. Lang. Soc. Psychol. **29**(1), 24–54 (2010). doi:10.1177/0261927X09351676

Thelwall, M., Buckley, K., Paltoglou, G., Cai, D., Kappas, A.: Sentiment in short strength detection informal text. J. Am. Soc. Inf. Sci. Technol. **61**(12), 2544–2558 (2010). doi:10.1002/asi.21416)

Tomkins, S.S., McCarter, R.: What and where are the primary affects? Some evidence for a theory. Percept. Mot. Skills **18**(1), 119–158 (1964). doi:10.2466/pms.1964.18.1.119

von Scheve, C., von Luede, R.: Emotion and social structures: towards an interdisciplinary approach. J. Theory Soc. Behav. **35**(3), 303–328 (2005). doi:10.1111/j.1468-5914.2005.00274.x

Chapter 4
The Social Sharing of Emotion in Interpersonal and in Collective Situations

Bernard Rimé

4.1 Introduction

Affective life consists of a variety of manifestations among which one can distinguish temporary variations in mood states, feelings or emotions one the one hand, and more permanent features of the individual such as temperamental traits on the other hand. Emotions can rightly be seen as the highlights of our affective life. They encompass five major characteristics. First, they mark a break in the course of our existence because of the sudden upsurge of a new element. Second, they encompass subjective experiences of a high intensity and with a definite shape such as anger, fear, sadness, or shame for instance. They are thus distinct from less intense affective experiences that only vary on an axis extending diffusely from positive to negative affects. Third, emotions manifest themselves both quickly and shortly, although as we will see, they generally entail significant extensions thereafter. Fourth, emotions have an episodic structure, with a beginning, an apex, and a denouement. As such, they are particularly suited to narration. Fifth, emotions consist of multimodal events that affect all aspects of the person, with changes at the neurological and physiological level, as well as at the cognitive, the behavioral and the subjective or phenomenal level. These complex manifestations take place in moments where the relationship of the individual to the environment changes abruptly. Emotions thus represent powerful bio-psychological signals that inform individuals of important changes occurring in their situation. They motivate people to adapt quickly. Through the action tendencies they include, they additionally formulate all at once proposals for responses to these changes.

In this chapter, we will review evidence that emotions are also almost inextricably linked to a process of social communication. When people go through an emotional

---

B. Rimé (✉)
University of Louvain, Louvain-la-Neuve, Belgium
e-mail: Bernard.Rime@uclouvain.be

J.A. Hołyst (ed.), *Cyberemotions*, Understanding Complex Systems,
DOI 10.1007/978-3-319-43639-5_4

experience, they immediately feel the need to talk with members of their entourage, and they actually do so in almost all cases. This is what we call "the social sharing of emotion". This process has only be studied systematically since two decades (Rimé et al. 1991a). It is of considerable importance for social life. It means in effect that the significant changes in the lives of individuals are systematically talked about and shared with the social network. Social communication thus gathers traces of almost all significant changes in the life of individuals. As long as these traces were exchanged via oral communication, they were fleeting and it was difficult to gauge the actual impact that shared experiences had for social life. Now that a considerable proportion of social communication is taking place in cyberspace, it leaves tangible marks. A large field of investigation has thus now been opened to the study of emotions, their impact, their social sharing and the consequences of that share to interpersonal relationships and to social cognitions.

The present chapter will give an overview of past work in the study of the social sharing of emotion. It is hoped that this overview could point to research avenues opened to future investigation in cyberspace. We will first examine basic observations about the social sharing of personal emotional experiences in interpersonal situations. We will then focus on the spread of emotional information that develops from such sharing situations. Third, we will discuss aspects of the social sharing of emotion that takes place during events affecting individuals collectively. In a final section, we will consider the collective sharing of emotions in collective emotional gatherings.

4.2 Social Sharing of Emotions: Basic Findings

Emotional episodes are subject to conversations in about 90 % of the cases and this is manifested most often repetitively—usually several times, with different people for a same emotional episode (for reviews Rimé 2009; Rimé et al. 1998, 1992). The more intense the emotion is, the higher the propensity to talk about it. The social sharing process is observed whatever the type of emotion involved (joy, fear, anger, sadness…) and whatever the valence of the experience (positive affect or negative affect). Only situations in which the person has felt shame and guilt elicit a restraint on the tendency to talk about it (Finkenauer and Rimé 1998). Manifestations of social sharing have also a strong cultural generality. They were observed at similar rates in Europe, Asia, and North America[1] (Singh-Manoux and Finkenauer 2001; Yogo and Onoe 1998[2]). From these observations, it can be concluded that the

[1]Mesquita, B.: Cultural variations in emotion: a comparative study of Dutch, Surinamese and Turkish people in the Netherlands, Unpublished Doctoral Dissertation, University of Amsterdam, The Netherlands (1993)

[2]Rimé, B., Yogo, M., Pennebaker, J.W.: [Social sharing of emotion across cultures]. Unpublished Raw Data (1996b)

process of talking about emotional experiences is a very general manifestation in such a way that it could be considered as an integral part of an emotional experience. Emotional sharing is started very early after the emotional event. A very reliable observation is that people first share their emotion on the day it happened in 60 % of the cases. In case of episodes of high emotional impact, the social sharing process, or at least the need to share the episode, may extend over weeks or even months, and sometimes over the entire life.

As the memory of the emotional episode fades away, sharing manifestations decline and the impact of the emotional experience is reduced and progressively becomes negligible. To illustrate, 1 week after a major academic exam, 100 % of students had talked about it within the 48 h preceding the survey. Two weeks after the exam, a rate of 94 % was still observed, whereas 3 weeks after the exam, the rate had fallen to 50 % (Rimé et al. 1998). The slope of the decline depends on the initial intensity of the experience. The higher the intensity of the initial emotion was, the lesser the steepness of the sharing extinction slope. Thus, compared with the university exam, the loss of a loved one obviously involves an emotion of a much higher intensity. Ten days after the death of a loved one, 97 % of respondents had shared their emotional experience in the 48 h preceding the survey. Four weeks after the death, the rate was still 86 %, and 12 weeks after the event, it still amounted to 79 %.[3]

4.3 Targets of Social Sharing

Who are the targets people select for sharing their emotions? Interesting trends about sharing addressees emerged from the comparison of age groups (Rimé et al. 1991a,b, 1992, 1996). Children aged between 6 and 8 who had been exposed to an emotion-eliciting narrative later manifested virtually no sharing toward peers of their classroom. Yet most of them shared the episode with their father and mother when back home. Other family members rarely were social sharing targets in this age group. Preadolescents (aged 8–12) were surveyed after a night game at a summer camp that, according to children's ratings, had induced a moderate intensity emotional state in them. They went back home on the day that followed the game. Three days later, parents' ratings showed that the night game had been shared by 97 % of the children. Parents clearly emerged as the privileged sharing partners—mother in 93 % and father in 89 %. Siblings served as recipients in 48 % of the cases, best friends in 33 %, peers in 37 % (peers generally were children who took part to the same summer camp), and grandparents in only 5 %. Among adolescents (aged 12–18), family members—predominantly parents—were by far the most often mentioned sharing target both among boys and girls. Friends collected about one-

[3]Zech, E.: La gestion du deuil et la gestion des émotions [Coping with grief and coping with emotions]. Unpublished Master Thesis, University of Louvain, Louvain-la-Neuve, Belgium (1994)

third of emotional sharing. Boyfriends and girlfriends were rarely mentioned, either because there was no, or because in this age group they were not yet eligible as sharing partners. But, as age cohorts got older, friends, including girl/boy friends and female best friends became increasingly important. Other people were rarely mentioned as sharing partners. Nonmembers of the immediate social network were simply absent from these communications. Among young adults (aged 18–33), the role of family members was considerably reduced, especially among males. In contrast, for both genders, spouses/partners emerged as major actors on the social sharing stage, whereas friends kept the same importance as in adolescents' data. The role of family members decreased again in middle-aged adults (aged 40–60), perhaps in part because parents are no longer available. Additionally, a considerable drop in the importance of friends occurred for male adults, but not for females. In this age group, spouses/partners predominated markedly as sharing targets. In men in particular, the spouse/partner was an exclusive target for more than three-quarters of respondents. Data collected on elderly people (aged 65–95) simply replicated this pattern.

To conclude, the social sharing of emotion is essentially addressed to members of one's close social network. From adulthood on, spouses and partners constitute the main sharing targets (over 75 %), followed by members of one's family (over 30 %) and friends (about 20 %). As previously mentioned, other categories of people, such as strangers or professionals, were rarely mentioned (less than in 5 % of the cases). These conclusions are somewhat qualified by observations regarding emotional episodes occurring in a professional context or in the framework of a specific emotional conditions such as illness. Soldiers in military operations, or hospital nurses working in emergency units overwhelmingly adopted their professional colleagues as the first sharing partners when they faced an emotional experience in their work. Cancer patients mentioned other cancer patients, their physician and psychologists as their most important sharing targets.

4.4 Emotional Reactivation and Motives to Share Emotions

That people systematically share a positive emotional experience comes as no surprise. But when it comes to negative experience, the propensity to talk about it as soon as it happened and to then share repetitively what happened with various members of one's network looks more puzzling. Our studies have addressed this puzzle in various ways. In a study conducted two decades ago (Rimé et al. 1991a), participants were instructed to recall and then to give a detailed description of an emotional episode of their recent past life. The episode had to be either one of joy, anger, fear or sadness, according to randomly distributed instructions. After the description, participants had to report what they felt during this task. Whatever the type of emotion they had described was, participants overwhelmingly reported vivid mental images of the recalled event together with related feelings and bodily sensations. This simply confirmed the known fact that accessing the memory of an emotional episode has the effect of reactivating the various components (i.e., physiological, sensory, experiential) of the emotion involved. Not surprisingly, participants who had to report an experience of joy rated their task as having been more pleasant than those who had to report an emotion of sadness, of fear, or of anger. More surprising was the fact that reporting fear, sadness or anger was rarely rated as unpleasant. Notwithstanding the reactivation of vivid images, feelings, and bodily sensations of a negative emotional experience, sharing such an experience was far from having elicited the aversion one would have expected. Even more striking was the fact that when participants were asked whether they would be willing to undertake the sharing of another emotional memory of the same type as the first one, virtually all of them (96 %) answered positively in all four emotional conditions. These data revealed the paradoxical character of the social sharing of emotion. Although sharing reactivates the various components of the emotion, it does seem to be a situation in which people engage much willingly, whatever the valence of the shared emotion.

Delfosse et al. (2004) investigated the reasons given by respondents for sharing their emotional experiences in current life. Here also, according to a random distribution, participants remembered an emotional event from their recent past having involved joy, anger, fear or sadness, and that they had shared with others. They were then asked about their motives for sharing the episode. The data collected in this study showed that the motives markedly differed depending on the valence of the event.

According to respondents' data, positive emotional memories were socially shared primarily in order to (1) recall the episode, (2) elicit the attention of the target person and to inform the latter of what happened. These findings are consistent with observations according to which talking about a past positive emotional experience elicits pleasurable emotional feelings both in the sender and in the receiver. In this regard, Langston (1994) introduced the concept of capitalization. Positive emotional episodes represent opportunities on which to seize or "capitalize". Whenever the memory of such an episode is reactivated, immediate benefits are taken under the form of a temporary boost of positive affect Seeking social contacts and letting others know about the event largely contributes to capitalizing on positive emotional experiences. Communicating positive events to others was indeed associated with an enhancement of positive affect far beyond the benefits due to the valence of the positive events themselves (for review see Gable and Reis 2010). Gable et al. (2004) observed that close relationships in which one's partner typically responds enthusiastically to capitalization were associated with higher relationship well-being (e.g., intimacy, daily marital satisfaction). Thus, sharing positive emotions not only boosts individuals' positive affect, it also enhances their social bonds.

As regarded negative emotional memories, Delfosse et al. (2004) found them associated with four major motives: (1) venting the emotion, (2) receiving understanding from the target for what happens, (3) enhancing social bonds and (4) receiving social support. "Venting the emotion" is a stereotypical response that really pops out in the population when the question of emotional expression comes to the floor. Common sense indeed quite willingly explains the propensity for social sharing of emotions by "liberatory" hypothesized effects of emotional expression. Thus, recommendations such as "talk about your emotional experience" or "get it off your chest" are adopted without hesitation by laypersons in Western countries as well as in Asia.[4] Psychological practitioners willingly subscribe to this thesis as well. To illustrate, after traumatic events, the so-called "psychological debriefing" techniques are much popular intervention methods in which victims are encouraged to express extensively the emotions they went through during the drama. Doing so is expected to ensure victims liberating effects and to prevent the development of psychological symptoms such as posttraumatic disorders. However, the meta-analysis of evaluation studies revealed that debriefings are not successful in reducing post-traumatic stress disorder and that in some cases, results are even going in the opposite direction (Arendt and Elklit 2001; Deahl 2000; Foa and Meadows 1997; Raphael and Wilson 2000; Rose and Bisson 1998; Rose et al. 2003; van Emmerik et al. 2002). These data therefore pleaded against the "discharge" hypothesis. Research on the effects of the social sharing of current life emotional episodes repetitively resulted in similar findings (Nils and Rimé 2012; for reviews Rimé 2009; Rimé et al. 1998). Contrary to the widespread expectations of common sense, simple sharing of emotion does not cause a reduction of the emotional and cognitive impact of memories of the shared emotional episode.

After the excluding the venting or discharge motive, we are left with three other motives alleged by respondents for sharing their negative emotional experiences: (1) receiving understanding from the target for what happens, (2) enhancing social bonds and (3) receiving social support. All three are social motives: addressees are expected to provide the sharing person with meaning, with social integration and with support. As will be seen in the section to follow, these observations fit nicely those collected in the investigation of the interpersonal dynamics that develop in the course of social sharing interactions.

4.5 The Interpersonal Dynamics of the Social Sharing of Emotion

A very special dynamic develops in person-to-person interaction focused on the social sharing of emotion by someone to a listening person.[5] First, listeners are found to exhibit a considerable interest for the narration of the emotional episode, whatever the valence of the episode—positive or negative. Actually, curiosity for emotionally negative scenes and information is quite common. When driving by a

[4]Zech, E.: The effects of the communication of emotional experiences. Unpublished Doctoral Dissertation. University of Louvain, Louvain-la-Neuve, Belgium (2000)

[5]Christophe, V.: Le partage social des émotions du point de vue de l'auditeur [Social sharing of emotion on the side of the target]. Unpublished Doctoral Dissertation, Université de Lille III, France (1997)

traffic accident, drivers slow down to watch. Pedestrians change their course to look at a building in flames. People are attracted by emotional stories in the media as well as in movies, novels, plays, drama, opera, songs, images, etc. Admittedly, a fascination for negative emotional materials literally permeates everyday life. This fascination is playing a critical role in the development and maintenance of the social sharing of emotion. By manifesting their interest to the sender, social sharing listeners stimulate the narrative process and become thus increasingly exposed to an emotional content. This, in turn, stimulates their own emotions. The more emotionally loaded the story, the higher the emotional arousal of listeners.

A second characteristic response of social sharing listeners takes the form of empathy. By manifesting emotions echoing those expressed by the sender, listeners contribute to the development of a climate of emotional communion or emotional fusion in the interaction. Such a climate stimulates a prosocial orientation among listeners: the more they are emotionally moved by the narration, the more they express warmth, support, understanding and validation to the sharing person.

A final manifestation resulting from the interpersonal sharing of emotion involves mutual attraction. When the sharing partners have preexisting ties, as is most often the case, one speaks of an enhancement of their mutual attraction. The social sharing of emotion modally ends up with an increase of positive affect on both sides and therefore leads to closer links between the interaction partners. A similar phenomenon was very consistently found in the study of situations of confidence and self-disclosure (for review Collins and Miller 1994). Those who are the receivers of more intimate confidences develop a higher level of affection for the one who engage in these confidences. And likewise, those who engage in intimate confidences enhance their affection for those who listen.

Observing the behavioral responses of social sharing listeners revealed additional traits of the latter. Christophe and Rimé (1997) observed that when intense emotions are shared, listeners reduce their use of verbal mediators in their responses. They engage less in verbal expression. As a substitute, they display nonverbal comforting behaviors, such as reducing the physical distance of the interaction, or even hugging, kissing, or touching the narrator. This suggests that sharing an intense emotional experience ends up decreasing the psychological distance between two persons, thus contributing to the maintenance and even to the improvement of their ties. Of course, the process has limits. As soon as the shared episode involves elements likely to represent a threat for the listener, the latter is likely to deny listening and to manifest avoidance. This is frequently observed when a person is experiencing serious health problems for instance. The threat that such a condition represents to those who are exposed to it leads to a reduction of their willingness to listen to the ill person (Cantisano et al. 2012; Herbette and Rimé 2004).

4.6 The Propagation of the Social Sharing of Emotion

If listening to an emotional story evokes emotion in the listener and if the emotion leads to social sharing of emotion, then we must expect that the listener shares in turn this story with others in a "secondary social sharing" (Christophe and Rimé 1997). Several studies confirmed that listeners indeed practice a secondary social sharing in about two-thirds of cases. The transmission of what they heard to members of their own social network is initiated in a majority of cases on the day listeners heard it. They usually share the story "three to four times" with "three to four people." As for the primary sharing, the frequency of secondary social sharing increases linearly with the intensity of the emotion felt when listening. Research conducted on large samples of people using diary research techniques by Curci and Bellelli (2004) fully confirmed the reality of secondary social sharing. In one of their studies, volunteer students completed a diary for 15 days in which they had to report daily an episode shared with them by someone who had experienced it. The collected data comprised 875 episodes (302 positives and 573 negatives). On the day they were heard, 54 % of these episodes were secondarily shared, with no difference as a function of valence—a result that virtually matches those collected for primary sharing (i.e., 60 % on average, see above). In addition, 55 % of events that were not shared on the day they were heard were shared on a later day. In this manner, 75 % of all episodes collected in this study were shared. This closely replicated previous findings by Christophe and Rimé (1997). According to the study, the rate of secondary social sharing recorded ranged from 66 % for the lowest estimate to 86 % for the highest estimate. This is a phenomenon of considerable importance for social life.

Do receivers of a secondary social sharing talk in their turn to third parties?[6] examined this question by asking respondents to search in their memories a situation where they had been the target of a secondary social sharing. They then rated how often they had shared thereafter what they heard. The findings showed that this occurred in 64 % of cases—several times in 31 % of cases, and only once in 33 % of cases. The tertiary sharing was initiated on the same day as the secondary sharing took place in 31 % of cases.

In sum, according to data from the study of primary social sharing, a person faced with an emotional experience of some intensity then talk about it with four or five persons. Each of the latter is likely to undertake a secondary social sharing with three or four people. These new targets will be talk in their turn to one, two or three other persons. If five people were exposed to the shared information in the first round, there will be approximately 18 in the second round and 30 in the third one. This means that 50–60 people will be informed of the event that has affected a single member of their community. The person at the source of the sharing process

[6]Christophe, V.: Le partage social des émotions du point de vue de l'auditeur [Social sharing of emotion on the side of the target]. Unpublished Doctoral Dissertation, Université de Lille III, France (1997)

is necessarily having close ties to those who disseminate information. Each relay has at least an indirect link with this person. Since the greater part of social sharing takes place on one same day, it can be speculated that the broadcast grows in a few hours.

The reality of this propagation process has been confirmed in a field study in which Harber and Cohen (2005) monitored the communications of 33 university students that their instructor had taken to visit a hospital mortuary. The intensity of the emotional reactions experienced by the students during the visit could predict not only the number of people that each student has shared the visit with others (primary division), but also the number people to whom receptors then told the story of the visit (secondary shares) and the number of people that the new receptors had then addressed in turn (tertiary sharing). In this manner, the study demonstrated that within a few days, nearly 900 people had heard of this event. Research on rumors and on urban legends has provided results that are very consistent with those of the study of social sharing of emotions (Heath et al. 2001). These studies showed that the circulation of stories is based on an emotional selection rather than on an informational selection. People are all the more willing to share the stories that evoke more emotion. In addition, the more stories are emotional, the wider their dissemination is. The speed of propagation of this type of information being a function of its emotional impact, a particularly fast social spread is to be expected in case of events with a high emotional impact. In addition, it can be expected that when the impact is very high, information by word of mouth will be relayed by mass media. This then enters the register of collective emotional episodes, which is the subject of the remainder of this chapter.

4.7 Social Sharing of Collective Emotional Events

The social impact of emotional events is particularly spectacular in the case of a collective emotional episode. This is what happens when a community is directly affected by an event such as a victory or defeat (in sports, in politics, etc.), a loss, a disaster or a common threat. Under such conditions, the direct experience of individuals is generally taken into relay by mass media. In case of media exposure, the number of individuals and communities who are concerned about the event can be extended considerably. Thus, the tsunami in the Indian Ocean in 2004 and the earthquake that struck Haiti in 2010 or Japan in 2011 first affected large communities directly. But the media coverage generated empathy across the entire planet. Similarly, the death of Princess Diana or of singer John Lennon affected individuals far beyond the communities directly affected due to the combined effect of the media coverage of the event and the prestige these personalities had in the world. In all cases that meet these examples, abundant collective sharing of emotion develops. It has spectacular features.

In case of individual emotional experiences, a single source disseminates information in direction of the periphery of the social network. However, the spread quickly reaches extinction, because empathy and emotion subside as one moves away from the initial source. The more the person to whom something happened is distant and unknown, the more a highly intense emotional story is needed to hold the interest of conversations. Instead, a collective emotional event will spark as many sources of social sharing of emotion as there are members in the social group concerned. In this manner, the social sharing of emotions can be expected to spread in all directions (Rimé 2007). In addition, as every moment of social sharing reactivates again in both the sender and the listener the emotion elicited by the event, the need for sharing is continuously reset for each of them.

Because of its repetitive aspect, the social sharing of a collective emotion contributes to the consolidation of the memory of the emotional episode, leading to a vivid memory of those events that caught people by surprise. Finkenauer and Rimé (1998) investigated the memory of the unexpected death of Belgium's king Baudouin in 1993 in a large sample of Belgian citizens. The data revealed that the news of the king's death had been widely socially shared. By talking about the event, people gradually constructed a social narrative and a collective memory of the emotional event. At the same time, they consolidated their own memory of the personal circumstances in which the event took place, an effect known as "flashbulb memory" (Conway 1995; Luminet and Curci 2009). The more an event is socially shared, the more it will be fixed in people's minds. Social sharing may in this way help to counteract some natural inclination people may have. Naturally, people should be driven to "forget" undesirable events. Thus, someone who just heard a bad news often inclines initially to deny what happened. The repetitive social sharing of the bad news contributes to realism.

As they elicit a spread of the social sharing in every direction, collective emotional events can cause chain reaction effects that are reminiscent of what occurs in a nuclear reactor. The emotional turmoil that ensues leads to a climate of mutual empathy and of collective emotional fusion. Whatever the emotional valence of a collective event, it is generally that start of a state of "honeymoon" in communities. People experience feelings such as "we feel the same, we are one, we are united." Pennebaker (1993) proposed a model of collective emotional responses to collective emotional events. The model first considers an initial period, or period of emergency, that takes place immediately after the event. It lasts about a month and is marked by intense emotional reactions of all members of the community. These reactions are accompanied by abundant mental rumination at the individual level. At the social level, intensive forms of communication develop (social sharing, media coverage), together with establishment of spontaneous connections and with numerous manifestations of generosity and of solidarity. We meet here the "honeymoon" situation that was mentioned above. Then, the model distinguishes a second phase, the so-called plateau, which also extends over approximately 1 month and represents an intermediate period. At this stage, the social sharing of emotion and the media coverage of the event disappear but mental rumination endures. Finally, 2 months after the starting point, a final period occurs. Social events of the "honeymoon" (spontaneous links and solidarity) decline, mental rumination disappears, and social life returns to normal.

A community cannot remain almost exclusively focused on the shared emotional episode and keep neglecting the routine economic and survival activities. Some natural slowing process is needed to avoid this pitfall. In a nuclear reactor, the fission is slowed down by the introduction of cadmium bars in the system. In an emotionally aroused community, the major cause of return to normal life lies in the saturation that gradually settles. Pennebaker (1993) reported a very illustrative example of this process while studying communications in an American community that had been struck by an earthquake. During the first days, manifestations of social sharing of emotion between the victims were plentiful. After a while, however, some of the victims showed a reversal of attitude and expressed it in a spectacular way. They would start wearing T-shirts that read, "Please do not share with me your experience of the earthquake." In fact, everyone was still so willing to tell things from one's own point of view, but had no more desire to listen to the experience of others. In the early days, the interest of each other for the stories of neighbors was considerable. But the repetition inherent in the collective situation had gradually eroded the interest.

The model of psychosocial reactions to a collective emotional event as proposed by Pennebaker (1993) was tested after the terrorist attacks in Madrid in March 2004 (Páez et al. 2007; Rimé et al. 2010). At that time, the popular emotion was considerable throughout Spain. Repetitive mass demonstrations were held with the purpose to condemn terrorism. A large sample of people who participated to varying degrees in these events was contacted three times: 1 week, 3 weeks and then 2 months after the event. These people were subjected to various measures, including their sense of belonging to the group, their position in relation to collective beliefs, as well as their personal well being and confidence in life. The data confirmed the model predictions regarding the emergence, plateau, and the extinction of psychosocial events. In addition, in accordance with the principle that sharing emotion causes the reactivation of the shared emotion, it was found that the reactivation of negative emotions generated by the dramatic events was even greater among respondents who had invested themselves heavily in the protests. But the data also highlighted important positive social effects resulting from the collective sharing process. On the one hand, the more people were involved in the protest demonstrations, the stronger was their sense of belonging to the group and the level of their cultural beliefs as assessed 8 weeks after the events. In addition, the importance of their participation in the collective social sharing was associated with feelings of well being and with positive feelings vis-á-vis their future life. These observations are particularly consistent with the theoretical model that will be examined in the next section.

4.8 The Collective Sharing of Emotions in Social Gatherings

The social sharing of emotions can take another type of collective form that is particularly common in social life. Members of a group or of a society gather deliberately in situations in which they experience emotions together. Such emotional gatherings can take many forms. The spectrum is wide, since at comprises at a time political rallies or protest demonstrations, funeral ceremonies, wedding celebrations, court sessions, music concerts, theater, sport events or religious rituals, and so forth. A century ago, in his classic book titled "The Elementary Forms of Religious Life", Emile Durkheim (1912) proposed an analysis of collective emotional gatherings primarily intended to account for religious rituals and ceremonies. Yet, his model is very likely valid for any type of situation in which people experience emotions in a crowd. Durkheim viewed such gatherings as a particularly effective way to periodically renew the membership of individuals to the group and reinvigorate in them the shared beliefs that underlie the life of any group. According to his analysis, in collective emotional gatherings, people generally gather in the presence of symbols that represent their membership group and evoking the beliefs shared by members. All the participants share a common concern and they focus their attention on a common object. The collective event then goes on involving abundant collective action and movements, with shared expressive gestures, dances, words, shouting or singing. These actions contribute to generate emotional states and an atmosphere of fervor. The shared focus and concern, the shared actions and movements, and the physical closeness of the participants contribute to favor emotional contagion. The elicited emotions echo and reinforce each other so that a climate of collective emotional fusion follows: individual feelings give way to shared feelings. For Durkheim, this generalized empathy is the action lever of collective rituals. It causes participants to experience a state of emotional fusion or communion: "we act the same way, we feel the same things, we are one". The feeling of belonging is thus revived and social cohesion follows in the group. Shared beliefs, diluted daily by the individual life, return to the forefront of the consciousness of each. They can then return to their individual occupations, inhabited again for a while by the strength of the group and shared beliefs. A strengthened faith in the existence enables them to cope with everyday life with a sense of meaning.

It can be stressed that the interpersonal process of sharing emotions, the social sharing of a collective event and the collective expression of emotions in emotional gatherings overlap almost completely with regard to their underlying social dynamic. The underlying social dynamic is actually one and the same, that of emotional fusion. In all three cases, participants reciprocally stimulate their emotions. Such a dynamic leads to a sense of unity and has consequences in terms of social relationships, social trust, individual well-being and confidence in the future. The essential difference between interpersonal and collective social sharing lies in the way emotions propagate. In person-to-person situations, the propagation develops in successive stages and the emotion vanishes from the one stage to the next one. By contrast, in the collective case, the emotional wave is instantaneous because it affects all members of a group at the same time.

Durkheim's model could be tested in the context of the "Gacaca" courts introduced all over Rwanda after the genocide that occurred in this country in 1994. These courts were inspired from the "Commissions for the Truth and Reconciliation" instituted for instance in South Africa after apartheid. The purpose of such procedures is to bring together victims and perpetrators in the presence of the members of the community. It is hoped that victims will find the opportunity to express their suffering and perpetrators will recognize their faults and will publicly express their repentances. In such collective situations, collective emotional expressions often reach paroxysmal dimension. Participants have very hard times in the situation, they leave it upset and sometimes retraumatized, but at the same time, they often feel a great personal and social benefit from their participation.

In two different studies (Kanyangara et al. 2007; Rimé et al. 2011), victims of the Rwandan genocide and those detained for their involvement in the genocide have completed questionnaires before and after their participation in the Gacaca court of the community they belong. In one of these studies, their responses could be compared to those of control groups composed of victims and prisoners who completed the questionnaires at the same time but belonged to communities where the Gacaca courts had not yet taken place. As was predicted by the model put forward by Durkheim, the data collected showed that participation in Gacaca courts had intensified the emotions of most of the participants in both groups. However, also in line with the model of Durkheim, indicators of social integration included in the questionnaires showed significant positive effects. Thus, after participation, there has been an increase in the level of positive stereotypes of the group of victims vis-á-vis the genocidal, and vice versa. Moreover, we know that in intergroup conflicts, opponents are typically perceived as forming a more homogeneous group than it actually is: "they are all alike". The results of both studies showed that the monolithic perception was significantly reduced in both groups of respondents after their participation in the Gacaca courts. Finally, in full agreement with the logic of Durkheim that emotions and emotional fusion is the lever action of collective rituals, data analysis evidenced the role played by the emotional arousal of the participants as a partial mediator of the effects of social integration just described.

4.9 Conclusions

The study of the social sharing of emotion reveals that emotion is hardly ever experienced in social isolation. Rather, an emotional experience triggers and consistently feeds up important social process. Every individual communicates his emotional experiences to others. Those who heard about these experiences shall in their turn inform people around them of what they heard. In this way, they propagate all at once the emotional information, the emotional impact of this information and the need to share the information. Particularly intensive exchanges are occurring between members of social groups when they cross a common emotional event. In this case, everyone is both a source and a target of emotional information and emotion reactivation. Finally, members of social groups gather regularly to experience emotions together or recall together common past emotional episodes. In such collective gatherings, the shared focus and concern, the shared actions and movements, and the physical closeness of the participants contribute to favor fast and powerful emotional contagion.

Why is emotion so closely associated with social orientation and social interactions? We have mentioned that common sense would favor an intra-individual explanation based upon the stereotype according to which expressing an emotion will end up "discharging" the associated emotional load. We saw that existing empirical data from the study of "psychological debriefings" after traumatic situations pleaded against the "discharge" hypothesis. Our own work on the effects of emotional sharing resulted in similar conclusions: contrary to the expectations of common sense, simple sharing of emotion does not cause a reduction of the emotional and cognitive impact of memories of the shared emotional episode. Thus, at odd with a very popular belief, emotional discharge is certainly not the primary function of the social sharing of emotion.

Rather, in this chapter, we have seen that the social sharing of emotion gives precedence to two well-documented processes. On the one hand, it rouses a specific socio-emotional process that (1) promotes emotional union between the sharing partners, (2) stimulates prosocial behaviors among targets, and (3) favors the social reintegration of individuals who lived a singular experience. On the other hand, the social sharing of emotion sparks a process of diffusion that allows the transmission of the individual experience to members of the social network. Those who receive this information are informed of what happened to one of them and how this one faced the situation. They will react to this information, spread it in turn, discuss it with others, and interpret it. Together, members of the social network will reflect upon the experience and they will derive lessons from it for the future of each of them. Through such a process, every significant experience of every single individual can enter the pool of shared knowledge, can impact on shared models of the world, shared worldviews and shared beliefs, and thus can engender changes into the systems of representations (concepts, beliefs, values, etc.) shared by the social milieu. In this sense, the social sharing of emotion is a tool for cultural transformation.

These findings allow us to get to the heart of what constitutes an emotional experience. An emotion necessarily reveals a mismatch between the events and the person's expectations, goals, models, values, and so forth. It should be reminded that the anticipation systems of individuals possess largely originate in socially shared knowledge, or cultural knowledge. An emotion thus signals a gap occurring between the current individual experience and the socially shared knowledge. From this perspective, it is less surprising that in emotional circumstances, individuals are quick to turn to the social network they are members of. And it is less surprising that this social network cares much about what happened to their individual members. The process of social sharing of emotion provides two functions that are essential with respect to survival: social integration of individuals and fine-grained adjustment of common knowledge on what can happen and how to face it. The model of collective emotional gatherings proposed by Emile Durkheim said nothing else. Their double function is to ensure social integration and to consolidate common belief systems.

In sum, emotional episodes have the effect of reviving a sense of unity among individuals and in social groups and this has important consequences for social cohesion, for social trust, for individual well-being and for confidence in the future.

4.10 Perspectives: The Social Sharing of Emotions in Cyberspace

The studies on which this chapter is based were conducted with respondents in small numbers and with methods limited to conventional data collection—most of the time using questionnaire techniques. This research has been able to highlight phenomena that were previously unknown and to open a number of promising avenues of investigation. However, it must be acknowledged that it is only in its infancy. From this point of view, the prospects offered by the study of emotions in cyberspace are simply gigantic. The study of online communications can spark a revolution in the field of investigation described in these pages. It will allow observing emotions, their expression and their social sharing in real time with massive data on communications flows, on the dynamics of exchanges, and on lexical indicators of underlying emotional, cognitive and social processes.

Events eliciting emotions of every possible types and of every level of intensity occur at every time in every parts of the world—would it be a bad act of purchase, the vision of a moving movie, attending a wedding or a funeral, or the sudden exposure to an earthquake. The study of online communications triggered by such events can offer an exceptionally fine-grained and reliable source of documentation on how people react to these events, how they talk about them, with whom, how many times, and for how long. Assessing reaction time, velocity of traffic, reaction time of target persons, duration of interactions, velocity of information propagation, extent of propagation wave, speed of extinction, indicators of reactivation or of remembrance and the like will provide a real time analysis of the dynamics of the social sharing of emotions. Such variables will be examined in function of the type of emotions involved, of the characteristics of the event, of its geographical location, of its socio-cultural context, of how distant witnesses are, of their socio-demographic and socio-political characteristics, and so forth.

Cyberspace data can powerfully feed the research about the management of emotion-eliciting experiences. Large scale data will help understanding how individuals and communities react to events, which are the factors that contribute to stress and trauma, and which are the ones promoting resilience and recovery. Such data will also document the impact the emotion communication dynamics has upon interpersonal relationships and social links, upon group integration and cohesion, and upon group members' assertiveness and confidence. Critical issues in the study of emotions are likely to receive responses from data with a magnitude never achieved before.

References

Arendt, M., Elklit, A.: Effectiveness of psychological debriefing. Acta Psychiatr. Scand. **104**(6), 423–437 (2001). doi:10.1034/j.1600-0447.2001.00155.x

Cantisano, N., Rimé, B., Munoz, T.: The social sharing of emotions in HIV/AIDS: a comparative study of HIV/AIDS, diabetes and cancer patients. J. Health Psychol. **18**(10), 1256–1267 (2012). doi:10.1177/1359105312462436

Christophe, V., Rimé, B.: Exposure to the social sharing of emotion: emotional impact, listener responses and the secondary social sharing. Eur. J. Soc. Psychol. **27**(1), 37–54 (1997). doi:10.1002/(SICI)1099-0992(199701)27:1<37::AID-EJSP806>3.0.CO;2-1

Collins, N.L., Miller, L.C.: Self-disclosure and liking: a meta-analytic review. Psychol. Bull. **116**(3), 457–475 (1994). doi:10.1037/0033-2909.116.3.457

Conway, M.A.: Flashbulb Memories. Lawrence Erlbaum, Mahaw, NJ (1995)

Curci, A., Bellelli, G.: Cognitive and social consequences of exposure to emotional narratives: two studies on secondary social sharing of emotions. Cognit. Emot. **18**(7), 881–900 (2004). doi:10.1080/02699930341000347

Deahl, M.P.: Psychological debriefing: controversy and challenge. Aust. N. Z. J. Psychiatry **34**(6), 929–939 (2000). doi:10.1080/000486700267

Delfosse, C., Nils, F., Lasserre, S., Rimé B.: Les motifs allégués du partage social et de la rumination mentale des émotions: comparaison des épisodes positifs et négatifs. Cahiers Internationaux de Psychologie Sociale **64**, 35–44 (2004)

Durkheim, E.: Les formes élémentaires de la vie religieuse [The Elementary Forms of Religious Life]. Alcan, Paris (1912)

Finkenauer, C., Rimé, B.: Socially shared emotional experiences vs. emotional experiences kept secret: differential characteristics and consequences. J. Soc. Clin. Psychol. **17**(3), 295–318 (1998). doi:10.1521/jscp.1998.17.3.29

Foa, E.B., Meadows, E.A.: Psychosocial treatments for posttraumatic stress disorder: a critical review. Annu. Rev. Psychol. **48**, 449–480 (1997). doi:10.1146/annurev.psych.48.1.449

Gable, S.L., Reis, H.T.: Good news! Capitalizing on positive events in an interper-sonal context. Adv. Exp. Soc. Psychol. **42**, 195–257 (2010). doi:10.1016/S0065-2601(10)42004-3

Gable, S.L., Reis, H.T., Impett, E.A., Asher, E.R.: What do you do when things go right? The intrapersonal and interpersonal benefits of sharing positive events. J. Pers. Soc. Psychol. **87**(2), 228–245 (2004). doi:10.1037/0022-3514.87.2.228

Harber, K.D., Cohen, D.J.: The emotional broadcaster theory of social sharing. J. Lang. Soc. Psychol. **24**(4), 382–400 (2005). doi:10.1177/0261927X05281426

Heath, C., Bell, C., Sternberg, E.: Emotional selection in memes: the case of urban legends. J. Pers. Soc. Psychol. **81**(6), 1028–1041 (2001). doi:10.1037/0022-3514.81.6.1028

Herbette, G., Rimé, B.: Verbalization of emotion in chronic pain patients and their psychological adjustments. J. Health Psychol. **9**(5), 661–676 (2004). doi:10.1177/1359105304045378

Kanyangara, P., Rimé, B., Philippot, P., Yzerbyt, V.: Collective rituals, intergroup perception and emotional climate: participation in "Gacaca" tribunals and assimilation of the Rwandan genocide. J. Soc. Issues **63**(2), 387–403 (2007). doi:10.1111/j.1540-4560.2007.00515.x

Langston, C.A.: Capitalizing on and coping with daily-life events: expressive responses to positive events. J. Pers. Soc. Psychol. **67**(6), 1112–1125 (1994). doi:10.1037/0022-3514.67.6.1112

Luminet, O., Curci, A. (eds.): Flashbulb Memories: New Issues and New Perspectives. Psychology Press, New York (2009)
Nils, F., Rimé, B.: Beyond the myth of venting: social sharing modes determine the benefits of emotional disclosure. Eur. J. Soc. Psychol. **42**(6), 672–681 (2012). doi:10.1002/ejsp.1880
Páez, D., Basabe, N., Ubillos, S., Gonzalez, J.L.: Social sharing, participation in demonstration, emotional climate, and coping with collective violence alter the March 11th Madrid bombings. J. Soc. Issues **63**(2), 323–338 (2007). doi:10.1111/j.1540-4560.2007.00511.x
Pennebaker, J.W.: Social mechanisms of constraint. In: Wegner, D.M., Pennebaker, J.W. (eds.) Handbook of Mental Control, pp. 200–219. Prentice Hall, Englewood Cliffs, NJ (1993)
Raphael, B., Wilson, J.P.: Psychological Debriefing: Theory, Practice and Evidence. Cambridge University Press, Cambridge (2000)
Rimé, B.: The social sharing of emotion as an interface between individual and collective processes in the construction of emotional climates. J. Soc. Issues **63**(2), 307–322 (2007). doi:10.1111/j.1540-4560.2007.00510.x
Rimé, B.: Emotion elicits the social sharing of emotion: theory and empirical review. Emot. Rev. **1**(1), 60–85 (2009). doi:10.1177/1754073908097189
Rimé, B., Noël, P., Philippot, P.: Episode émotionnel, réminiscences cognitives et ré-miniscences sociales. Cahiers Internationaux de Psychologie Sociale **11**, 93–104 (1991a)
Rimé, B., Mesquita, B., Philippot, P., Boca, S.: Beyond the emotional event: six studies on the social sharing of emotion. Cognit. Emot. **5**(5–6), 435–465 (1991b). doi:10.1080/02699939108411052
Rimé, B., Philippot, P., Boca, S., Mesquita, B.: Long-lasting cognitive and social consequences of emotion: social sharing and rumination. In: Stroebe, W., Hewstone, M. (eds.) European Review of Social Psychology, vol. 3, pp. 225–258. Wiley, Chichester (1992)
Rimé, B., Dozier, S., Vandenplas, C., Declercq, M.: Social sharing of emotion in children. In: Frijda, N. (ed.) ISRE 96. Proceedings of the IXth Conference of the International Society for Research in Emotion, ISRE, Toronto, ON, pp. 161–163 (1996)
Rimé, B., Finkenauer, C., Luminet, O., Zech, E., Philippot, P.: Social sharing of emotion: new evidence and new questions. In: Stroebe, W., Hewstone, M. (eds.) European Review of Social Psychology, vol. 9, pp. 145–189. Wiley, Chichester (1998)
Rimé, B., Paez, D., Basabe, N., Martinez, F.: Social sharing of emotion, post-traumatic growth, and emotional climate: follow-up of Spanish citizen's response to the collective trauma of March 11th terrorist attacks in Madrid. Eur. J. Soc. Psychol. **40**(6), 1029–1045 (2010). doi:10.1002/ejsp.700
Rimé, B., Kanyangara, P., Yzerbyt, V., Páez, D.: The impact of Gacaca tribunals in Rwanda: psychosocial effects of participation in a truth and reconciliation process after a genocide. Eur. J. Soc. Psychol. **41**(6), 695–706 (2011). doi:10.1002/ejsp.822
Rose, S., Bisson, J.: Brief early psychological interventions following trauma: a systematic review of the literature. J. Trauma Stress **11**(4), 697–710 (1998). doi:10.1023/A:1024441315913
Rose, S., Bisson, J., Wessely, S.: A systematic review of single-session psychological interventions ('debriefing') following trauma. Psychother. Psychosom. **72**(4), 176–184 (2003). doi:10.1159/000070781
Singh-Manoux, A., Finkenauer, C.: Cultural variations in social sharing of emotions: an intercultural perspective on a universal phenomenon. J. Cross-Cult. Psychol. **32**(6), 647–661 (2001). doi:10.1177/0022022101032006001
van Emmerik, A.A., Kamphuis, J.H., Hulsbosch, A.M., Emmelkamp, P.M.: Single session debriefing after psychological trauma: a meta-analysis. Lancet **360**(9335), 766–771 (2002). doi:10.1016/S0140-6736(02)09897-5
Yogo, M., Onoe, K.: The social sharing of emotion among Japanese students. Poster session presented at ISRE'98. The Biannual conference of the International Society for Research on Emotion, Wuerzburg (1998)

Chapter 5
Measuring Emotions Online: Expression and Physiology

Dennis Küster and Arvid Kappas

5.1 Introduction

Emotions are embodied mental processes. They are considered *mental*, because emotions are elicited by information processing in the brain. They are considered *embodied*, because in consequence of this information processing, many changes happen in the brain and in many other parts of the body. It is assumed that these changes are evolutionary selected adaptations to classes of challenges that provide functional benefits for the procreation of the genes that shape this concerted action of systems over the life span (see also Chap. 3).

While the development of human emotions lies much in our ancestral past, even modern day situations, such as reading an insult in a traditional letter or in a digital blog post on a computer screen, trigger event cascades that are not only shaped by experiences made throughout our life-time, but also using systemic constraints that are ancient. It is because of this heritage, that pixels on a computer screen can change the frequency with which our heart beats change the distribution of blood from the core of the body to the periphery, or trigger metabolic cascades that ultimately will change the availability of energy for cells, providing conditions for adaptive action. It does not matter that these physiological changes are not necessary to prepare the individual in front of a computer screen for action. Typically, all it might take to respond is to write a message, change the channel, or switch off the device—none of these requires much physical action, however, the emotional reaction is not scaled to whatever is necessary in the specific situation in the here-and-now. Instead, the response might be beneficial if the opponent would have to be physically threatened, or we would have to make a quick physical exit out of the situation.

D. Küster (✉) • A. Kappas
Department of Psychology and Methods, Jacobs University Bremen, Campus Ring 1, 28759 Bremen, Germany
e-mail: d.kuester@jacobs-university.de; a.kappas@jacobs-university.de

J.A. Hołyst (ed.), *Cyberemotions*, Understanding Complex Systems,
DOI 10.1007/978-3-319-43639-5_5

Because of the complexity with which these bodily systems interact, understanding emotions involves an assessment of many different systems, ranging from self-report of how people feel, to expressive behavior, and changes that are linked to affective influences on the peripheral and central nervous system of individuals. This is necessary because the cohesion between these systems is moderate to low and not a single measure is seen as an absolute criterion for diagnosing a particular emotional state (Mauss and Robinson 2009; Hollenstein and Lanteigne 2014). In other words, someone might report being angry but not show any measurable change in their expression, or physiology. Inversely, somebody might report not being touched by the statement that "climate change is nonsense", but does so with the eyebrows pulled down, and a marked increase in sweat on the palms of their hands. Emotion scientists would consider both of these scenarios affectively relevant and any understanding of emotions on the Internet can also be seen in such a multi-level context.

In recent years, the Internet as a source of increasingly comprehensive and rich data about human emotional behavior at a large scale (Golder and Macy 2011) has attracted a dramatic increase of interest from researchers across disciplinary boundaries (Mayer-Schönberger and Cukier 2013). Thus, online measurement of emotional contents in text, such as those obtained with **sentiment**-mining tools like *SentiStrength* (see Chap. 7) have been at the core of many publications within the CYBEREMOTIONS project (see Chap. 6; Chmiel et al. 2011; Garas et al. 2012; Paltoglou et al. 2013; Thelwall et al. 2010) any beyond (Alvarez et al. 2015; Tanase et al. 2015). However, it is not really known how emotions contained in text relate to how a reader or writer feels, and to what extent they may be associated with a full-blown affective response. In this chapter, we will consider the emotion detected by sentiment mining as a property of the text and we will investigate how emotion-in-text, relates to emotion in the person.

In contemporary laboratory research, due to the evolutionary heritage of emotions outlined earlier, components of the emotional response are frequently collected at more than one *level*. Bodily and expressive changes, or other variables, such as differences in response times, are given weight in addition to what subjects report verbally if they are asked or presented specific response scales—because emotions are understood to take place not only in the spotlight of conscious subjective experience but also at more automatic levels involving the entire body (see Cacioppo et al. 1996; Scherer 1984, 2009). In fact, emotion theories generally posit the presence of a synchronized emotional response across several levels, yet empirical support for this assertion has often been weak or inconsistent (Hollenstein and Lanteigne 2014). In addition, there are social pressures that might lead participants in laboratory research to downplay, or exaggerate their responses. Because of this context, we suggest in this chapter that some of these time tested laboratory measures for emotion research can be fruitful also in the study of cyberemotions, even if much of the visible emotional online content initially appears to take the shape of plain text. Because we, as psychologists, assume that it is ultimately the people behind the keyboard who are emotional, we need to investigate critically to what extent text-based analysis tools and self-report relate to other measures of human emotions. In other words, we need to discuss the issue of cohesion between different levels of the emotional response, such as what people write when they are emotional, what they can report about their emotional feelings, and what bodily indicators can tell us about emotional states associated with cyberemotions. We will repeatedly return to this issue as we discuss the different measures and examples presented in this chapter.

A basic design decision relates to the choice of theoretical framework to operate in. This is important because it has immediate implications for the question which indicators will be most suitable. One of the most important distinctions here is whether emotions are conceived of as discrete states, such as happiness, or anger, or at a more abstract level as points in a two- or three-dimensional space. For example, sadness would be a negative state that is typically characterized by low to medium activation, whereas anger might be just as negative, but much more active. Happiness might be as active as anger, but clearly positive, and we can conceive of states in such a two-dimensional space, such as very slightly positive, and very slightly activated that do not have a clear categorical label that would correspond to such a state (e.g., Yik et al. 2011). Related to the basic decision about the theoretical framework is the question of which measures should be used. Ideally, whatever measure or measures we use to detect cyberemotions should be economic to use, highly reliable, not interfering with the ongoing situation, and in agreement with other indicators across a wide range of experimental contexts. Unfortunately, apart from the fact that no reliable gold standard is available, there likewise is no single measure or set of measures that would be the optimal fit for every situation. Rather, each method and each individual indicator has its own strengths and weaknesses that may or may not be critical for any individual study. In consequence, emotion researchers have to make a number of careful design decisions.

In many cases, adaptation of laboratory research for the purposes of validation and comparison with computer-based assessments of emotional online texts is facilitated by the use of dimensional models rather than discrete models of basic and qualitatively different emotional states (e.g., Barrett and Russel 1999; Russell 2009). In comparison to discrete emotion perspectives, dimensional models do not have to preserve all of the assumed qualitative differences between discrete states, and this can simplify the measurement considerably. Dimensional models further have a strong aim to organize emotional states in terms of a limited number of underlying dimensions (Mauss and Robinson 2009), a property which makes it easier for dimensional models to be "understood" by a computer, and thereby minimizes the risks of cross-category misclassifications based on highly contextualized context-dependent information. For these reasons, a dimensional framework was chosen for the majority of research in the CYBEREMOTIONS project. For the purposes of research within a framework focusing on discrete emotional states, some of the considerations and examples presented in this chapter would need to be modified. However, the overarching issue of limited cohesion would remain.

One of the most reliable indicators of negative affect relates to changes in activation of the *Corrugator Supercilii* muscles that are involved in pulling the eyebrows together and down. In addition to being strongly activated, such as when a subject is frowning in response to exposure to a negative stimulus, these muscles can also relax beyond what would be observed for a neutral baseline. Thus, when a subject is in a positive emotional state, the *Corrugator Supercilii* muscles are likely to be relaxed. In consequence, the state of relative relaxation vs. activation of these muscles can provide a mapping of emotional valence that has been shown to be largely linear (Mauss and Robinson 2009). However, even this measure is not always correlated with other behavioral or physiological changes. These limits of cohesion should therefore be considered seriously, and factors that might moderate cohesion need to be understood to improve affect detection (Kappas 2010). Because no single "marker" is yet in sight that could reliably detect emotions on its own, emotion researchers from all disciplines have to make a number of informed decisions about their measurement paradigms.

Our primary concern regarding the measurement of cyberemotions is the question to what extent it can be shown that the indicators that we use to measure emotions indeed measure emotions in a way that is meaningful. Arguably, if behavior of individuals on the Internet could be used to reliably infer their emotional states, even in the wild, i.e., outside of the lab, then this would open up entirely new horizons for applied research. A basic step in this direction is the validation of our measures by tools that have been carefully tested in laboratory research on emotions. For example, in the case of sentiment analysis, automatic classifiers can be validated by ratings from independent human judges who were not involved in the initial training of the classifier. Thus, in a recent study together with Paltoglou, Thelwall and colleagues (Paltoglou et al. 2013, see Chap. 7), we compared evaluations by human judges with various automatic prediction methods. Surprisingly, the results of this study vastly exceeded our a-priori expectations for the strength of this relationship. In fact, the results of this small validation study found correlations of up to 0.89 for hedonic valence (positive to negative), and 0.42 for arousal (calm to excited). Therefore, while these results are not directly transferable to other samples and domains of cyberemotions on the Internet, it reinforces the argument that the degree of observed cohesion varies between measures and experimental context. As new measures are tested, and empirical designs are improved, further research may uncover more successful ways of detecting emotions online. For example, new techniques, such as infrared thermal imaging, have recently been tested successfully in the study of guilt in children (Ioannou et al. 2013), as well as in the context of facial responses toward social ostracism (Paolini et al. 2016). Yet to what extent such technological advances will indeed lead to systematic increases of cohesion with other components of the emotional response remains an empirical question. In the remainder of this chapter, we will go into more detail on different aspects of experimental measurement and design, and the question of cohesion will reappear for every measure that we are going to discuss.

5.2 Self-report Measures of Cyberemotions as Beliefs vs. Feelings

Self-report measures of emotion capture what we can verbalize about our emotional feelings (Barrett 2004), as well as our beliefs about emotions (Robinson 2002a). While people have been shown to differ in how much they focus on different dimensions of feeling states, such as pleasantness and intensity (Barrett 2004), self-report data is generally seen as an indispensable tool. Sometimes, it is even seen as the only tool worth using. An essential observation, however, is that people often have difficulties to cleanly separate emotional feelings from generalized beliefs about emotions, such as how someone is supposed to feel at a funeral (see Robinson 2002a). The study of such biases and other kinds of contextual influences on human perception is a bread and butter topic of social psychology.

Research has shown that generalized statements about emotions can sometimes differ dramatically from reports made during, or directly subsequent to an event. For example, Barrett et al. (1998) have shown that stereotypical sex-related differences, such as the widely held belief that women are more emotional than men, only emerge when participants are asked to provide relatively global self-descriptions. In one study, males would agree more to items such as "I rarely experience strong emotions" (Barrett et al. 1998, p. 561). Yet when they were instead asked to provide momentary ratings of their actual emotional experience during a 10-min interaction, or when they recorded their emotional responses to all of their emotional interactions during a 1-week period, none of these stereotypical sex differences could be shown. Thus, there are conditions where self-report can be assumed to reflect emotional experience relatively well, such as when we evaluate our "online" emotional experience when it happens (Robinson 2002a). However, when we have to recall emotions after the fact, we increasingly rely on more generalized beliefs rather than on episodic memory (Robinson 2002a,b). In other words, "offline" post-hoc emotional self-reports describe relatively stable beliefs about how we usually feel, whereas event-specific or time-dependent knowledge about a specific emotional experience is more likely to be retrieved if we can focus on our current feelings. What does this imply for the offline study of online cyberemotions?

For cyberemotions, current emotional feelings are likely of particular interest to researchers who, e.g., aim to improve dynamic models of the emergence of emotional states across an online community (see Chaps. 8, 10 and 11). In other cases, emotion dynamics may be of lesser interest, such as when a static corpus of text is to be analyzed for emotional content in general. The important point here, however, is that emotional self-report data may vary in the extent to which it reflects episodic emotional experiences such as feelings vs. more generalized stereotypes about what people may *typically* feel (Robinson 2002a; Mauss and Robinson 2009). This has implications for what types of self-report data are suitable for different types of research on cyberemotions, and how it should be collected.

We concur with Mauss and Robinson (2009) in suggesting to measure subjective experience of emotions as closely as possible to when and where it actually occurs. Similarly, when we expose subjects to emotional statements collected from the Internet, we may either ask them to report about their own emotional response to just having read these statements—or we might ask them to annotate the emotional content of the statements for us. This renders the phrasing of each question particularly sensitive. For example, if a researcher is interested in current emotional states, rather than generalized beliefs, this should be clearly reflected in the phrasing of each relevant item. To give a very simple example, we might ask a participant in the laboratory "How do you feel right now?" on a rating scale for valence—instead of asking "How positive was this picture?" What is most important here, however, is to be aware of how such differences may influence participants, and adapt our designs and methods accordingly. Often, the answer will be that self-report alone is insufficient as the only indicator of emotional state of the individual.

Once the idea of using self-report as the uncontested gold standard is abandoned, the issue of consistency between different measures rises to the fore. For example, when we (Paltoglou et al. 2013) compared human valence ratings to *SentiStrength* metrics on the same corpus of forum posts, participants were asked: "How did the thread you just read make you feel?" This question was presented repeatedly after each individual forum discussion so that participants would be able to base their answers on current emotional experiences instead of memory or generalized stereotypes. As already discussed, we were rather surprised to find substantial correlations between human ratings and *SentiStrength*, suggesting that the algorithms were indeed able to predict episodic feeling states associated with human emotional responses to online content. Nevertheless, caution is still advised because subjects in this online study might have found it socially more desirable to provide stereotypical answers based on the content of the threads rather their immediate emotional experience as such. This is an example where data from other components of the emotional response, such as facial activity, could have been informative. If, for example, facial responses were to show strong evidence for coherence with self-report, social desirability might be ruled out as an alternative explanation. Of course, additional measures at the same time increase complexity to the point that the entire pattern of data should be interpreted rather than any individual measure on its own. Thus, data from other levels, such as bodily indicators, themselves need to be analyzed and integrated within an appropriate multi-level analysis framework (see Cacioppo et al. 2000).

5.3 From Feelings to Bodily Responses: Psychophysiological Measures

The next two sections of this chapter are devoted to a selection of psychophysiological measures of emotion. In general, psychophysiological measures offer opportunities to extend the study of cyberemotions beyond self-report and automated text analyses, yet they still have to face certain technical limitations when taken outside of the laboratory. For example, it is generally not yet possible to record online physiological responses at a large scale across all individual members of an online community (but see Kappas et al. 2013). Such, however, are primarily technical limitations at the level of large scale or community-wide measurement that may be bridged by technological advancement of the tools. It may even become possible to record at least some of these measures reliably at a large scale in the foreseeable future using mobile devices or webcams (Picard 2010; Poh et al. 2011). Of greater importance to the present discussion, however, is how they, i.e., the activity they record, relate to the other components of the emotional response.

Psychophysiological measures are indicators of bodily responses that are assumed to be associated with psychological phenomena. In comparison to emotional self-report, they promise to be more objective and less influenced by certain confounding factors such as socially desirable responding (Ravaja 2004; see also Paulhus 1991, 2002). Certain peripheral bodily responses like facial activity and skin conductance have furthermore been shown to be associated with emotional self-report in the laboratory (Mauss and Robinson 2009), and they can be recorded at a considerably lower cost than most physiological measures of central nervous system (CNS) activity. Thus, while CNS measures such as electroencephalography (EEG) continue to be of great interest to laboratory research on emotions, other measures may be more likely to bridge the gap to larger scale online emotions.

Unfortunately, a comprehensive discussion of all psychophysiological measures that have been shown to be associated with emotion is well beyond the scope of this chapter. Instead, we will have to focus on a few examples, and refer the interested reader to an overview chapter (e.g., Larsen et al. 2008), or one of the standard reference volumes on psychophysiology (e.g., Andreassi 2007; Cacioppo et al. 2000, 2007). While we will include a few basic technical considerations for the measures that are discussed for the following examples, our main aims in the present discussion are to illustrate some of the basic principles, and to show how measures of bodily responses might be used to complement and validate self-report data.

5.3.1 Measuring Valence: Facial Electromyography

Perhaps the single most important bodily component of emotions is what we show, or fail to show, on our faces (see Darwin 1872/2005). In the psychophysiological laboratory, there are two principal ways of assessing movements of the face. First, visible facial muscle activity can be classified by trained and certified coders of highly standardized anatomically based systems, such as the Facial Action Coding System (FACS, Ekman and Friesen 1978). Second, activation of the facial musculature, such as those associated with smiling and frowning, can be recorded by means of facial electromyography (EMG) using electrodes glued on the skin. While both methods have their individual strengths and weaknesses, we will focus on the latter because a particular strength of facial EMG is that even very subtle responses below the visual detection threshold can be recorded, including muscular relaxation (van Boxtel 2010). Facial EMG has further been of particular relevance for the validation of cyberemotions because it has frequently been mapped onto a valence-arousal dimensional space based on circumplex models of affect (see Russell 1980; Yik et al. 2011). As discussed above, the use of a valence-arousal model was one of the early design decisions in the CYBEREMOTIONS project. If, discrete emotion states such as anger, fear, and happiness were to be measured instead, FACS-coding might be more appropriate. Facial EMG, however, is particularly suitable for the measurement of how pleasant vs. unpleasant certain emotional stimuli are perceived—i.e., the valence of an emotional response (Mauss and Robinson 2009).

Facial EMG is a technique that can be used to record muscle activity with the help of small electrodes attached to the face. These specialized, re-useable, electrodes are filled with conducive gel and attached to specific locations on the face following highly standardized recording procedures (Fridlund and Cacioppo 1986). While a certain degree of cleaning of the skin is required to reduce electrical resistance, this procedure is otherwise non-invasive for the participants. However, it can be perceived as relatively obtrusive (van Boxtel 2010) when compared to being filmed. Standardized recording procedures are particularly important because some of the most diagnostic information derived from facial EMG can be relatively small changes in the range of a few microvolts (Tassinary and Cacioppo 2000).

Substantial research has shown that episodic positive and negative affect can reliably be distinguished on the basis of activation at the sites of the *Zygomaticus Major* (smiling) and *Corrugator Supercilii* (frowning) muscles (Brown and Schwartz 1980; Larsen et al. 2003; van Boxtel 2010), and this property is essential for the validation of other valence-based measures, such as text-based instruments (see Küster and Kappas in press). Nevertheless, considerable caution is still needed because the magnitude of convergence with subjective report is limited (Hollenstein and Lanteigne 2014), even for the best measures of facial activity (Mauss and Robinson 2009), and the usefulness of facial activity as a readout of emotional states has been hotly debated on both empirical and theoretical grounds (Fridlund 1991, 1994; Kappas 2003). Furthermore, the most successful laboratory research on the relationship between self-reported emotions and facial EMG has typically used highly standardized emotional images or individual words (e.g., Bradley et al. 2001; Lang et al. 1993; Larsen et al. 2003) that can be judged rapidly on their general emotional content. This, however, is substantially different from evaluating full sentences, or even paragraphs of text that people read on the Internet.

In one simple study, we (Kappas et al. 2010) recently compared emotional responses to images with responses to reading threads taken from online discussion forums. While we observed some surprisingly large correlations between facial EMG activity and subjective report that were generally on par with the coherence observed for the images (Bradley et al. 2001), there were rather large differences in reading times between participants (Fig. 5.1b). Thus, while many participants took somewhere between 10 and 20 min to read all of the texts, others took 30 min or more. Clearly, some participants will have read the forum posts more thoroughly

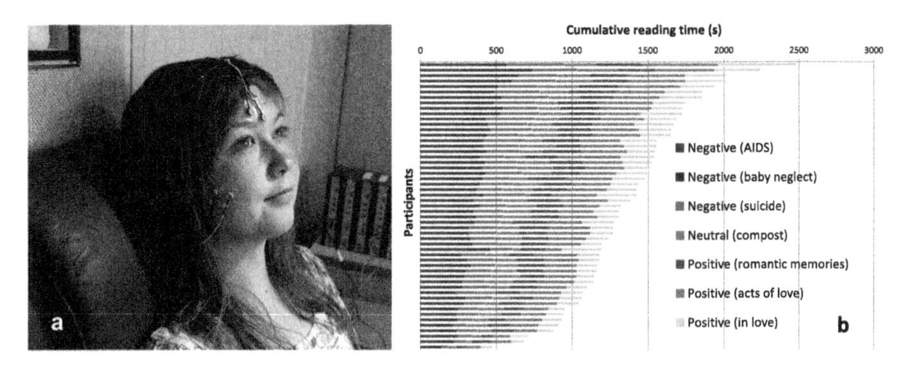

Fig. 5.1 (**a**) A participant in the laboratory while recording facial EMG from two sites (*Corrugator Supercilii, Zygomaticus Major*; photo: Jacobs University); (**b**) cumulative reading times of 53 individual participants expressed in seconds

than others, and for some participants, reading has likely been associated with emotional feelings. However, in other cases, participants may have based their emotional self-report more on generalized stereotypes as discussed by Robinson (2002b) for emotional reports made after the fact.

As this example illustrates, there is often a tradeoff between the ecological validity of responses and maintaining a level of experimental control typical for psychological laboratory research on emotions. Do we, for example, enforce equal reading times for all participants by showing each text for a predefined number of seconds? In standard paradigms of experimental psychology, such intervals are usually fixed precisely—yet in the case of longer texts, this practice would be of questionable value if some participants indeed need twice as long to process any given stimulus. Worse, even if we knew where participants were looking at the moment that a particular facial response occurred, the type of emotional story presented in a given forum thread is already so deeply embedded in contextual information that it quickly becomes surprisingly complicated to unambiguously associate the individual emotional responses with smaller units such as individual words or statements. Likewise, it can be difficult to draw conclusions from subjective report data in such cases because we cannot reasonably ask participants to rate each and every word individually.

The example of facial EMG in a reading study illustrates another more general issue associated with a multi-level measurement of cyberemotions. Psychophysiological measures, such as EMG, can provide quasi-continuous data, whereas there will usually only be a limited number of discrete data points for any measure of self-report. Thus, while continuous psychophysiological data can potentially be very useful for the study of emotion dynamics, there is a mismatch in resolution between measures such as facial EMG and self-report that precludes certain kinds of comparisons, and that may reduce the observed level of coherence in others. Unfortunately, there is no easy solution. That is, while participants could, for example, be given a control device such as a slider or a joystick to continually adjust their perceived emotional state, this is not very feasible when participants have to simultaneously process complex material such as forum posts.

Just for the sake of completeness two variants of assessing facial activity shall be mentioned. Because many researchers do not possess the means of a psychophysiology laboratory, but also do not have access to trained coders, for example of FACS, they might try to use lay-people as judges of emotional expressions. This cannot be recommended, as these measures are typically not very reliable. In a related vein, there is the promise of using *automatic coding* via software. This is indeed something to look forward to in the next few years. However, at the time of writing, none of the systems available are able to measure the full set of movements that a FACS coder would include. Systems that try to identify discrete emotions without providing output linked to specific facial regions are also not recommended because a) these are based on rigid patterns describing stereotypical expressions, and b) much of the action of facial movements appears to be associated with blends of Action Units, and partial displays—hence a dimensional framework might be the best choice. Once the technical problems of computer coding have been solved, this might be a preferred way of assessing expressive behavior as laptops, portable devices, and increasingly stand-alone monitors tend to have cameras built in. While the spatial and temporal resolution of EMG will always be superior, no preparation of the skin is required for visual coding. Nevertheless, the possibility of large-scale remote measurement is clearly on the horizon; and the integration of such emerging technologies with advances in modeling of complex data, such as dynamic patterns in facial temperature (Jarlier et al. 2011), will likely contribute to a more reliable (remote) measurement of facial activity in a multi-disciplinary approach to the study of cyberemotions.

5.3.2 Measuring Physiological Arousal: Electrodermal Activity (EDA)

In many cases, research on cyberemotions will not succeed to satisfactorily represent ongoing emotional processes in users unless arousal is considered in addition to hedonic valence (positive vs. negative). Thus, while it may sometimes appear to be sufficient to determine if someone experienced a particular online situation as unpleasant, a pure valence categorization would be entirely blind to what type of unpleasant emotional state has been elicited. For example, moderately unpleasant states associated with comparatively low arousal such as sadness or boredom may differ dramatically from other unpleasant emotional states such as fear or anger. To distinguish these types of fundamentally different cases, arousal will typically have to be measured alongside valence. Valence and arousal together have been shown to account for a large portion of the variance across a wide range of emotional rating situations (Russell 2003).

A comparatively simple to use physiological measure of arousal is Electrodermal activity (EDA). EDA refers to small changes in electrical conductivity of the skin that are associated with variation in the production of sweat by the eccrine sweat glands. Apart from eccrine sweat glands, there are the aprocrine sweat glands, however, these have remained relatively unstudied in respect to skin conductance (Dawson et al. 2000). Changes in EDA have been the subject of study for well over 100 years, and they have been observed in response to a large and varied number of stimuli. Importantly, EDA has been shown to be associated with activation of the sympathetic nervous system (Wallin 1981), and has been widely used as an indicator of sympathetic arousal (Dawson et al. 2000; Boucsein et al. 2012). For the study of cyberemotions, this implies that physiological arousal can in principle be measured continuously by means of attaching two small electrodes to a suitable place on the skin. However, as we have already seen for the measurement of valence via facial EMG, there are still a number of limitations to be considered.

One major limitation of EDA data is the fact that electrodermal activity cannot tell us what precisely the participant is responding to at any given moment. Instead, this information has to be inferred from the experimental design in which EDA was elicited. Thus, a participant may show very similar responses when there is a sudden noise in the environment, when she is reading an emotionally activating online forum discussion, or when she suddenly remembers that she still has to make an important phone call later in the afternoon. The use of control conditions can reduce, but not eliminate, this problem.

Another characteristic of EDA is that changes can be both slow (tonic) and fast (phasic), both of which can overlap substantially but should be considered separately (Dawson et al. 2000; Boucsein 2012). For the study of emotions, the phasic changes are often of particular interest because they may be tied to a specific experimental event, e.g., the presentation of a particularly activating emotional statement in an ongoing discussion. Such phasic electrodermal responses are called skin conductance responses (SCRs), and they are typically assumed to be associated with a significant experimental event if they occur within a specific time window, such as 1–4 s after stimulus onset (Boucsein et al. 2012). This shift in time between stimulus and response has to do with the physical properties of the relevant sweat glands, and is discussed in more detail in the respective guideline literature (Boucsein 2012; Boucsein et al. 2012). However, tonic changes of skin conductance across a longer time window can likewise be informative about physiological arousal.

While EDA can, in principle, be measured relatively easily and inexpensively from a variety of different locations on the body, not all of these recording sites are equally reliable (van Dooren et al. 2012). The most frequently recommended placement for the EDA electrodes is on the distal phalanges of two fingers on the non-dominant hand (Boucsein et al. 2012). However, for the study of cyberemotions, this location is typically compromised by substantial movement artefacts when participants require both hands to control input devices such as a keyboard. Fortunately, official guidelines exist for an alternative, and comparably reliable recording site at the arches of the feet (Boucsein et al. 2012; van Dooren et al. 2012).

While this location might at first glance appear inconvenient when considering the "sweaty feet", the feet's propensity to produce sweat that is comparable to the sweat produced at the palms of the hands is precisely the reason why the feet are generally suggested as the next best alternative to the hands (see van Dooren et al. 2012). In our laboratory, we have repeatedly used specifically this recording site because it maximizes the freedom of movement of the hands when participants have to type. By resting their feet on a comfortable footstool, for example, most typing-induced movement artefacts can be completely avoided.

Apart from the recording site, interindividual differences between subjects are another important consideration. The potentially most worrisome issue relates to the finding that up to about 25 % of subjects can be classified as electrodermal non-responders (Venables and Mitchell 1996), i.e., people who never or only rarely show significant skin conductance responses. This limitation naturally reduces cohesion with subjective report or text-analyses that can be obtained for all subjects, if only self-report and EDA are measured. Interindividual differences are furthermore not limited to the case of non-responders. Thus, a substantial body of literature has shown that there are significant effects of basic demographic factors such as age, gender, and ethnicity on electrodermal activity (Boucsein 2012). One illustrative example in this context is the observation that self-report of arousal and electrodermal activity may increasingly diverge as people get older. Specifically, Gavazzeni et al. (2008) found that older adults rated the intensity of negative images higher than younger adults—whereas the level of electrodermal activity was reduced for the older adults. In another example study, Ketterer and Smith (1977) observed significant interactions between gender and electrodermal responses to music vs. a series of advertisement paragraphs (verbal stimulus). In this study, females responded more frequently than males to the verbal stimulus—whereas the reverse was found for the musical stimulus. For the measurement of online emotions, these findings imply that researchers should aim to obtain at least minimal demographic data on participants' age, gender, ethnicity, and handedness. None of this means that the collection of EDA cannot contribute to an estimation, or model of, subjects' arousal. However, it serves to emphasize our point that EDA, unfortunately, cannot serve as a simple objective gold standard because it is itself known to be influenced by a large number of factors. Rather, additional measures might be used to fill in the blanks and control some of the variance.

How then might EDA, despite its deficiencies, still contribute to the evaluation of other measures? In an example related to the production of emotional texts we could, e.g., aim to test the assumption that females, on the basis of gender-based stereotypes about themselves, should care more about correctly and tactfully responding to a controversial negative online topic than males. To address such a question, a researcher might begin by repeatedly asking participants about their self-perceived arousal while they are composing forum posts in an experiment. However, as discussed above, self-report questions can only be asked at a limited frequency without creating too much interference with the ongoing task. This means that there will usually only be a few discrete points of measurement for self-report data. In

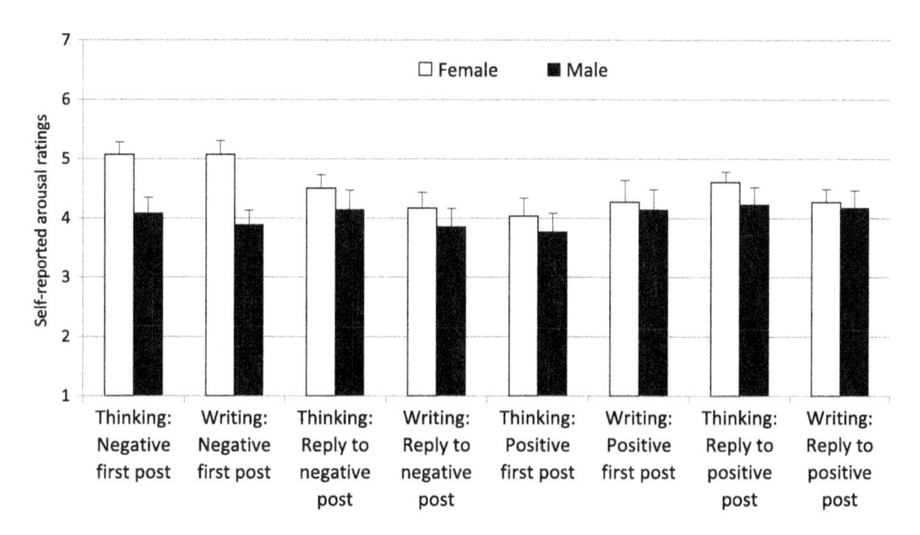

Fig. 5.2 Self-reported emotional arousal of male and female participants across different stages of composing online forum posts. Participants had to either reply to a positive vs. negative topic, or write the first post of a topic. The figure distinguishes intervals where participants were thinking about what to write and intervals where they were actually writing. Higher values on the 7-point scale represent greater intensity of self-rated arousal. Error bars denote standard error of the means

many cases, researchers will even decide to have only a single moment of self-report measurement to minimize distractions during the task. EDA, however, can be measured continuously throughout the entire experiment.

At the same time, experimental design decisions can be used to make self-report data more useful even for questions pertaining to emotion dynamics. We could, for example, ask subjects about their emotional experience when they are starting to think about what they are going to write—and then ask them again after they have actually written it. This is, in fact, what we have done in a recent writing study with 59 right-handed participants (30 female, 29 male) in our laboratory (Küster et al. 2011). These participants were asked to write online forum posts while physiological measures such as EDA and EMG were recorded, and they were asked to intermittently report about their perceived emotional state (valence, arousal).

In our writing study (Fig. 5.2), it became evident that participants felt at least moderately activated by the writing task, and that female participants tended to report feeling slightly more excited than male participants—in particular when they had to compose an opening forum post on a negatively valenced topic. Likewise, when we showed the same group of participants a selection of emotional images from a set of standardized pictures taken from the International Affective Picture System (IAPS; Lang et al. 2008; Fig. 5.3), female participants generally reported somewhat higher subjective arousal than males. In comparison to the writing task, only the most extreme negative images used in this study reached roughly the same level of arousal (see the rightmost three stimuli in Fig. 5.3: A duck dying from an oil spill; a painful dental operation; a gun pointed at the subject). This suggests

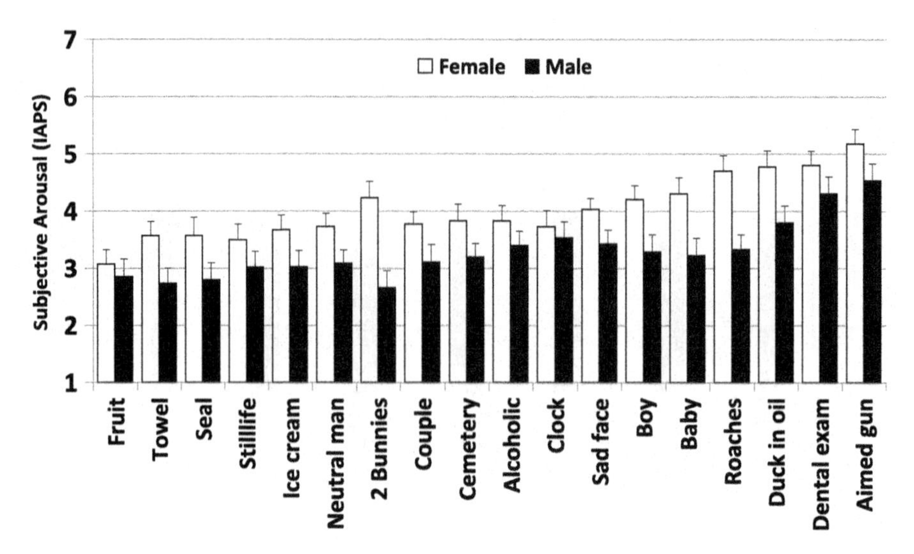

Fig. 5.3 Self-reported emotional arousal of male and female participants in response to a standardized set of frequently used emotional images (IAPS; Lang et al. 2008). Higher values on the 7-point scale represent greater intensity of self-rated arousal. Error bars denote standard error of the means

that participants typically perceived the writing tasks as at least somewhat more activating than watching emotional images, and that females appeared generally more responsive to both tasks than male participants who may have "played it cool".

Next, we turn to consider the electrodermal activity shown by the same group of participants while looking at the images, and while engaging in the online writing task (Fig. 5.4). Here, a rather complex pattern can be observed. On first glance, engaging in the different stages of the writing task appears to have elicited a generally higher level of arousal than looking at emotional images. This suggests a certain level of cohesion that could also be observed when EDA at the feet was correlated with self-reported arousal in response to the images. Depending on gender (male vs. female) and location of measurement (left foot vs. right foot) these correlations were of medium ($r = 0.44$; female left foot) to large ($r = 0.60$; male right foot) magnitude, suggesting a substantial albeit not perfect level of cohesion between both measures for a standard picture viewing task.

On closer inspection of the data of the writing study, however, it appears that there was substantially more variance in the levels of EDA across the different phases of the task—and between both genders. Thus, while female participants appeared to have been subjectively more excited, this was not evident from the level of electrodermal activity. We can interpret this result in the context of earlier findings on gender differences in tonic EDA (Ketterer and Smith 1977) discussed above—however, it cannot be taken at face value. Rather we should remind ourselves of the issue of timing of episodic emotional self-reports discussed, among others, by Mauss and Robinson (2009). In the case of standardized images presentation,

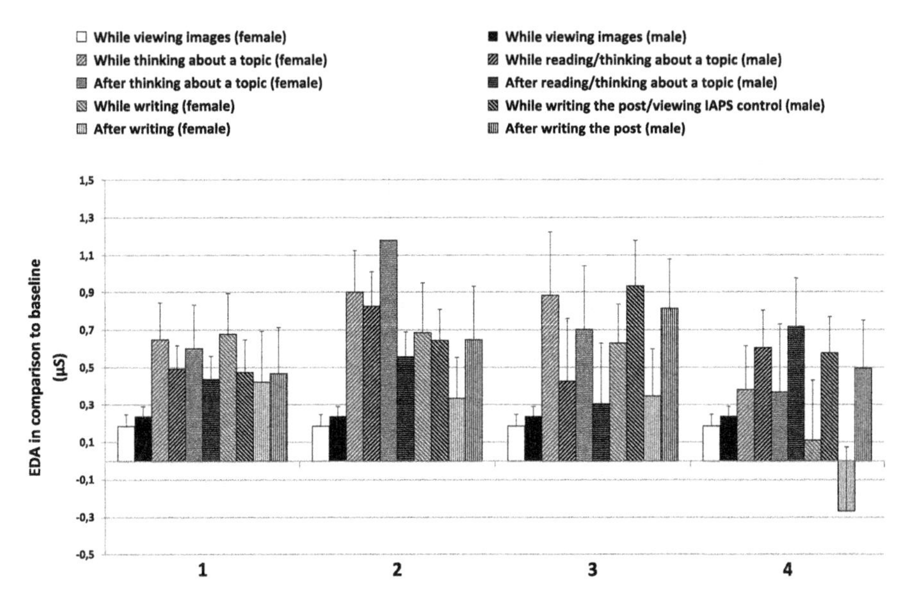

Fig. 5.4 Electrodermal activity of 59 participants (30 female, 29 male) expressed as a difference to the baseline activity level across different stages of composing online forum posts. Male and female participants had to either reply to a positive vs. negative topic, or write the first post of a topic. The figure distinguishes intervals where participants were thinking about what to write, intervals where they were actually writing, and intervals where they were merely looking at a set of widely used emotional images. Error bars denote standard error of the means

such as the intervals of 6 s per image that have been customary and time-tested in a large volume of psychophysiological research, the object of the emotional self-report is quite clearly defined in the mind of participants. For a complex writing task spanning minutes, however, the precise object of the rating may have been much less clear. At this point, we might wonder if participants actually report some sort of mean of their emotional experience during the writing task that would be equivalent to an averaged level of electrodermal activity—or if they rather focus on a peak experiences at the beginning or end of a task. While research from other areas of psychology suggests the latter, this is a question about the emotion dynamics of emotional writing that would likely best be studied with the aid of continuous measures.

In the writing study, thinking and writing phases were experimentally separated even though they may usually take place nearly simultaneously in everyday life. The advantage was that another point of self-report could be gained. In addition, further tentative conclusions could be drawn. First, it appears that pondering what to write online may often be associated with a greater intensity of emotional excitement (arousal) than the actual act of writing itself. Second, some of the results suggest that males may have benefitted less from the writing activity than females. I.e., at least in some cases, females may have enjoyed a greater post-writing relaxation effect than males. Some of these theoretically more interesting findings would require further research and replication. However, for the present purposes, they serve as an illustration of both the potential and the limitations of adding physiological measures such as EDA to the mix of measures of online emotions. Psychophysiological measures provide the study of cyberemotions with tools to address the complexity of dynamically unfolding online emotions—and at the same time they force researchers to deal with a greater level of complexity that brings along its own problems and potential confounds as part of the deal.

The interpretation of EDA as a sign of arousal is based on the notion that relevant stimuli trigger a response of the sympathetic branch of the autonomic nervous system. For completeness sake, it should be mentioned that here too there is potentially an alternative to EDA assessment. For example, pupil dilation is associated with arousal in a very similar way (Bradley et al. 2008). The problem is that pupil dilation is also dependent of the brightness of the visual object the eye is focusing on. This makes it very difficult to use this measure when the stimuli (e.g., web sites) are quite heterogeneous with regard to brightness and contrast. Cardiac correlates of sympathetic arousal, such as the pre-ejection period of the heart require rather complicated recording set-ups that mean that EDA, even if recorded at the feet, is the best available measure for arousal in the context of the assessment of arousal in cyberemotions (for a recent study employing both PEP and EDA, see, Kreibig et al. 2013).

In the CYBEREMOTIONS project, as discussed above, data collected in the laboratory has been used to validate sentiment classifiers, as well as for the validation and development of models of emotion dynamics. To give a more concrete example of how this kind of data might be used within a modeling context, we will conclude this section with a brief look at the role of empirical data in an agent-based framework such as the models for online emotions developed at ETH Zurich (see Chap. 10). First, the validity of a model can be tested by simulations of the behavior of agents driven by the parameters and equations used in a model. For example, a simulation of 100,000 agents per type of thread was used at ETH Zurich for comparison with the subjective report data collected at our laboratory (Fig. 5.5). Even though certain differences between the empirical data and the model simulations might still remain, this strategy can lead to a good level of agreement. Second, validated models, such as the ETH framework, can be used to study emotion dynamics such as those evident in the facial EMG and EDA. E.g., one of the central assumptions of the ETH model has been that arousal drives user participation in online discussions: Such assumptions can be tested at more than one level when data from both subjective report and physiological response parameters are considered.

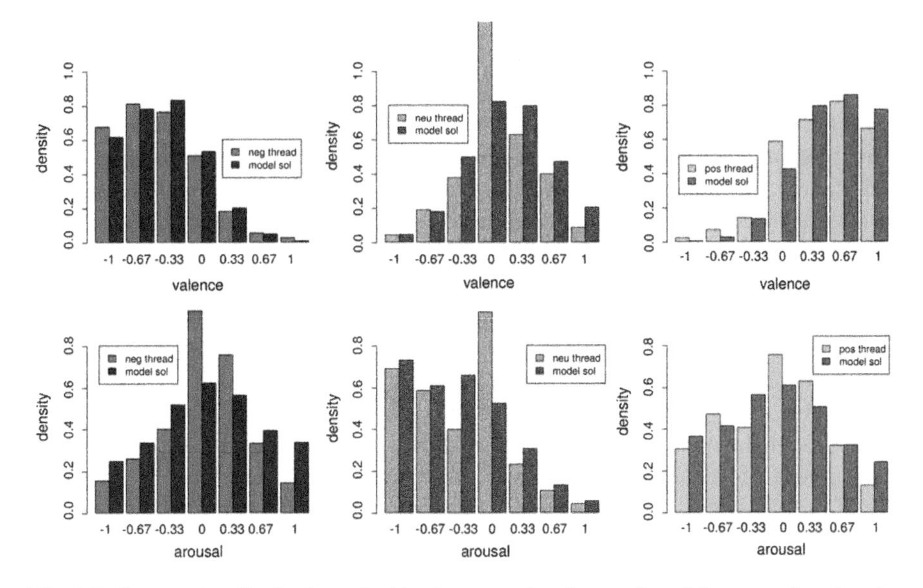

Fig. 5.5 Comparative distributions of subjective report data from real participants reading threads in our laboratory (*light bars*) vs. simulated agents based on the parameters and models developed at ETH Zurich (*dark bars*). Valence distributions are shown in the *top row*, arousal distributions in the *bottom row*

5.4 Conclusions

In this chapter, we have argued for a multi-level approach in the study of cyber-emotions that combines self-report data with information from other levels of emotional processing such as bodily responses. We have discussed strengths and limitations of a selection of indicators, with a particular focus on emotional self-report, facial electromyography, and electrodermal activity. Emotional self-report remains an important corner stone in the study of online emotions that ties directly into sentiment analysis and the computational treatment of expressions of emotional states in written text (Chap. 6). However, neither self-report nor any other individual measure has to date been identified that could reliably identify one of the other components of the emotional response by a single criterion. In some cases, computational sentiment analysis corresponds surprisingly well with emotional self-report, and this is quite encouraging for further research that is based on software such as *SentiStrength* (see also Chap. 6). However, self-report often correlates only weakly with other emotion measures. From a psychophysiological perspective, this is not very surprising because measures aiming at different levels of emotional functioning are understood as in fact measuring different, albeit partially overlapping, aspects of emotional processing.

We have argued that psychophysiological measures such as facial EMG or EDA can make a valuable contribution to research on cyberemotions. This applies to the continued need for basic research in the laboratory as well as the future application of large-scale remote measurement devices. It is tempting to think of these types of objective measures as a solution to the problem of obtaining a simple and continuous readout of the private emotional states of online participants. However, while each of these measures can be informative about different aspects of emotions, we emphasize that they cannot provide a readout of emotional feelings as such. That said, some physiological measures, such as a relaxation of the *Corrugator Supercilii* muscles associated with frowning, have been identified that correlate fairly reliably with perceived emotional valence. Nevertheless, we argue that psychophysiological measures can often be more useful if they are not primarily intended as physiological markers for emotional feeling states. As we have discussed for the example of electrodermal activity during the composition of online forum posts, an important strength of physiological measures is that emotional responses can be recorded continuously without participants having to pay special attention to their ongoing emotional feeling states. This often allows a more meaningful interpretation of the associated subjective data as well as conclusions about the dynamics of emotions while participants are engaged with an emotional online task. In the context of the interdisciplinary project from which this book emerged, we have seen how physiological data and subjective report can be used to validate and improve agent-based models (see Chap. 10).

The lack of cohesion that we have discussed throughout this chapter has, in some ways, more complex implications than the technical limitations it appears to impose upon the predictive value of any one measure of emotion. Is this entirely a matter of imprecise measurement instruments, movement artefacts, placement of electrodes, or social desirability? It appears likely that certain issues of cohesion can be reduced, or even eliminated by technical improvements on the side of recording sensors. However, we have to keep in mind that participants' self-report about their emotional feelings is itself likely to be influenced by bodily states (see Allen et al. 2001; Strack et al. 1988). Humans certainly do not perceive bodily responses such as facial expressions, sweating, or heartbeats in quite the same way as electrodes and amplifiers do. And yet, bodily sensations clearly play an important role in how we experience, report, and think about them. For example, intercultural research has shown that people from very different cultures like Belgium, Mexico, and Indonesia associate emotions with bodily sensations, such as to feel the heart beating faster (Breugelmans et al. 2005). Most people nevertheless have surprising difficulties to correctly discriminate and monitor their own heartbeats, unless trained (Katkin et al. 1982; Wiens et al. 2000).

While our main focus in this chapter has been on measures of emotion that have become established in experimental psychology and psychophysiology, it should also be noted that sentiment analysis (Chap. 7), the emerging corner stone of such a large proportion of research on cyberemotions, is still occupying a niche role in conventional emotion research. This is surprising given the potential of this measure, yet perhaps more understandable as an example of the interdisciplinary boundaries that have to be overcome in this field. An exception has been the recent development of the LIWC (*Linguistic Inquiry and Word Count*) by Pennebaker and colleagues (e.g., Pennebaker et al. 2007) and their previous work on therapeutic writing (e.g., Pennebaker 1993) that has captured sustained interest from clinical and social psychologists. Thus, there has been a clinical interest in text-analyses involving therapeutic essays and longitudinal studies (Danner et al. 2001) but surprisingly little basic research involving these measures in the psychological laboratory. We, by no means, intend to downplay the partially still untapped potential of automated text-analyses for emotion research. However, this exciting new measure is presented in more detail in the next chapter of this volume, and we will instead focus on a few more general issues, as well as examples of widely used measures in psychology laboratories.

Ideally, smart affective sensors and dynamic emotion models would be able to on-the-fly interpret how emotional responses from different levels of measurement interact to produce distinct emotional states. In addition, they would be context-sensitive. Some progress on the context-sensitivity capabilities of automated systems has already been made throughout the lifetime of the CYBEREMOTIONS project (Thelwall et al. 2013; cf. Chap. 7). However, to truly understand intriguing aspects of emotions, such as the interplay of subjective experience and bodily responses, future models would have to develop and test further assumptions on what aspects of bodily activation directly or indirectly influence other components, such as subjective experience. While some of these questions have been of interest to psychologists and psychophysiologists since the days of William James, it is only since recently that some of the more complicated issues and their inherent complexity can be addressed by computational modeling. Clearly, further research is needed here. One outcome of this process may eventually be better physiological markers for subjective emotional states. However, as we have argued throughout this chapter, researchers have to be prepared for the complexity of this endeavor. For this purpose, we have to be aware that there still is no gold standard, and that emotions involve the entire body, rather than "just feelings".

References

Allen, J.J.B., Harmon-Jones, E., Cavender, J.H.: Manipulation of frontal EEG asymmetry through biofeedback alters self-reported emotional responses and facial EMG. Psychophysiology **38**(4), 685–693 (2001). doi:10.1111/1469-8986.3840685

Alvarez, R., García, D., Moreno, Y., Schweitzer, F.: Sentiment cascades in the 15M movement. EPJ Data Sci. **4**(1), 1–13 (2015). doi:10.1140/epjds/s13688-015-0042-4

Andreassi, J.L.: Psychophysiology: Human Behavior and Physiological Response, 5th edn. Erlbaum, Mahwah, NJ (2007)

Barrett, L.F.: Feelings or words? Understanding the content in self-report ratings of experienced emotion. J. Pers. Soc. Psychol. **87**(2), 266–281 (2004). doi:10.1037/0022-3514.87.2.266

Barrett, L.F., Russell, J.A.: The structure of current affect: controversies and emerging consensus. Curr. Dir. Psychol. Sci. **8**(1), 10–14 (1999). doi:10.1111/1467-8721.00003

Barrett, L.F., Robin, L., Pietromonaco, P.R., Eyssell, K.M.: Are women the "more emotional" sex? Evidence from emotional experiences in social context. Cognit. Emot. **12**(4), 555–578 (1998). doi:10.1080/026999398379565

Boucsein, W.: Electrodermal Activity. Springer, New York (2012). doi:10.1007/978-1-4614-1126-0

Boucsein, W., Fowles, D.C., Grimnes, S., Ben-Shakhar, G., Roth, W.T., Dawson, M.E., Filion, D.L.: Publication recommendations for electrodermal measurements. Psychophysiology **49**(8), 1017–1034 (2012). doi:10.1111/j.1469-8986.2012.01384.x

Bradley, M.M., Codispoti, M., Cuthbert, B.N., Lang, P.J.: Emotion and motivation I: defensive and appetitive reactions in picture processing. Emotion **1**(3), 276–298 (2001). doi:10.1037//1528-3542.1.3.276

Bradley, M.M., Miccoli, L., Escrig, M.A., Lang, P.J.: The pupil as a measure of emotional arousal and autonomic activation. Psychophysiology **45**(4), 602–607 (2008). doi:10.1111/j.1469-8986.2008.00654.x

Breugelmans, S.M., Poortinga, Y.H., Ambadar, Z., Setiadi, B., Vaca, J.B., Widiyanto, P.: Body sensations associated with emotions in Rarámuri Indians, rural Javanese, and three student samples. Emotion **5**(2), 166–174 (2005). doi:10.1037/1528-3542.5.2.166

Brown, S.-L., Schwartz, G.E.: Relationships between facial electromyography and subjective experience during affective imagery. Biol. Psychol. **11**(1), 49–62 (1980). doi:10.1016/0301-0511(80)90026-5

Cacioppo, J.T., Berntson, G.G., Crites, S.L. Jr.: Social neuroscience: principles of psychophysiological arousal and response. In: Higgins, E.T., Kruglanski, A.W. (eds.) Social Psychology: Handbook of Basic Principles, pp. 72–101. Guilford, New York (1996)

Cacioppo, J.T., Tassinary, L.G., Berntson, G.G.: Psychophysiological science. In: Cacioppo, J.T., Tassinary, L.G., Berntson, G.G. (eds.) Handbook of Psychophysiology, 2nd edn., pp. 3–23. Cambridge University Press, Cambridge (2000)

Cacioppo, J.T., Tassinary, L.G., Berntson, G.G.: Handbook of Psychophysiology, 3rd edn. Cambridge University Press, Cambridge (2007)

Chmiel, A., Sienkiewicz, J., Thelwall, M., Paltoglou, G., Buckley, K., Kappas, A., Hołyst, J.A.: Collective emotions online and their influence on community life. PLoS ONE **6**(7), e22207 (2011). doi:10.1371/journal.pone.0022207

Danner, D.D., Snowdon, D.A., Friesen, W.V.: Positive emotions in early life and longevity: findings from the nun study. J. Pers. Soc. Psychol. **80**(5), 804–813 (2001). doi:10.1037/0022-3514.80.5.804

Darwin, C.: The Expression of the Emotions in Man and Animals. Digireads.com, Stilwell, KS (2005) (Original work published 1872)

Dawson, M.E., Schell, A.M., Filion, D.L.: The electrodermal system. In: Cacioppo, J.T., Tassinary, L.G., Berntson, G.G. (eds.) Handbook of Psychophysiology, 2nd edn., pp. 200–223. Cambridge University Press, Cambridge (2000)

Ekman, P., Friesen, W.V.: The Facial Action Coding System. Consulting Psychologists Press, Paolo Alto, CA (1978)

Fridlund, A.J.: Sociality of solitary smiling: potentiation by an implicit audience. J. Pers. Soc. Psychol. **60**(2), 229–240 (1991). doi:10.1037/0022-3514.60.2.229

Fridlund, A.J.: Human Facial Expression: An Evolutionary View. Academic, San Diego, CA (1994)

Fridlund, A.J., Cacioppo, J.T.: Guidelines for human electromyographic research. Psychophysiology **23**(5), 567–589 (1986). doi:10.1111/j.1469-8986.1986.tb00676.x

Garas, A., García, D., Skowron, M., Schweitzer, F.: Emotional persistence in online chatting communities. Sci. Rep. **2**, 402 (2012). doi:10.1038/srep00402

Gavazzeni, J., Wiens, S., Fischer, H.: Age effects to negative arousal differ for self-report and electrodermal activity. Psychophysiology **45**(1), 148–151 (2008). doi:10.1111/j.1469-8986.2007.00596.x

Golder, S.A., Macy, M.W.: Diurnal and seasonal mood vary with work, sleep, and daylength across diverse cultures. Science **333**(6051), 1878 (2011). doi:10.1126/science.1202775

Hollenstein, T., Lanteigne, D.: Models and methods of emotional concordance. Biol. Psychol. **98**, 1–5 (2014). doi:10.1016/j.biopsycho.2013.12.012

Ioannou, S., Ebisch, S., Aureli, T., Bafunno, D., Ioannides, H.A., Cardone, D., Manini, B., Romani, G.L., Gallese, V., Merla, A.: The autonomic signature of guilt in children: a thermal infrared imaging study. PLoS ONE **8**(11), e79440 (2013). doi:10.1371/journal.pone.0079440

Jarlier, S., Grandjean, D., Delplanque, S., N'Diaye, K., Cayeux, I., Velazco, M.I., Sander, D., Vuilleumier, P., Scherer, K.R.: Thermal analysis of facial muscles contractions. IEEE Trans. Affect. Comput. **2**(1), 2–9 (2011). doi:10.1109/T-AFFC.2011.3

Kappas, A.: What facial activity can and cannot tell us about emotions. In: Katsikitis, M. (ed.) The Human Face: Measurement and Meaning, pp. 215–234. Kluwer, Dordrecht (2003)

Kappas, A.: Smile when you read this, whether you like it or not: conceptual challenges to affect detection. IEEE Trans. Affect. Comput. **1**(1), 38–41 (2010). doi:10.1109/T-AFFC.2010.6

Kappas, A., Küster, D., Theunis, M., Tsankova, E.: Cyberemotions: subjective and physiological responses to reading online discussion forums. Poster Presented at the 50th Annual Meeting of the Society for Psychophysiological Research, Portland, OR (2010)

Kappas, A., Krumhuber, E., Küster, D.: Facial behavior. In: Hall, J.A., Knapp, M.L. (eds.) Nonverbal Communication, pp. 131–166. de Gruyter/Mouton, Berlin, New York (2013)

Katkin, E.S., Morell, M.A., Goldband, S., Bernstein, G.L., Wise, J.A.: Individual differences in heartbeat discrimination. Psychophysiology **19**(2), 160–166 (1982). doi:10.1111/j.1469-8986.1982.tb02538.x

Ketterer, M.W., Smith, B.D.: Bilateral electrodermal activity, lateralized cerebral processing and sex. Psychophysiology **14**(6), 513–516 (1977). doi:10.1111/j.1469-8986.1977.tb01190.x

Kreibig, S.D., Samson, A.C., Gross, J.J.: The psychophysiology of mixed emotional states. Psychophysiology **50**(8), 799–811 (2013). doi:10.1111/psyp.12064

Küster, D., Kappas, A.: Measuring emotions in individuals and internet communities. In: Benski, T., Fisher, E. (eds.) Internet and Emotions, pp. 48–61. Routledge (2013)

Küster, D., Kappas, A., Theunis, M., Tsankova, E.: Cyberemotions: subjective and physiological responses elicited by contributing to online discussion forums. Poster Presented at the 51st Annual Meeting of the Society for Psychophysiological Research, Boston, MA (2011)

Lang, P.J., Greenwald, M.K., Bradley, M.M., Hamm, A.O.: Looking at pictures: affective, facial, visceral, and behavioral reactions. Psychophysiology **30**(3), 261–273 (1993). doi:10.1111/j.1469-8986.1993.tb03352.x

Lang, P.J., Bradley, M.M., Cuthbert, B.N.: International affective picture system (IAPS): affective ratings of pictures and instruction manual. Technical Report A-8. University of Florida, Gainesville, FL (2008)

Larsen, J.T., Norris, C.J., Cacioppo, J.T.: Effects of positive and negative affect on electromyographic activity over zygomaticus major and corrugator supercilii. Psychophysiology **40**(5), 776–785 (2003). doi:10.1111/1469-8986.00078

Larsen, J.T., Berntson, G.G., Poehlmann, K.M., Ito, T.A., Cacioppo, J.T.: The psychophysiology of emotion. In: Lewis, R., Haviland-Jones, J.M., Barrett, L.F. (eds.) The Handbook of Emotions, 3rd edn., pp. 180–195. Guilford, New York (2008)

Mauss, I.B., Robinson, M.D.: Measures of emotion: a review. Cognit. Emot. **23**(2), 209–237 (2009). doi:10.1080/02699930802204677

Mayer-Schönberger, V., Cukier, K.: Big Data: A Revolution That Will Transform How We Live, Work, and Think. Houghton Mifflin Harcourt, New York (2013)

Paltoglou, G., Theunis, M., Kappas, A., Thelwall, M.: Predicting emotional responses to long informal text. IEEE Trans. Affect. Comput. **4**(1), 107–115 (2013). doi:10.1109/T-AFFC.2012.26

Paolini, D., Alparone, F.R., Cardone, D., van Beest, I., Merla, A.: "The face of ostracism": the impact of the social categorization on the thermal facial responses of the target and the observer. Acta Psychol. **163**, 65–73 (2016). doi:10.1016/j.actpsy.2015.11.001

Paulhus, D.L.: Measurement and control of response bias. In: Robinson, J.P., Shaver, P.R., Wrightsman, L.S. (eds.) Measures of Personality and Social Psychological Attitudes. Academic, San Diego, CA (1991)

Paulhus, D.L.: Socially desirable responding: the evolution of a construct. In: Braun, H.I., Jackson, D.N., Wiley, D.E. (eds.) The Role of Constructs in Psychological and Educational Measurement, pp. 49–46. Erlbaum, Mahwah, NJ (2002)

Pennebaker, J.W.: Putting stress into words: health, linguistic and therapeutic implications. Behav. Res. Ther. **31**(6), 539–548 (1993). doi:10.1016/0005-7967(93)90105-4

Pennebaker, J.W., Booth, R.J., Francis, M.E.: Linguistic Inquiry and Word Count: LIWC 2007, Austin, TX (2007). LIWC www.liwc.net

Picard, R.W.: Emotion research by the people, for the people. Emot. Rev. **2**(3), 250–254 (2010). doi:10.1177/1754073910364256

Poh, M.-Z., McDuff, D.J., Picard, R.W.: Advancements in noncontact, multiparameter physiological measurements using a webcam. IEEE Trans. Biomed. Eng. **58**(1), 7–11 (2011). doi:10.1109/TBME.2010.2086456

Ravaja, N.: Contributions of psychophysiology to media research: review and recommendations. Media Psychol. **6**(2), 193–235 (2004). doi:10.1207/s1532785xmep0602_4

Robinson, M.D., Clore, G.L.: Episodic and semantic knowledge in emotional self-report: evidence for two judgment processes. J. Pers. Soc. Psychol. **83**(1), 198–215 (2002a). doi:10.1037//0022-3514.83.1.198

Robinson, M.D., Clore, G.L.: Belief and feeling: evidence for an accessibility model of emotional self-report. Psychol. Bull. **128**(6), 934–960 (2002b). doi:10.1037//0033-2909.128.6.934

Russell, J.A.: A circumplex model of affect. J. Pers. Soc. Psychol. **39**(6), 1161–1178 (1980). doi:10.1037/h0077714

Russell, J.A.: Core affect and the psychological construction of emotion. Psychol. Rev. **110**(1), 145–172 (2003). doi:10.1037/0033-295X.110.1.145

Russell, J.A.: Emotion, core affect, and psychological construction. Cognit. Emot. **23**(7), 1259–1283 (2009). doi:10.1080/02699930902809375

Scherer, K.R.: On the nature and function of emotion: a component process approach. In: Scherer, K.R., Ekman, P. (eds.) Approaches to Emotion, pp. 293–317. Lawrence Erlbaum Associates, Hillsdale, NJ (1984)

Scherer, K.R.: The dynamic architecture of emotion: evidence for the component process model. Cognit. Emot. **23**(7), 1307–1351 (2009). doi:10.1080/02699930902928969

Strack, F., Martin, L.L., Stepper, S.: Inhibiting and facilitating conditions of the human smile: a nonobtrusive test of the facial feedback hypothesis. J. Pers. Soc. Psychol. **54**(5), 768–777 (1988). doi:10.1037/0022-3514.54.5.768

Tanase, D., Garcia, D., Garas, A., Schweitzer, F.: Emotions and activity profiles of influential users in product reviews communities. Front. Phys. **3**, 87 (2015). doi:10.3389/fphy.2015.00087

Tassinary, L.G., Cacioppo, J.T.: The skeletomotor system: surface electromyography. In: Cacioppo, J.T., Tassinary, L.G., Berntson, G.G. (eds.) Handbook of Psychophysiology, 2nd edn., pp. 3–23. Cambridge University Press, Cambridge (2000)

Thelwall, M., Buckley, K., Paltoglou, G., Cai, D., Kappas, A.: Sentiment in short strength detection informal text. J. Am. Soc. Inf. Sci. Technol. **61**(12), 2544–2558 (2010). doi:10.1002/asi.21416

Thelwall, M., Buckley, K., Paltoglou, G., Skowron, M., García, D., Gobron, S., Ahn, J., Kappas, A., Küster, D., Hołyst, J.A.: Damping sentiment analysis in online communication: discussions, monologs and dialogs. In: Computational Linguistics and Intelligent Text Processing. Lecture Notes in Computer Science, vol. 7817, pp. 1–12 (2013). doi:10.1007/978-3-642-37256-8_1

van Boxtel, A.: Facial EMG as a tool for inferring affective states. In: Spink, A.J., Grieco, F., Krips, O.E., Loijens, L.W.W., Noldus, L.P.J.J., Zimmerman, P.H. (eds.) Proceedings of Measuring Behavior 2010, pp. 104–108. Eindhoven, The Netherlands (2010). Retrieved from Measuring Behavior 2012. http://www.measuringbehavior.org/files/ProceedingsPDF(website)/Boxtel_Symposium6.4.pdf

van Dooren, M., de Vries J.J.G., Janssen, J.H.: Emotional sweating across the body: comparing 16 different skin conductance measurement locations. Physiol. Behav. **106**, 298–304 (2012). doi:10.1016/j.physbeh.2012.01.020

Venables, P.H., Mitchell, D.A.: The effects of age, sex and time of testing on skin conductance activity. Biol. Psychol. **43**(2), 87–101 (1996). doi:10.1016/0301-0511(96)05183-6

Wallin, B.G.: Sympathetic nerve activity underlying electrodermal and cardiovascular reactions in man. Psychophysiology **18**(4), 470–476 (1981). doi:10.1111/j.1469-8986.1981.tb02483.x
Wiens, S., Mezzacappa, E.S., Katkin, E.S.: Heartbeat detection and the experience of emotions. Cognit. Emot. **14**(3), 417–427 (2000). doi:10.1080/026999300378905
Yik, M., Russell, J.A., Steiger, J.H.: A 12-point circumplex structure of core affect. Emotion **11**(4), 705–731 (2011). doi:10.1037/a0023980

Part II
Sentiment Analysis

Chapter 6
Sensing Social Media: A Range of Approaches for Sentiment Analysis

Georgios Paltoglou and Mike Thelwall

6.1 Introduction

Sentiment analysis deals with the *computational* treatment of expressions of *private states* in written text, that is, human states that are not open to objective observation or verification (Quirk et al. 1985). Examples of such states include opinions, sentiments, emotions, beliefs and speculations (Wiebe et al. 1999). An example of an expression of a private state is the statement "I am feeling very positive today", as it reflects an internal evaluation by the subject. In contrast, expressions of non-private states such as "I am going to work" can be objectively observed and verified. Simply put, sentiment analysis deals with designing algorithms and implementing software (i.e., computer programs) that are able to detect and analyse expressions of private states in an automatic or semi-automatic manner. The field is also known as *opinion mining*[1] and in this chapter we use both definitions interchangeably.

Research in the field had been considerably popular in recent years, both in academia and industry. The reasons for this phenomenon can be found in the diverse set of applications that it has been utilized: from forecasting box office movie revenues (Asur and Huberman 2010) to estimating countries' gross happiness indexes (Kramer 2010) and tracking the affective responses of social media users to emerging news stories (Thelwall et al. 2011). In essence, opinion mining offers researchers and practitioners the possibility of analysing large amounts of data in

[1]Refer to Chap. 1.5 of Pang and Lee (2008) for a detailed discussion about the terminology.

G. Paltoglou (✉) • M. Thelwall
School of Mathematics and Computer Science, University of Wolverhampton, Wulfruna Street, Wolverhampton WV1 1LY, UK
e-mail: g.paltoglou@wlv.ac.uk; m.thelwall@wlv.ac.uk

J.A. Hołyst (ed.), *Cyberemotions*, Understanding Complex Systems,
DOI 10.1007/978-3-319-43639-5_6

an efficient (i.e., timely) and effective (i.e., precise) manner to extract their affective content.

Such an analysis is non-trivial and often very challenging as studies have shown that even humans tend to disagree on the affective content of online communication (Paltoglou et al. 2010; Paltoglou and Buckley 2013). Simplistic approaches, such as comparing the occurrences of positive and negative terms in text, are particularly inadequate (Pang et al. 2002). The reason for this is that, in contrast to keyword-based information systems (e.g., search engines) where the occurrence of a term often provides significant evidence about the topicality of a document,[2] the same cannot be said about affective analysis of documents. This is because internet users can be particularly creative when expressing opinions and emotions. For example, lexical phenomena that increase the inherent difficulty in predicting the affective content of online text include:

- *Thwarted expectations* (Pang et al. 2002) which occur when the final sentences of a text changes its overall affective appraisal:

  This film should be brilliant. It sounds like a great plot, the actors are first grade, and the supporting cast is good as well, and Stallone is attempting to deliver a good performance. However, it can't hold up.

- *Irony*, which occurs when words or expressions with a typical positive affective content are figuratively used for expressing a negative opinion (Carvalho et al. 2009):

  Oh, that's beautiful!

  Let's pray that the human race never escapes from Earth to spread its iniquity elsewhere.

- *Mixed emotions*, when more than one diametrically-opposed opinions are expressed in a short text segment:

  And I think you are a slanderous and ignorant fool,mostly because Rush doesn't use drugs any longer. Great how free speech works, isn't it? I mean, where else can all the people who don't know anything get to talk and then I can show up and show them how ignorant they are? I love it!

- *Context*, when it's not the explicit textual content of a communication that contains an expression of a private state, but rather the context in which it is embedded. For example, the statement:

  Go read the book!

  when referring to a movie, clearly indicates a strong negative bias towards it, although its textual content doesn't explicitly contain any affective information.

[2]Most information retrieval systems and algorithms explicitly use this information in estimating the relevancy of a document in regard to a user query (Robertson et al. 2004).

The type of affective analysis that can be conducted and the output of sentiment analysis algorithms also significantly differs, based on the environment of the analysis. Examples include, but are not limited, to:

- A *ternary* prediction on whether the examined text contains positive, negative affective content or is neutral/objective (i.e., doesn't contain expressions of private states). Examples include online reviews praising or criticising products (Pang et al. 2002) or opinions in support or against proposed legislation (Whitelaw et al. 2005).
- A *categorical* extraction of affective content where the output can be one of several potential states such as nervousness, anxiety, fear, fatigue and tension (Mishne 2005; Strapparava and Mihalcea 2008).
- A *numeric* prediction in specific affective dimensions, such as valence or arousal (González-Bailón et al. 2012; Dodds and Danforth 2010), that indicate the level or positivity and mobilization respectively.[3]

As previously explained, the type of analysis that is most appropriate for a specific environment and application requirement is based on a number of factors. Typical factors include the typical number of words per communication unit (e.g., blog posts vs. tweets), the type of language that is prevalent in the particular medium (e.g., the amount of informal abbreviations used) and the desired subsequent analysis; for example, a simple ternary classification may be sufficient for product reviews but for online political discussions, the detection of extreme states of negativity and mobility may be more useful in taking preventive action (Paltoglou 2014).

In this chapter, we discuss a range of different approaches to solve the problem of accurately predicting the nature of private states expressed in social media. Section 6.2 will focus on machine learning solutions, i.e., solutions that require some pre-annotated data to automatically extract the underlying patterns that characterise different affective content. Section 6.3 will present the lexicon-based solutions that were investigated within the project, that is, algorithms that rely on sentiment dictionaries. Lastly, the chapter concludes with a summary and a discussion of the potential future directions of the field in Sect. 6.4. A detailed analysis of SentiStrength (Thelwall et al. 2010), one of the most successful algorithms for sentiment strength detection in social media, and its extensions is given in Chap. 7 and it is often used in this chapter as a baseline for evaluation.

6.2 Machine Learning Solutions

In this section, we briefly present the general principles of machine learning and subsequently focus on a range of relevant approaches that have been used for sentiment analysis.

[3] Valence and arousal are discussed in more depth in Sect. 6.2.2.

Generally, machine learning is based on the 'learn by example' paradigm according to which software programs *learn* to perform a specific task from *data*, rather than being explicitly programmed for the task by humans (Sebastiani 2002). The specific task at hand here is the detection and identification of affective content in text. The data that is used comprises textual segments (e.g., product reviews, tweets, blog posts) that have been read by humans and manually annotated with a category, such as positive, negative or neutral. Due to the utilization of manually annotated data, the machine learning approaches that we discuss here fall within the 'supervised' category.[4]

Machine learning algorithms function in two stages; during the *training phase*, an inductive process learns the general characteristics of a category (e.g., positive, negative) by observing the properties of documents that belong to the category. Typical properties that are being observed include the terms that comprise the document, their frequency and particular syntactic patterns. The details of the inductive process give rise to a number of machine learning algorithms, such as Naive Bayes (John and Langley 1995), Maximum Entropy (Le Cessie and Van Houwelingen 1992), Support Vector Machines (Platt 1999; Joachims 1999) and others. The acquired knowledge is stored and subsequently applied to unseen text, during the *testing* or *application* stage to determine the most appropriate category. A detailed analysis of machine learning is beyond the score of this chapter, but the interested reader is encouraged to seek the books of Bishop (2006) and Mitchell (1997).

6.2.1 Hierarchical Language Models

6.2.1.1 Background

In this initial approach, we address the problem of extracting the emotions from text segments through an hierarchical Language Model (h-LM) classifier. The classifier functions in a two-tier fashion; the first stage classification determines whether or not the examined text is objective or subjective and the second determines the polarity of the document (e.g., positive or negative) if it was classified as subjective in the first tier classification. The final output of the classifier is therefore one of $\{objective, negative, positive\}$.

We utilize language model classifiers (Manning et al. 2008; Manning and Schuetze 1999) for both classification tasks. As the classifiers are probabilistic in nature, that is, they not only output the most appropriate category for a text segment but also the probability that the text segment belongs to that category, the final output includes the probabilities $P(objective|D)$, $P(subjective|D)$ for the first stage classification and $P(negative|D)$, $P(positive|D)$ for the latter, where $P(c|D)$

[4]In contrast, 'unsupervised' machine learning algorithms do not require annotated data.

is defined as the posterior probability of text segment D belonging to class c. By default, if $P(objective|D) \geq P(subjective|D)$ (respectively, $P(negative|D) \geq P(positive|D)$) then the document is classified as *objective* (similarly, *negative*). This is equivalent to setting a threshold value $thres_{obj} = 0.5$ and classifying a document as objective if $P(objective|D) \geq thres_{obj}$.

Language models were initially used for speech recognition (Jelinek et al. 1991) tasks, but have also been extensively used in other application areas, such as Information Retrieval (Manning et al. 2008) and classification (Manning and Schuetze 1999). In our implementation we use $n - gram$ token models employing Witten-Bell smoothing (Witten and Bell 1991). For brevity, we omit the details of the implementation and the algorithm here. The full details can be found in Mitrović et al. (2010).

6.2.1.2 Experiment Setup and Results

We experiment with unigrams and bigrams ($n = 1, 2$ respectively) which provide a good compromise between effectiveness (i.e., predictive capability) and efficiency (i.e., storage and memory requirements). We train the language model classifiers on the BLOGS06 dataset (MacDonald and Ounis 2006; MacDonald et al. 2008). The training dataset consists of 16,481 *objective* documents, 7930 *negative* documents and 9968 *positive* documents. Negativity and positivity in the dataset are defined as negative or positive opinions towards specific entities (e.g., people, companies, films). Refer to Mitrović et al. (2010) for the full details of the dataset extraction process.

For the first level classification (i.e. objective vs. subjective) we use the documents from the objective category and the union of positive and negative as subjective. For the second stage classifier (i.e. positive vs. negative), we use the respective positive and negative documents.

To evaluate the predictive accuracy of the produced classifiers we use two external datasets; the first is extracted from the BBC Message Boards,[5] where registered users are allowed to start discussions and post comments on existing discussions on a variety of topics, ranging from UK/World news to religion. The second dataset is extracted from the social news website Digg,[6] where people share and discuss news and ideas. The difference in language between the two datasets[7] and the diversity of topics that they contain offer a unique opportunity to explore the effectiveness of emotional classification in both controlled and unrestrained forum-like environments.

In order to create the *testing* dataset, a sample of comments is extracted from both datasets and given to three human annotators to manually annotate their emotional

[5]http://www.bbc.co.uk/messageboards/.

[6]http://www.digg.com.

[7]The BBC message boards are closely administered and moderated, while Digg posts aren't.

Table 6.1 Number of documents per category for the BBC and Digg datasets

Data set	Number of documents		
	Neutral	Positive	Negative
BBC	96	41	457
Digg	144	107	221

content. Both datasets and the annotation process are described in detail in Paltoglou et al. (2010) and are freely available. Table 6.1 presents some relevant statistics about their content.

Due to the fact that the resulting dataset is highly uneven, in reference to the number of documents in each category, we use the average value of the $F1$ measure for both classes to quantify classification quality, instead of the more typically-used accuracy. The $F1$ for class c is defined as:

$$F1_c = \frac{2P_cR_c}{R_c + P_c} \tag{6.1}$$

where P and R are the precision and recall that the classifier attains for class c respectively, defined as:

$$Precision_c = \frac{tp}{tp + fp}, \quad Recall_c = \frac{tp}{tp + fn} \tag{6.2}$$

where tp is the number of documents correctly classified as belonging to class c ("true positive"), fp is the number of documents falsely classified as belonging to class c ("false positive") and fn is the number of documents falsely classified as not belonging to class c ("false negative"). The final average $F1$ measure is calculated as $F1 = \frac{1}{|c|}\sum_c F1_c$. *Perfect* classification would give an F1 value of 1, because both precision and recall would be equal to 1 for all classes, as $fp = fn = 0$.

In order to present a thorough examination of the performance of Language Models, we report results with different probability thresholds. Specifically, for the objective vs. subjective classification we vary the objectivity threshold ($thres_{obj}$) in the $\{0, 1\}$ range using intervals of 0.005; subsequently, a document is classified as objective if $P(objective|D) > thres_{obj}$. We also do the same for the positive vs. negative classification varying the positive threshold ($thres_{pos}$), therefore a document is classified as positive if $P(positive|D) > thres_{pos}$.

For comparison, we use the Lexicon-Based algorithm by Paltoglou and Thelwall (2012), that will be analyzed in Sect. 6.3.1 and also include the results for *random* prediction (where one of the two classes is chosen by chance) and *majority*, where the class with the highest number of documents is always selected. Figures 6.1 and 6.2 report the effectiveness of the Language Models on the BBC and Digg datasets on both objective vs. subjective and positive vs. negative classification tasks respectively.

The results indicate that although the Lexicon-Based algorithm consistently outperforms language models, with appropriate threshold tuning (i.e., $thres_{obj} \approx 0.85$,

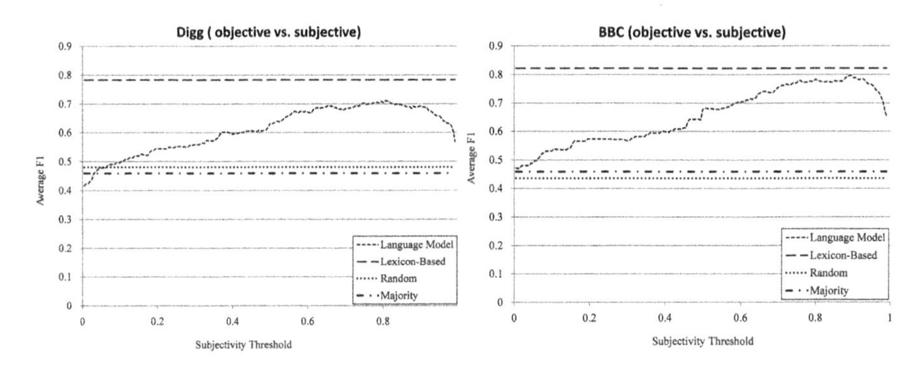

Fig. 6.1 Average F1 value on the objective vs. subjective classification task

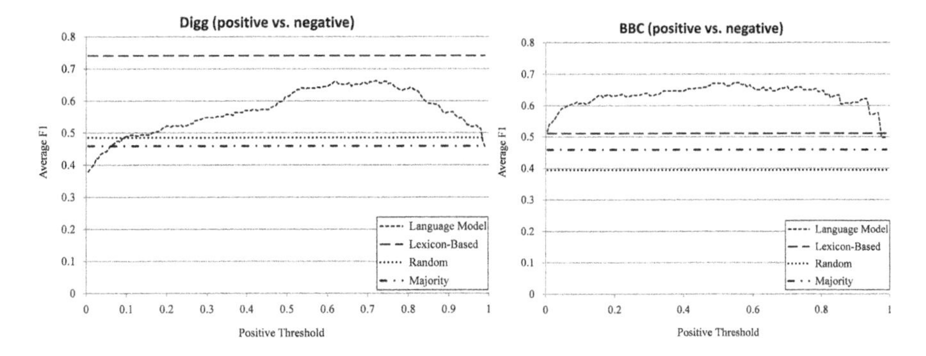

Fig. 6.2 Average F1 value on the positive vs. negative classification task

$thres_{pos}$ ≈0.75), their performance can be optimized. Considering that the Lexicon-based classifier uses extensive syntactic signals to detect negation, exclamation marks, intensifiers and other prose elements, while the language model uses only word tokens without prior knowledge of positive or negative terms and performs no syntactic analysis, it can be concluded that the observed performance is very good.

6.2.2 Predicting Levels of Valence and Arousal

6.2.2.1 Background

Although the ternary classification scheme that was examined above, according to which a text segment is classified as containing either positive, negative or no opinions, is adequate in most environments, there are scenarios and applications where a more fine-grained analysis of emotional content is required. For example, an analysis of online blogs may require the identification of the level of emotional arousal of the authors, so that a prediction of the possibility of an action (e.g., demonstration) is made.

In this section, we investigate the problem of assigning a text segment *two* separate scores; each one is related to the affective intensity in the two psychological dimensions of *valence* and *arousal*, which have been shown to be a vital aspect of emotions[8] (Russell 1980). The former can be defined as the hedonic dimension of experience (i.e., pleasure and displeasure) and the latter describes the level of mobilization or energy (Barrett and Russell 1999). Therefore, a high-arousal state denotes a highly energetic state (e.g., arousement, alertness) while a low-arousal state generally describes a state of tiredness and immobility.

Overall, the problem can be defined as follows; given a text segment, is it possible to determine the level of valence and arousal that is expressed within it? We use a categorical scale for both dimensions; specifically, for arousal the possible values are *{very low, low, moderate, high, very high}* and for valence they are *{very negative, negative, neutral, positive, very positive}*. For example, a blog post could be identified as expressing a very high level of arousal and be very negative.

We approach the problem through a machine learning point-of-view and utilize two state-of-the-art, supervised algorithms: ϵ-Support Vector Regression (ϵ-SVR) and One-versus-All, multi-class Support Vector Machines (*OvA* SVMs). In the former case, the aim is to find a function $f(x)$ that has at most ϵ deviation from the correct targets y_i for all the training data and at the same time is as *flat* as possible. ϵ-SVR has been used extensively in research (Lee and Pang 2005; Shimada and Endo 2008) in relation to the problem of predicting a real-value for a given text segment, a problem similar to the one that we are investigating here. We use the implementation of ϵ-Support Vector Regression by Chang and Lin (2001) and convert its output into an ordinal value with half adjust.

One-vs-All Support Vector Machines classification is a typical extension of the binary classification functionality of SVMs for multi-class settings. During training, M binary classifiers are built, where M is the number of available classes, each one trained to distinguish between a single class and the rest of the classes put together. During testing, all M classifiers are utilized and the one which outputs the largest value (greatest confidence) is chosen as the correct prediction. Despite its conceptual simplicity, *OvA* classification has been shown to be both theoretically valid (Rifkin and Klautau 2004) and generally have good classification accuracy (Lee and Pang 2005). We use the implementation of *OvA* SVMs by Fan et al. (2008) and Keerthi et al. (2008).

In both cases we represent text segments as feature vectors in N-dimensional space, where N is the size of the vocabulary (i.e. the number of unique terms) in the training dataset and a document D is represented as $\{w_1, w_2, \ldots, w_N\}$. We utilize a binary feature weighting scheme according to which the weight of a term is 1 if the term is present in the document and 0 otherwise.

[8]Studies (Bradley and Lang 1999) have shown that emotions can be characterised as "coincidence of values on a number of different strategic dimensions", such as valence and arousal.

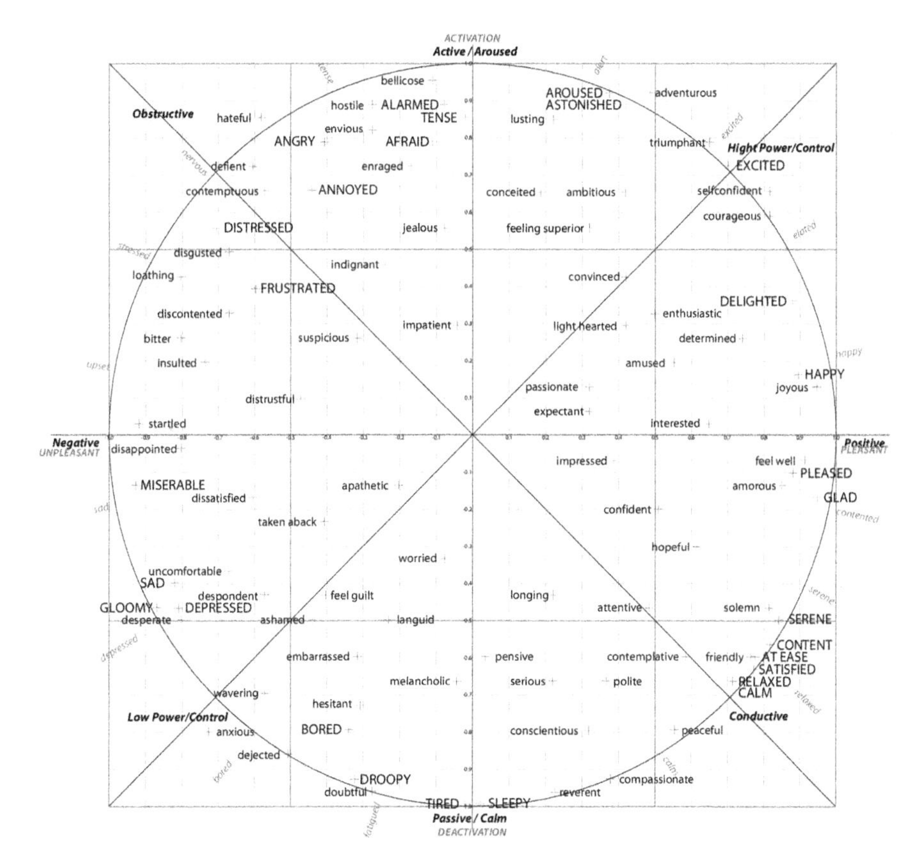

Fig. 6.3 Two dimensional circumplex space model. Figure taken from Ahn et al. (2010) and adapted from Scherer (2005). *Upper-case notation* denotes the terms that were used by Russell (1983)

6.2.2.2 Experiment Setup and Results

For training and testing dataset, we use a corpus of blogs posts from LiveJournal[9] by Mishne (2005). The collection contains 815,494 individual diary-like blog posts authored by users of the web service, each post annotated with either a free-form mood or a mood from a predetermined list. Based on Russell's circumplex space model (Russell 1980) we map a subset of the original moods to a two-dimensional cartesian coordinate system, where the affective states are represented as points and the x and y axis represent intensity of *valence* and *arousal* respectively (Scherer 2005). A representation of the circumplex model can be seen in Fig. 6.3 (see also Ahn et al. 2010 and Russell 1980). Table 6.2 presents the subset of moods from

[9]http://www.livejournal.com.

Table 6.2 Extracted "moods" from the LiveJournal dataset that are used to signify different levels of arousal and valence based on their location on Russell's circumplex model (see Fig. 6.3)

Arousal	Moods	# posts	Valence	Moods	# posts
Very low	Sleepy, tired	31,857	Very negative	Sad, depressed	12,502
Low	Content, depressed	17,560	Negative	Annoyed, bored	21,031
Moderate	Disappointed, good	8390	Neutral	Tired, sleepy	31,857
High	Distressed, annoyed	10,480	Positive	Contemplative, amused, peaceful	37,949
Very high	Enraged, afraid, angry	4188	Very positive	Happy, good	20,035
Total		72,475	Total		123,374

Table 6.3 Results for the 5-scale valence and arousal dimensions

	Valence		Arousal	
Algorithm	Accuracy	Accuracy (± 1)	Accuracy	Accuracy (± 1)
OvA SVM	0.482	0.747	0.516	0.761
ϵ-SVR	0.343	0.816	0.346	0.761
SentiStrength.C4.5	0.306	0.712	–	–

the dataset that is used for training and testing, as well as their respective levels of valence and arousal. For example, according to the figure, the emotional state of *depressed* expresses a *low* level of arousal and is *very negative* in terms of its valence. A complete description of the process and the setup can be found in Paltoglou and Thelwall (2013).

For the experiments conducted, we use 10-fold stratified[10] cross-validation to train and evaluate the algorithms. As predictive metrics, we use accuracy (i.e., the percentage of correctly classified text segments) and accuracy ±1, which considers any prediction within a distance of 1 of the correct class as correct, that is, if the actual arousal of a text segment is *very low* and the algorithm predicts *low*, then we consider this a correct prediction. As baseline for the dimension of valence we use SentiStrength (Thelwall et al. 2010), which is covered in detail in Chap. 7. We map its output to the 5-point ordinal scale that is being investigated here by training a pruned C4.5 decision tree (Quinlan 1993), using the implementation in the Weka software (Hall et al. 2009) and default parameter values. For more details, see Paltoglou and Thelwall (2013).

Table 6.3 presents the results for both dimensions of valence and arousal. An initial observation is that the results for both dimensions and both approaches are better than a random baseline (20 % accuracy), indicating the potential for predicting valence and arousal intensity in heterogeneous, large-scale social media

[10]Stratification ensures that the splits for training and testing are as equal as possible for every class.

environments. The results also show that intensity levels of both valence and arousal can be predicted with some success, despite the complexity of the task. In both cases, OvA SVM classification seems to be more accurate, obtaining an accuracy of 0.482 and 0.516 for valence and arousal respectively using only token features, while ϵ-SV regression performs worse, having an accuracy of 0.343 and 0.346 respectively (statistically significant at $p < 0.01$). Both approaches also perform significantly better than SentiStrength ($p < 0.01$).

The results from the looser ± 1 accuracy show that regression produces better results in both dimensions of valence and arousal (0.816 vs. 0.747 and 0.811 vs. 0.761, respectively). The results indicate that although regression provides lower exact intensity prediction accuracy, its effectiveness outperforms OvA SVM if a metric that doesn't penalize small-scale errors is adopted.

6.3 Lexicon-Based Solutions

Despite the success of machine learning solutions for sentiment analysis, there are environments where such approaches are inappropriate or unfeasible. Typical examples include settings where training data is very difficult or expensive to find and/or create or environments where the heterogeneity of content is so pronounced that different models have to be build for each specific sub-domain. An example of such an environment is *product reviews*, where machine learning models trained on one category (e.g., electronics) attain much lower performance on another type, such as DVDs or movies. The standard solution to this problem is to *adapt* previously trained models to the new domain (Ponomareva and Thelwall 2012), but this process still requires some training data to be created, which may not always be feasible.

In these types of scenarios, lexicon-based solutions are often preferable. These are based on *emotional dictionaries*; lexicons, where lemmas are annotated with affective semantics, for example the level of positivity or negativity they typically convey. There is a significant number of such lexicons available that have been produced either automatically or semi-automatically (Jijkoun et al. 2010). They usually extend the WordNet (Miller 1995) lexical database with additional annotations. Examples include WordNet-Affect (Strapparava and Valitutti 2004) and SentiWordNet (Baccianella et al. 2010).

Complementary to the aforementioned lexicons that were developed mainly within the field of computational linguistics, is a set of affective dictionaries that were motivated and produced by psychological studies, typically *manually* annotated by human participants. Those include the "Linguistic Inquiry and Word Count" (LIWC) (Pennebaker and Francis 1999) and the "Affective Norms for English words" (ANEW) (Bradley and Lang 1999) lexicons. The former classifies words in one or several, not necessarily affective, categories, such as social, family, time, positive, anger, etc. while the latter provides for each word three values of valence, arousal and dominance on a $[1, 9]$ scale. Both have been used in a number of large scale studies (Owsley et al. 2006; González-Bailón et al. 2012; Dodds and Danforth 2010).

In the next two sections, we will present the design, implementation and testing of two such lexicon-based solutions. Both were conceived to answer different research questions and needs and are applicable in different environments and scenarios.

6.3.1 Ternary Classification

6.3.1.1 Background

The classifier is a typical example of an *unsupervised* approach, because it doesn't require any training (i.e., can be applied "off-the-shelf"). It is based on estimating the intensity of negative and positive emotion in text in order to make a ternary prediction for subjectivity and polarity, that is, its output is one of $\{neutral, positive, negative\}$. The notion that both negative and positive emotion is present in a text may seem somewhat peculiar, but is in accordance with a number of psychological studies (i.e., Schimmack 2001; Cornelius 1996; Fox 2008) and is therefore adopted as the underlying premise of the approach. The level of valence in each scale is measured in two independent ratings $\{C_{pos}, C_{neg}\}$; one for the positive dimension ($C_{pos} = \{1, 2 \ldots, 5\}$) and one for the negative ($C_{neg} = \{-1, \ldots, -5\}$), where higher absolute values indicate stronger emotion and values $\{1, -1\}$ indicate lack of (i.e., objective text).

For example, a score like $\{+3, -1\}$ would indicate the presence of only positive emotion, $\{+1, -4\}$ would indicate the presence of (quite strong) negative emotion and $\{+4, -5\}$ would indicate the presence of both negative and positive emotion. In order to make a ternary prediction, the most prevalent emotion, i.e., the one with the highest absolute value, is returned as the final judgement, e.g. positive in the first example above and negative in the other two. For example, the sentence

I hate the fact that I missed the bus, but at least I am glad I made it on time:-)

expresses both negative and positive emotion, where the latter is considered dominant. We solve conflicts of equality (e.g. $\{+3, -3\}$) by taking into consideration the number of positive and negative tokens and giving preference to the class with the largest number of tokens. A document is classified as objective if its scores are $\{+1, -1\}$. Note that the $\{C_{pos}, C_{neg}\}$ ratings are only used as an intermediate step in making the final prediction. That is in contrast to SentiStrength which outputs both values.

The algorithm uses a variation of the LIWC affective dictionary, described above, as its basis. It functions in the following manner: given a text segment, it detects all words that belong to LIWC and extracts their polarity and intensity from it. The initially produced scores are subsequently modified if various prose signals are detected within the neighborhood of the extracted affective tokens. Such signals include negation, capitalization, exclamation marks, emoticons, intensifiers and diminishers. For example the phrase

I am happy

would be given a score of $\{+3, -1\}$ while

I am very happy

would be given a score of $\{+4, -1\}$, because the intensifier "very" is detected near the emotional word "happy". Similarly,

I am very happy!!!

would be assigned a score of $\{+5, -1\}$ because, in addition to the intensifier, exclamation marks are also detected in the text, both of which are considered as signals that increase the emotional intensity of the basic affect expressed by the token "happy". The rest of the prose signals function in a similar fashion. The full details of the algorithm are provided in Paltoglou and Thelwall (2012). It is publicly available for research purposes, both as a *.jar* file with an open API and a command-line interface, as well as a C++ *.dll* at http://www.CyberEmotions.eu.

6.3.1.2 Experiment Setup and Results

We use three different and diverse datasets to evaluate the effectiveness of the classifier. The first data set is extracted from Twitter[11] and it comprises two subsets. The first one (henceforth referred to as *Train*) is collected through the Twitter API, based on the existence of particular emoticons, which are used to provide an indication of the polarity of the text: positive for tweets that contain ':)' or ':-)' and negative for tweets that contain ':(' or ':-('. The second subset (henceforth referred to as *Test*) is humanly annotated for objective, positive and negative emotion. More information about both subsets is provided by Pak and Paroubek (2010).

The second dataset is extracted from Digg and is described in detail in Sect. 6.2.1.2. The third dataset is extracted from the social website MySpace[12] and comprises a sample of comments exchanged between friends in each other's public profile. The dataset is described in detail by Thelwall and Wilkinson (2010). Table 6.4 presents the distribution of objective, positive and negative documents in the datasets.

We use three machine learning approaches in order to compare the effectiveness of the proposed lexicon-based classifier: Naive Bayes (NB) (John and Langley 1995), Maximum Entropy (MaxEnt) (Le Cessie and Van Houwelingen 1992) and Support Vector Machines (SVM) (Platt 1999; Joachims 1999), using binary

[11]http://www.twitter.com.

[12]The extraction took place in 2009, before the website changed its focus to promotion of music.

Table 6.4 Number of documents per class for each data set used

Data set	Number of documents		
	Neutral	Positive	Negative
Digg	144	107	221
MySpace	153	400	105
Twitter *train*	–	232,442	151,955
Twitter (test)	33	108	75

Table 6.5 Subjectivity and polarity classification on the Twitter, Digg and MySpace datasets

Method	Twitter		Digg		MySpace	
	Subjectivity	Polarity	Subjectivity	Polarity	Subjectivity	Polarity
Majority	45.8	37.1	41.0	40.3	43.4	44.2
Random	43.1	49.6	48.0	48.4	46.1	45.3
Lexicon	70.9	**86.5**	**77.0**	**76.2**	**79.9**	**80.6**
SVM	**75.3**	71.0	71.9	72.7	78.6	73.2
NB	71.7	75.0	66.6	69.2	77.0	72.6
MaxEnt	63.9	80.7	59.9	70.1	71.9	63.6

The reported metric is the average $F1$ value. Results should be read in columns, as each one presents the results on a specific dataset and task. Bold results indicate the best performance for the particular dataset and classification task. All results, with the exception of the polarity classification task in Twitter (second column) are based on 10-fold cross-validation

unigrams as features, similarly to Sect. 6.2.2.1. All machine learning algorithms are implemented using the Weka Toolkit (Hall et al. 2009).[13]

Results are based on 10-fold stratified cross-validation for the machine learning approaches for the Digg and MySpace datasets for two classification tasks: objective vs. subjective (i.e., subjectivity detection) and positive vs. negative (i.e., polarity detection). For the Twitter dataset, for the latter task, we use the *Train* subset to train the classifiers and test them on the *Test* subset. For the subjectivity detection task, we use 10-fold stratified cross-validation as previously. The lexicon classifier is tested on the complete datasets, since it doesn't require any reference corpus. We also present results using two baselines: majority and random. In the former case, the class with the highest number of documents is always predicted and in the latter case, a class is selected at random. Reported analyses are based on the F1 metric (refer to Sect. 6.2.1.2).

The results of the experiments are presented in Table 6.5. Overall, it can be observed that in the majority of environments and tasks the lexicon-based solution outperforms machine learning approaches. The results provide a strong indication of the robustness of the lexicon-based solution and its wide applicability in social media settings, without any need for parameter setting or tuning.

[13]All the Weka *.arff* files for all datasets are available upon request.

6.3.2 *ANEW-Based Approach*

6.3.2.1 Background

As discussed in Sect. 6.2.2.1, there is a range of application scenarios where a simple ternary classification scheme isn't sufficient. In these settings, the automatic extraction of the level of emotional arousal, in additional to valence, can be particularly useful. An example of analysis of this type is reported in Sect. 6.2.2.1. Although the presented solutions provide a good initial attempt, it is often the case that a greater level of granularity of emotional analysis (e.g., a real-value prediction on a specific range) can be more insightful. In addition, in contrast to most approaches that focus on the emotional content from the perspective of the author, it is often very important to distill the interpretation of emotion by the *reader*.

In this section, we present a method for extracting the level of valence and arousal in long, informal text segments (i.e., whole forum discussion threads) from the perspective of the reader as a real value on a $\{1, 7\}$ range. Attention should be paid to the specific environment where the analyses is applied as it is very probable that the proposal solutions wouldn't be as effective in shorter text segments (e.g., individual forum posts) as they gather more evidence about the affective content of the text segment, the longer it is.

The solution is lexicon-based and, in contrast to the previously discussed approach, it is based on the ANEW lexicon (Bradley and Lang 1999). As discussed in Sect. 6.3, it provides a value of valence, arousal and dominance on a $[1, 9]$ scale for each token. The algorithm functions in the following manner: given a text segment, it detects all the words in it that also appear in ANEW and extracts their respective values in the dimensions of valence and arousal. Subsequently, and without applying any sort of syntactical or prose analysis, it estimates the overall affective content of the text segment by calculating either the weighted arithmetic (*wAM*) or geometric mean (*wGM*) of the tokens in the text, as follows:

$$wAM_d = \frac{\sum_{t=1}^{N}(tf_t * \mu_{t,d})}{\sum_{t=1}^{N} tf_t} \tag{6.3}$$

$$wGM_d = \sqrt[\prod_{t=1}^{N} tf_t]{(\prod_{t=1}^{N} \mu_{t,d}^{tf_t})} \tag{6.4}$$

where N is the number of unique ANEW tokens found in the text, $\mu_{t,d}$ is the mean value given to term t in ANEW in dimension d, where $d = \{valence, arousal\}$ and tf_t is the term frequency of the term in the text. Alternatively, a Gaussian Mixture Model (*GMM*) is built using the mean values and standard deviations of ANEW tokens. More information about the algorithms is provide by Paltoglou et al. (2013).

The simplicity of the approach should be noted as, in contrast to the solution that is presented in Sect. 6.3.1, it makes no use of syntactic or prose elements, such as exclamation marks, emoticons, etc. Despite those limitations, experiments show

Table 6.6 Results for predicting valence and arousal on a real-value $\{1, 9\}$ range

	Valence			Arousal		
Algorithm	*MSE*	*MAE*	r^a	*MSE*	*MAE*	r^a
wAM	2.39	1.20	0.77	0.73	0.72	0.27
wGM	1.64	**1.07**	0.81	0.63	0.65	0.29
wGMM	3.88	1.51	0.63	0.79	0.74	0.12
bAM	2.34	1.27	0.85	0.66	0.69	0.41
bGM	**1.61**	1.12	**0.87**	**0.56**	**0.62**	**0.42**
bGMM	3.28	1.34	0.61	0.72	0.71	0.18
SentiStrength	4.75	1.83	0.67	–	–	–

Approaches with the *w* prefix denote weighted mean-based solutions, while approaches prefixed with *b* denote Boolean term frequency weights ($tf_t = 1$ in Eqs. (6.3) and (6.4) for all ANEW terms found). Bold results indicate the best performance for the particular metric
[a] Pearson's correlation coefficient

that it is surprisingly effective in predicting the affective response of readers of forum discussion threads. Examples where this type of sentiment analysis has been used include longitudinal analyses of public opinion in online political discussion fora (González-Bailón et al. 2012) or the quantification of the emotional state of populations by examination of representative text segments, such as popular song lyrics, weblogs and others (Dodds and Danforth 2010). Nonetheless, the effectiveness of the approach was evaluated for the first time by Paltoglou et al. (2013).

6.3.2.2 Experiment Setup and Results

As a basis for the comparison and validation of the classifier, 91 human judges were asked to rate on a 7-point Likert-type scale the emotional impact of reading 20 forum discussions in reference to how positive or negative and calm or energetic they made them feel. The forum discussions used as stimuli were selected by four psychologists who identified discussions, with perfect inter-rater agreement, that encompass the whole range of the emotional valence and arousal continua.[14] We compare the output of the algorithm with the average of the human annotations on the dimensions of valence and arousal. Because the predictions are real-valued, previously described metrics, like accuracy or F1 are inappropriate. Subsequently, we evaluate the results using Pearson's correlation coefficient *r*, Mean Square Error (MSE) and Mean Absolute Error (MAE). For valence, we utilize SentiStrength as a strong baseline.

Results are presented in Table 6.6. We focus on Pearson's *r* metric as a basis, since it is the most indicative of the correlation between human judgement and

[14]The full list of discussion threads can be found in the Appendix material as http://doi.ieeecomputersociety.org/10.1109/T-AFFC.2012.26.

automatic prediction. Overall, it can be observed that despite the simplicity of the approaches, valence can be predicted on a very good level (i.e., maximum of 0.87). Arousal, in contrast, is more difficult to predict with automatic methods, with the majority of approaches attaining a Pearson's r correlation of 0.12 to 0.29. The use of Boolean weights results in a significant increase in predictive power, to a maximum of 0.41 for *bAM* and 0.42 for *bGM*, indicating the potential of successfully predicting levels of arousal.

Comparing the different solutions, the Geometric Mean approach seems to always outperform all other mean-based approaches.[15] It is notable that in accordance with machine learning approaches, the use of Boolean features instead of term frequency-based (e.g., *bAM* vs. *wAM*) consistently results in better predictive power. SentiStrength also has a good level of correlation (i.e., 0.67) despite the fact that it wasn't designed for the particular task. Overall, it can be concluded the given long segments of text from social media, valence can be predicted automatically with a good level of accuracy while arousal generally is more challenging.

6.4 Summary

In this chapter, we presented a number of opinion mining solutions for detecting and analyzing the affective content of online social media. Depending on the kind of analysis that is desirable and the particular environment that a solution would be applied to, different approaches were explored and evaluated.

We discussed machine learning approaches in Sect. 6.2, which depend on the availability of training data. In Sect. 6.2.1 we presented an hierarchical Language Model classifier that classifies a text segment on a ternary basis as positive, negative or neutral. The classifier was evaluated on two social media platforms and the results indicated the despite the lack of syntactic analysis or detection of social-media specific features, it can perform very adequately, especially with appropriate parameter tuning. Solutions that detect the level of arousal, as well as valence using an ordinal 5-level scale were explored in Sect. 6.2.2. The results showed that such a fine-grained analysis is possible, especially when small-scale errors are ignored.

In environments where training data is difficult to find or different subdomains are significantly diverse, lexicon-based solutions are typically more appropriate. Those were introduced in Sect. 6.3. We focused on two different scenarios; the first deals with ternary classification of short social media communication and uses extensive prose signals in order to extract the affective content of text. The latter is more appropriate for longer text segments, such as whole forum thread discussions and outputs a numeric value on a $\{1, 9\}$ range of the level of valence and arousal. They were respectively introduced and described in Sects. 6.3.1 and 6.3.2.

[15] That may be due to the fact that by definition the geometric mean is less susceptible to outliers than the arithmetic mean.

Both approaches showed very good results in automatically extracting the affective content of text segments, indicating that despite the popularity of machine learning solutions, lexicon-based ones are still a viable and reliable option.

Overall, we can conclude that despite the challenging nature of designing algorithms and software that are able to automatically detect the expression of privates states in social media, significant progress has been made in recent years. All presented algorithms and datasets are freely available for research purposes from the CyberEmotions website.[16]

Acknowledgements This work was supported by a European Union grant by the 7th Framework Programme, Theme 3: Science of complex systems for socially intelligent ICT. It is part of the CyberEmotions project (contract 231323).

References

Ahn, J., Gobron, S., Silvestre, Q., Thalmann, D.: Asymmetrical facial expressions based on an advanced interpretation of two-dimensional russell's emotional model. In: ENGAGE 2010, pp. 1–12 (2010)

Asur, S., Huberman, B.A.: Predicting the future with social media. In: Huang, X.J., King, I., Raghavan, V., Rueger, S. (eds.) Proceedings of the 2010 IEEE/WIC/ACM International Conference on Web Intelligence and Intelligent Agent Technology, vol. 01, pp. 492–499. IEEE Computer Society, Washington (2010). doi:10.1109/WI-IAT.2010.63

Baccianella, S., Esuli, A., Sebastiani, F.: Sentiwordnet 3.0. In: Calzolari, N., Choukri, K., Maegaard, B., Mariani, J., Odijk, J., Piperidis, S., Rosner, M., Tapias, D. (eds.) Proceedings of the 7th International Conference on Language Resources and Evaluation (LREC'10), Valetta, pp. 2200–2204 (2010)

Barrett, L.F., Russell, J.A.: The structure of current affect: controversies and emerging consensus. Curr. Dir. Psychol. Sci. **8**(1), 10–14 (1999). doi:10.1111/1467-8721.00003

Bishop, C.M.: Pattern Recognition and Machine Learning. Information Science and Statistics. Springer, New York (2006)

Bradley, M.M., Lang, P.J.: Affective norms for English words (ANEW): instruction manual and affective ratings. Tech. Rep. C-1, University of Florida: Center for Research in Psychophysiology (1999)

Carvalho, P., Sarmento, L., Silva, M.J., de Oliveira, E.: Clues for detecting irony in user-generated contents: oh…!! it's "so easy" ;-). In: Jiang, M., Yu, B. (eds.) Proceedings of the 1st International CIKM Workshop on Topic-Sentiment Analysis for Mass Opinion, pp. 53–56. ACM, New York (2009). doi:10.1145/1651461.1651471

Chang, C.C., Lin, C.J.: LIBSVM: a library for support vector machines (2001). Software available at http://www.csie.ntu.edu.tw/~cjlin/libsvm

Cornelius, R.R.: The Science of Emotion. Prentice Hall, Upper Saddle River (1996)

Dodds, P., Danforth, C.: Measuring the happiness of large-scale written expression: songs, blogs, and presidents. J. Happiness Stud. **11**(4), 441–456 (2010). doi:10.1007/s10902-009-9150-9

Fan, R.E., Chang, K.W., Hsieh, C.J., Wang, X.R., Lin, C.J.: LIBLINEAR: a library for large linear classification. J Mach Learn Res **9**(August), 1871–1874 (2008)

Fox, E.: Emotion Science. Palgrave Macmillan, London (2008)

[16]http://www.cyberemotions.eu.

González-Bailón, S., Banchs, R.E., Kaltenbrunner, A.: Emotions, public opinion, and U.S. presidential approval rates: a 5-year analysis of online political discussions. Hum. Commun. Res. **38**(2), 121–143 (2012). doi:10.1111/j.1468-2958.2011.01423.x
Hall, M., Frank, E., Holmes, G., Pfahringer, B., Reutemann, P., Witten, I.H.: Theweka data mining software: an update. SIGKDD Explor. Newsl. **11**(1), 10–18 (2009). doi:10.1145/1656274.1656278
Jelinek, F., Merialdo, B., Roukos, S., Strauss, M.: A dynamic language model for speech recognition. In: Marcus, M.P. (ed.) Proceedings of the Workshop on Speech and Natural Language, pp. 293–295. Association for Computational Linguistics, Stroudsburg (1991). doi:10.3115/112405.112464
Jijkoun, V., de Rijke, M., Weerkamp, W. (2010) Generating focused topic-specific sentiment lexicons. In: Hajic, J., Carberry, S., Clark, S. (eds.) ACL 2010, Proceedings of the 48th Annual Meeting of the Association for Computational Linguistics, pp. 585–594. Association for Computational Linguistics, Stroudsburg
Joachims, T.: Making large-scale SVM learning practical. In: Schoelkopf, B., Burges, C.J.C., Smola, A.J. (eds.) Advances in Kernel Methods: Support Vector Learning, pp. 169–184. MIT Press, Cambridge (1999)
John, G.H., Langley, P.: Estimating continuous distributions in bayesian classifiers. In: Besnard, P., Hanks, S. (eds.) Proceedings of the 11th Conference on Uncertainty in Artificial Intelligence, pp. 338–345. Morgan Kaufmann Publishers, San Francisco (1995)
Keerthi, S.S., Sundararajan, S., Chang, K.W., Hsieh, C.J., Lin, C.J.: A sequential dual method for large scale multi-class linear svms. In: Li, Y., Liu, B., Sarawagi, S. (eds.) Proceedings of the 14th ACM SIGKDD International Conference on Knowledge Discovery and Data Mining, pp. 408–416. ACM, New York (2008). doi:10.1145/1401890.1401942
Kramer, A.D.: An unobtrusive behavioral model of "gross national happiness". In: Mynatt, E., Fitzpatrick, G., Hudson, S., Edwards, K., Rodden, T. (eds.) CHI' 10 Proceedings of the SIGCHI Conference on Human Factors in Computing Systems, pp. 287–290. ACM, New York (2010). doi:10.1145/1753326.1753369
Le Cessie, S., Van Houwelingen, J.C.: Ridge estimators in logistic regression. Appl. Stat. J. R. Stat. Soc. C **41**(1), 191–201 (1992). doi:10.2307/2347628
Lee, L., Pang, B.: Seeing stars: exploiting class relationships for sentiment categorization with respect to rating scales. In: Knight, K., Ng, H.T., Oflazer, K. (eds.) ACL 2005 Proceedings of the 43rd Annual Meeting on Association for Computational Linguistics, pp. 115–124. Association for Computational Linguistics, Stroudsburg (2005)
MacDonald, C., Ounis, I.: The TREC Blogs06 collection: creating and analysing a blog test collection. Tech. Rep. TR-2006-24, Department of Computer Science, University of Glasgow (2006)
MacDonald, C., Ounis, I., Soboroff, I.: Overview of the TREC 2008 Blog Track. In: The Sixteenth Text REtrieval Conference (TREC 2008) Proceedings, NIST Special Publication SP 500-277, p. 1 (2008)
Manning, C.D., Schuetze, H.: Foundations of Statistical Natural Language Processing, 1st edn. MIT Press, Cambridge (1999)
Manning, C.D., Raghavan, P., Schütze, H.: Introduction to Information Retrieval, 1st edn. Cambridge University Press, Cambridge (2008)
Miller, G.A.: Wordnet: a lexical database for English. Commun. ACM **38**(11), 39–41 (1995). doi:10.1145/219717.219748
Mishne, G.: Experiments with mood classification in blog posts. In: Proceedings of ACM SIGIR 2005 Workshop on Stylistic Analysis of Text for Information Access (2005)
Mitchell, T.M.: Machine Learning, 1st edn. McGraw-Hill, New York (1997)
Mitrović, M., Paltoglou, G., Tadić, B.: Networks and emotion-driven user communities at popular blogs. Eur. Phys. J. B **77**(4), 597–609 (2010). doi:10.1140/epjb/e2010-00279-x
Owsley, S., Sood, S., Hammond, K.J.: Domain specific affective classification of documents. In: Computational Approaches to Analyzing Weblogs, Papers from the 2006 AAAI Spring Symposium, Technical Report SS-06-03, pp. 181–183. AAAI Press, Menlo Park (2006)

Pak, A., Paroubek, P.: Twitter as a corpus for sentiment analysis and opinion mining. In: Calzolari, N., Choukri, K., Maegaard, B., Mariani, J., Odijk, J., Piperidis, S., Rosner, M., Tapias, D. (eds.) Proceedings of the 7th International Conference on Language Resources and Evaluation (LREC'10), Valetta, pp. 1320–1326 (2010)

Paltoglou, G.: Sentiment analysis in social media. In: Agarwal, N., Wigand, R.T., Lim, M. (eds.) Online Collective Action: Dynamics of the Crowd in Social Media. Lecture Notes in Social Networks, pp. 3–18. Springer, Wien (2014). doi:10.1007/978-3-7091-1340-0_1

Paltoglou, G., Buckley, K.: Subjectivity annotation of the microblog 2011 realtime adhoc relevance judgments. In: Serdyukov, P., Braslavski, P., Kuznetsov, S.O., Kamps, J., Rüger, S., Agichtein, E., Segalovich, I., Yilmaz, E. (eds.) Advances in Information Retrieval, 35th European Conference on IR Research, ECIR 2013, Moscow, March 2013. Proceedings, pp. 344–355. Springer, Berlin/Heidelberg (2013). doi:10.1007/978-3-642-36973-5_29

Paltoglou, G., Thelwall, M.: Twitter, Myspace, Digg: unsupervised sentiment analysis in social media. ACM Trans. Intell. Syst. Technol. **3**(4), 66:1–66:19 (2012). doi:10.1145/2337542.2337551

Paltoglou, G., Thelwall, M.: Seeing stars of valence and arousal in blog posts. IEEE Trans. Affect. Comput. **4**(1), 116–123 (2013). doi:10.1109/T-AFFC.2012.36

Paltoglou, G., Thelwall, M., Buckely, K.: Online textual communication annotated with grades of emotion strength. In: Proc. EMOTION, pp. 25–31 (2010)

Paltoglou, G., Theunis, M., Kappas, A., Thelwall, M.: Predicting emotional responses to long informal text. IEEE Trans. Affect. Comput. **4**(1), 106–115 (2013). doi:10.1109/T-AFFC.2012.26

Pang, B., Lee, L.: Opinion mining and sentiment analysis. Found. Trends Inf. Retr. **2**(1–2), 1–135 (2008). doi:10.1561/1500000011

Pang, B., Lee, L., Vaithyanathan, S.: Thumbs up?: sentiment classification using machine learning techniques. In: Proceedings of the ACL-02 Conference on Empirical Methods in Natural Language Processing, EMNLP '02, vol. 10, pp. 79–86. ACL, Stroudsburg (2002)

Pennebaker, J.W., Francis, M.E.: Linguistic Inquiry and Word Count, 1st edn. Lawrence Erlbaum, Mahwah (1999)

Platt, J.C.: Fast training of support vector machines using sequential minimal optimization. In: Schoelkopf, B., Burges, C.J.C., Smola, A.J. (eds.) Advances in Kernel Methods: Support Vector Learning, pp. 185–208. MIT Press, Cambridge (1999)

Ponomareva, N., Thelwall, M.: Do neighbours help?: an exploration of graph-based algorithms for cross-domain sentiment classification. In: Proceedings of the 2012 Joint Conference on Empirical Methods in Natural Language Processing and Computational Natural Language Learning, Association for Computational Linguistics, Jeju Island, pp. 655–665. ACL, Stroudsburg (2012)

Quinlan, J.R.: C4.5: Programs for Machine Learning. Morgan Kaufmann Series in Machine Learning, 1st edn. Morgan Kaufmann, San Francisco (1993)

Quirk, R., Greenbaum, S., Leech, G., Svartvik, J.: A Comprehensive Grammar of the English Language. Longman, New York (1985)

Rifkin, R., Klautau, A.: In defense of one-vs-all classification. J. Mach. Learn. Res. **5**, 101–141 (2004)

Robertson, S., Zaragoza, H., Taylor, M.: Simple bm25 extension to multiple weighted fields. In: Proceedings of the 13th ACM International Conference on Information and Knowledge Management, pp. 42–49. ACM, New York (2004)

Russell, J.A.: A circumplex model of affect. J. Pers. Soc. Psychol. **39**(6), 1161–1178 (1980). doi:10.1037/h0077714

Russell, J.A.: Pancultural aspects of the human conceptual organization of emotions. J. Pers. Soc. Psychol. **45**(6), 1281–1288 (1983). doi:10.1037/0022-3514.45.6.1281

Scherer, K.R.: What are emotions? And how can they be measured? Soc. Sci. Inf. **44**(4), 695–729 (2005). doi:10.1177/0539018405058216

Schimmack, U.: Pleasure, displeasure, and mixed feelings: are semantic opposites mutually exclusive? Cognit. Emot. **15**(1), 81–97 (2001). doi:10.1080/02699930126097

Sebastiani, F.: Machine learning in automated text categorization. ACM Comput. Surv. **34**(1), 1–47 (2002). doi:10.1145/505282.505283

Shimada, K., Endo, T.: Seeing several stars: a rating inference task for a document containing several evaluation criteria. In: Washio, T., Suzuki, E., Ting, K.M., Inokuchi, A. (eds.) Advances in Knowledge Discovery and Data Mining, 12th Pacific-Asia Conference, PAKDD 2008 Osaka, May 2008 Proceedings. Lecture Notes in Computer Science (Lecture Notes in Artificial Intelligence), vol. 5012, pp. 1006–1014. Springer, Berlin/Heidelberg (2008). doi:10.1007/978-3-540-68125-0_106

Strapparava, C., Mihalcea, R.: Learning to identify emotions in text. In: Wainwright, R.L., Haddad, H. (eds.) Proceedings of the 2008 ACM Symposium on Applied Computing (SAC), pp. 1556–1560. ACM, New York (2008). doi:10.1145/1363686.1364052

Strapparava, C., Valitutti, A.: WordNet-Affect: an affective extension of WordNet. In: Lino, M.T., Xavier, M.F., Ferreira, F., Costa, R., Silva, R. (eds.) Proceedings of the 4th International Conference on Language Resources and Evaluation (LREC'04), pp. 1083–1086. European Language Resources Association, Paris (2004)

Thelwall, M., Wilkinson, D.: Public dialogs in social network sites: What is their purpose? J. Am. Soc. Inf. Sci. Technol. **61**(2), 392–404 (2010). doi:10.1002/asi.21241

Thelwall, M., Buckley, K., Paltoglou, G., Cai, D., Kappas, A.: Sentiment strength detection in short informal text. J. Am. Soc. Inf. Sci. Technol. **61**(12), 2544–2558 (2010). doi:10.1002/asi.21416

Thelwall, M., Buckley, K., Paltoglou, G.: Sentiment in twitter events. J. Am. Soc. Inf. Sci. Technol. **62**(2), 406–418 (2011). doi:10.1002/asi.21462

Whitelaw, C., Garg, N., Argamon, S.: Using appraisal groups for sentiment analysis. In: Herzog, O., Scheck, H.J., Fuhr, N., Chowdhury, A., Teiken, W. (eds.) Proceedings of the 14th ACM International Conference on Information and Knowledge Management, pp. 625–631. ACM, New York (2005). doi:10.1145/1099554.1099714

Wiebe, J.M., Bruce, R.F., O'Hara, T.P.: Development and use of a gold-standard data set for subjectivity classifications. In: Dale, R., Church, K.W. (eds.) Proceedings of the 37th Annual Meeting of the Association for Computational Linguistics on Computational Linguistics, pp. 246–253. Association for Computational Linguistics, Stroudsburg (1999). doi:10.3115/1034678.1034721

Witten, I.H., Bell, T.C.: The zero-frequency problem: estimating the probabilities of novel events in adaptive text compression. IEEE Trans. Inf. Theory **37**(4), 1085–1094 (1991). doi:10.1109/18.87000

Chapter 7
The Heart and Soul of the Web? Sentiment Strength Detection in the Social Web with SentiStrength

Mike Thelwall

7.1 Introduction

Emotions and sentiments are critical to many human activities, including communication. People not only engage in social communication because they enjoy it or because it helps to fulfil emotional needs but also use sentiment to help convey meaning and react to sentiments expressed towards them or others. Hence, those seeking to model or understand communication patterns on a large scale need to account for the emotions of the participants or at least the sentiments expressed in their messages, whether the emotions arise before the communication or during it.

The importance of emotions applies not only to real time face-to-face communication. It is well documented that people can feel and express emotions through computer mediated communication (CMC) even if it is asynchronous and text-based (Walther and Parks 2002). For example emoticons arose as a partial solution to the lack of body language and intonation to express emotion in informal types of text-based CMC (Derks et al. 2008). Hence, to effectively analyse any area of the social web, emotion should be taken into account for all except the simplest models. If using real data for such analyses, it is necessary to have an automatic method to extract sentiment from text and this is the sentiment analysis task.

Sentiment analysis software reads text and uses an algorithm to produce an estimate of its sentiment content. This estimate can be in several different forms: binary—either positive/negative or objective/subjective; trinary—positive/neutral/negative; scale—e.g., −5 (strongly negative) to +5 (strongly positive); dual scale—e.g., +1 (no positivity) to +5 (strong positivity) and −1 (no negativity) to −5 (strong negativity); and multiple—e.g., happiness (0–100),

M. Thelwall (✉)
School of Mathematics and Computer Science, University of Wolverhampton, Wulfruna Street, Wolverhampton WV1 1LY, UK
e-mail: m.thelwall@wlv.ac.uk

J.A. Hołyst (ed.), *Cyberemotions*, Understanding Complex Systems,
DOI 10.1007/978-3-319-43639-5_7

sadness (0–100), fear (0–100). Sentiment analysis algorithms tend to use either machine learning or lexical methods. A machine learning approach may start by converting each text into a list of words, consecutive word pairs and consecutive word triples (i.e., 1- to 3-grams) and then, based upon a human coded set of texts, learn which of these features tend to associate with sentiment scores, using this information to classify new cases. In contrast, a lexical approach may start with some language information, such as a list of sentiment words and their polarities, and use this together with grammatical structure knowledge, such as the role of negation, to estimate the sentiment of texts. To illustrate the difference between the two, a machine learning approach may classify "I am not happy" as negative because the bigram "not happy" occurs almost always in texts in the training set coded as negative by humans whereas the lexical approach may choose negative because "happy" is a known positive word and "not" is a known negating word that occurs immediately before it. The two approaches seem to have similar levels of accuracy (however measured) depending upon the types of texts classified and the amount of human classified training data available. Nevertheless, lexical sentiment analysis seems to be superior from a pragmatic perspective for many social research applications because it is less likely to pick up indirect indicators of sentiment that will generate spurious sentiment patterns. For instance a machine learning approach might extract unpopular politicians' names as negative features because they tend to occur in negative texts but this would result in even objective or neutral texts about them being classified as negative, undermining any derived analysis of sentiment in political communication.

This chapter describes SentiStrength, a free sentiment analysis program that uses a lexical approach to classify social web texts. It uses the dual positive and negative scales because psychological research reports that humans can experience positive and negative emotions simultaneously and to some extent independently (Norman et al. 2011). It also uses the lexical approach for the pragmatic reasons given above and harnesses CMC conventions for expressing sentiment to capture non-standard expressive text. As the results below show, it works well without any training data on a wide range of social web texts and approaches human-level accuracy in most tested cases. The exceptions where it performs less well are sets of texts with widespread irony or sarcasm, such as informal political discussions, and narrowly-focused topics with frequently used sentiment terms that are either rare in other topics or tend to have a specialist meaning within the narrow topic examined.

This chapter explains in detail how SentiStrength works, reports evaluations of it on Twitter, YouTube and other data sets, describes how to customise it for specific topics or topics with a negative mood, introduces an extension to enhance its accuracy within sets of related texts, explains how to customise it for languages other than English and reports analyses of Twitter and YouTube.

7.2 SentiStrength

SentiStrength is available in two versions, Java and Windows. The Windows version can be downloaded free from the website http://sentistrength.wlv.ac.uk/ and the (commercial) Java version is available free from the author for researchers and educational users. There is also an interface on the SentiStrength web site to try out SentiStrength live online. The web site includes the main English version as well as several other language variants. SentiStrength's commercial users include Yahoo! (Kucuktunc et al. 2012; Weber et al. 2012) and a range of online information management companies around the world. It was also used to power a light display on the EDF Energy London Eye during the London 2012 Olympic Games by continually monitoring the average sentiment of Olympic-related tweets. The Java version can process 16,000 tweets per second on a standard PC and can be configured to output dual, scale, binary and trinary results (as described above).

The SentiStrength resources, such as the sentiment lexicon and emoticon list, are stored as separate text files and SentiStrength must be pointed to the location when started. It can process text in various ways (depending upon the version), including: single texts submitted via the command line, batches of texts in a single or multiple plain text files with each line of a file classified separately; listening on an internet port; and reading stdin.

7.3 The Core SentiStrength Algorithm

The heart of SentiStrength is a lexicon of 2310 sentiment words and word stems obtained from the Linguistic Inquiry and Word Count (LIWC) program (Pennebaker et al. 2003), the General Inquirer list of sentiment terms (Stone et al. 1966) and ad-hoc additions made during testing, particularly for new CMC words. The stemming used is simple and indicated in the lexicon with a wildcard at the end of a word. For instance *amaz** matches all words starting with amaz, such as amazed and amazing. For each text, SentiStrength outputs a positive sentiment score from 1 to 5 and a negative score from −1 to −5. Matching this, each word or stem in the dictionary is given a positive or negative score within one of these two ranges. These scores were initially human assigned based upon a development corpus of 2600 comments from the social network site MySpace, and subsequently updated through additional testing. The weights for the terms in the sentiment lexicon have been tested against several data sets and can be fine-tuned by SentiStrength using a machine learning approach, as discussed below. The reason for primarily relying upon human input for the sentiment weights is that many of the terms occur rarely in texts and so a machine learning approach would need a huge number of classified texts to give sufficient coverage to assign lexicon weights well. This is a long tail effect because even though many individual terms in the lexicon are rare, collectively the rare terms occur often enough to affect the performance of SentiStrength.

The lexicon is used in a simple way. When SentiStrength reads a text, it splits it into words and separates out emoticons and punctuation. Each word is then checked against the lexicon for matching any of the sentiment terms. If a match is found then the associated sentiment score is retained. The overall score for a sentence is the highest positive and negative score for its constituent words and for multiple sentences and the maximum scores of the individual sentences is taken. For example, the text "Mike is horrible and nasty but I am lovely. I am fantastic." would be classified as follows, "Mike is horrible[-4] and nasty[-3] but I am lovely[2]. <sentence score: 2,-4> I am fantastic[3]. <sentence: 3,-1>" with numbers in square brackets indicating sentiment strength of the preceding word, and angle brackets indicating sentence scores. The overall classification for this text is the maximum positive and negative strength of each sentiment, which is 3 and −4.

The above scores are kept unless they are modified by any of SentiStrength's additional rules. An odd feature of the lexicon is that it contains some non-sentiment terms with a score of 1 (no positivity) or −1 (no negativity). There are two reasons for these terms. Some are included to match non-sentiment variants of sentiment stems. For instance *amaz** is a positive stem but *amazon* is added as a neutral term to catch this non-sentiment term that would otherwise match *amaz**. This works because SentiStrength returns the longest matching term in cases of multiple matches. Some neutral terms are also included as reminders that they have been assessed for inclusion in the lexicon but rejected.

In addition to the lexicon, SentiStrength includes a list of emoticons together with human-assigned sentiment scores. Emoticons are somewhat tricky to automatically extract because although they are typically constructed from lists of punctuation characters and surrounded by spaces, some contain numbers or letters and they may be followed by punctuation that is not part of the emoticon. Hence emoticon extraction is imperfect. SentiStrength also has a list of idioms with sentiment strength weights. These are all multiple word phrases with a meaning that is different from their component word. These idiom scores override the lexicon scores. For example, the stock phrase "shock horror" has an idiom score of −2 for mildly negative and overrides the strong negative scores for shock (−3) and horror (−4).

A weakness of SentiStrength is that it does not attempt to use grammatical parsing (e.g., part of speech tagging) to disambiguate between different word senses. This is because it is designed to process very informal text from the social web and so, unlike typical linguistic parsers, does not rely upon standard grammar for optimal performance. Some grammatical information is used by SentiStrength, however, as the rules below show, and the idiom table can also be used for a brute force approach. To illustrate this, the word "like" can express positive sentiment (e.g., "I like you") or can be used as a comparator (e.g., "Mike looks like an idiot"). SentiStrength gives a neutral score to *like* but has phrases containing *like* in its idiom list with a positive score to override the neutral score for *like* when it is used in a common positive way (e.g. "he likes", "I like", "we like", "she likes").

7.4 Additional Sentiment Rules

In addition to the sentiment term strength lexicon, the idiom list and the emoticon list, SentiStrength incorporates a number of rules to cope with special cases. These were mainly derived from testing on the 2600 My Space comments development data set by examining cases of wrong scores given by early visions of SentiStrength and formulating general rules to cope with them. The following rules are incorporated into SentiStrength (Thelwall et al. 2012a).

1. An **idiom list** is used to identify the sentiment of a few common phrases. This overrides individual sentiment word strengths. The idiom list is extended with phrases indicating word senses for common sentiment words, as described above.
2. The word "**miss**" is a special case with a positive strength of 2 and a negative strength of −2. It is frequently used to express sadness and love simultaneously, as in the common phrase, "I miss you".
3. A **spelling correction algorithm** deletes repeated letters in a word when the letters are more frequently repeated than normal for English or, if a word is not found in an English dictionary, when deleting repeated letters creates a dictionary word (e.g., hellp − > help).
4. **At least two repeated letters** added to words give a strength boost sentiment words by 1. For instance haaaappy is more positive than happy. Neutral words are given a positive sentiment strength of 2 instead.
5. A **booster word list** is used to strengthen (e.g., very +1) or weaken (e.g., somewhat −1) the sentiment of any immediately following sentiment words.
6. A **negating word list** is used to neutralise any following sentiment words (skipping any intervening booster words). (e.g., "I do not hate him", is not negative).
7. An **emoticon list with polarities** is used to identify additional sentiment (e.g., :) scores +2).
8. Sentences with **exclamation marks** have a minimum positive strength of 2, unless negative (e.g., "hello Pardeep!!!").
9. **Repeated punctuation** with one or more exclamation marks boost the strength of the immediately preceding sentiment word by 1.
10. **Two consecutive moderate or strong negative terms** with strength at least −3 increase the strength of the second word by 1 (e.g., "He is a nasty[−3] hateful[−4] person" scores −5 for negativity due to this boost.

There are also a number of additional rules in SentiStrength that have been tested but do not improve its performance. These are disabled in the default configuration but can be enabled by users if they are likely to work on a particular type of data.

1. Sentiment terms in **CAPITAL letters** receive a strength increase of 1.
2. **Two consecutive moderate or strong positive terms** with strength at least 3 increase the strength of the second word by 1.

3. **Sentences containing irony** have their positive sentiment reduced to 1 and their negative sentiment equal to 1 less than their positive sentiment. Irony is operationalized by having positive sentiment in conjunction with the presence of a term or phrase from a user-defined list (e.g., politicians' names or derogatory terms for politicians).

Many of the additional rules can be disabled or modified in SentiStrength, if desired. For instance the booster words feature can be disabled by emptying the booster word list and the number of words allowed between a negating word and a sentiment word can be user defined. Caution should be used when modifying the defaults: whilst a change may improve scores on some texts it may reduce overall accuracy by giving worse scores on other, perhaps unexpected cases.

Some of the rules also need to be modified for non-English versions of SentiStrength and there are some options for this. For example, in Germanic languages negating words are typically placed after sentiment words and this aspect of the rule is a modification available in SentiStrength. If a test data set is used to evaluate SentiStrength then this test data should not also be used to evaluate SentiStrength rule modifications (or any other SentiStrength modifications) because this would invalidate the test results due to the potential for over-fitting the algorithm—i.e., tailoring it too much to the test data so that it is more accurate on the test data than on other similar data.

7.5 Supervised and Unsupervised Modes

SentiStrength has the capability to optimise its lexicon term weights for a specific set of human-coded texts (i.e., a collection of texts with human-assigned sentiment scores for each one). It does this by repeatedly increasing or decreasing the term weights by 1, one term at a time, and then assessing whether this change increases, decreases or does not affect the overall classification accuracy for the human coded texts. Changes that improve accuracy are kept and the process is repeated until no term strength change improves the overall classification accuracy (i.e., it is a hill climbing algorithm). This process can easily lead to *over-fitting* because only one occurrence of a term can be used to change its lexicon strength, although it is possible to increase the threshold required for a change in the algorithm. This means requiring a bigger increase in accuracy for a change in term strength in order to retain the change.

If the above process is used to optimise the SentiStrength weights then this is its *supervised* mode; without training is the *unsupervised* mode. As the results below show, supervised mode has similar overall accuracy to that of unsupervised mode, but it should logically outperform unsupervised mode if enough training data is used.

As a final point on the lexical term strength optimisation process, the reason why stemmed terms are included in the lexicon rather than a complete list of matching terms is to improve the power of the term strength optimisation algorithm because the stemmed terms can occur more often than each individual matching complete word.

7.6 Evaluating SentiStrength

SentiStrength can be evaluated by applying it to a set of texts that have been coded for sentiment by humans and comparing the SentiStrength scores with the human scores. For the best results, the average of at least three different human coders should be used for the texts. This is because coding is subjective and one coder is more likely to give unusual results than the average of three or more. The coders should be chosen and assessed for accuracy and consistency because a large number (>1000) of texts need to be coded for a reliable assessment. One way to select coders is to give them the same 100 texts to classify as a pilot study and then choose the three coders that agree most with each other. Experience with SentiStrength suggests that only 1 in 5 coders give accurate enough results to be useful. The best metric to assess the degree of agreement between the coders is Krippendorff's inter-coder weighted alpha (Krippendorff 2004) with the weight for a mismatch being the difference between the two scores. This metric is one of the standard options for social sciences content analysis studies and is available as a menu option in the Windows version of SentiStrength. This metric can be used for the initial pilot and also to report the level of agreement on the complete data set for the selected coders once they have finished.

The human-coded corpus is not likely to be definitive in the sense that any reasonable person would agree with the classifications because emotions can be expressed in different ways by different groups of people. For example, the phrase "the film was bad" could be interpreted as indicating a positive sentiment by people who routinely use the word *bad* as a strong positive sentiment term. Hence the objective for creating a human-coded corpus should be to create a consistent and reasonable set of classifications rather than one that would be universally recognized as correct.

Once the human-coded corpus is ready, SentiStrength can be applied to it and its results compared with the human coder average. The best metric for the comparison is the Pearson correlation because this is one of the few standard performance metrics for sentiment analysis that takes into account how close an estimation is to the correct value when they are not identical. It is also superior in practice to the alternatives, such as Mean Absolute Deviation (MAD) and Mean Squared Error (MSE) in that it gives a result that is more easily interpreted by researchers outside the sentiment analysis field since the Pearson correlation is simple and well known.

The SentiStrength-human comparison gives two separate correlations, one for positive sentiment strength and one for negative sentiment strength. If these are significantly positive then this is evidence that SentiStrength works better than random guessing. Higher positive correlations indicate better performance and can

Table 7.1 Unsupervised and supervised SentiStrength against the baseline measure (predicting the most common class) and the standard machine learning algorithm (from a set of nine) and feature set size (from 100, 200 to 1000) having the highest correlation with the human-coded values

	Positive correct (%)	Negative correct (%)	Positive correlation	Negative correlation
BBC Forums[a]				
Unsupervised SentiStrength	51.3	46.0	0.296	**0.591**
Supervised SentiStrength	60.9	48.4	0.286	0.573
Best machine learning	**76.7**	**51.1**	**0.508**	0.519
Digg				
Unsupervised SentiStrength	53.9	46.7	0.352	0.552
Supervised SentiStrength	57.9	50.5	**0.380**	**0.569**
Best machine learning	**63.1**	**55.2**	0.339	0.498
MySpace				
Unsupervised SentiStrength	62.1	70.9	**0.647**	0.599
Supervised SentiStrength	62.1	72.4	0.625	**0.615**
Best machine learning	**63.0**	**77.3**	0.638	0.563
Runners world				
Unsupervised SentiStrength	53.5	50.9	0.567	0.541
Supervised SentiStrength	53.9	55.8	0.593	0.537
Best machine learning	**61.5**	**65.3**	**0.597**	**0.542**
Twitter				
Unsupervised SentiStrength	59.2	66.1	0.541	0.499
Supervised SentiStrength	63.7	67.8	0.548	0.480
Best machine learning	**70.7**	**75.4**	**0.615**	**0.519**
YouTube				
Unsupervised SentiStrength	44.3	56.1	0.589	0.521
Supervised SentiStrength	46.5	57.8	0.621	0.541
Best machine learning	**52.8**	**64.3**	**0.644**	**0.573**
All 6				
Unsupervised SentiStrength	53.5	58.8	0.556	0.565
Supervised SentiStrength	56.3	61.7	0.594	**0.573**
Best machine learning	**60.7**	**64.3**	**0.642**	0.547

Correlation is the most important metric

[a] The metrics used are: accuracy (% correct) and correlation. Best values on each data set and each metric are in bold. *Source*: extracted from Thelwall et al. (2012a)

be used to compare different versions or settings for SentiStrength and to compare its performance on different corpora.

Table 7.1 reports the correlations between SentiStrength and the human coder average for a range of different types of social web text. The positive correlations in all cases together indicate its general applicability to social web texts, even in unsupervised mode. In other words it would be reasonable to apply it to any new source of social web texts, even in the absence of training data. The table shows that the supervised mode, with lexicon term weights automatically adjusted based upon the training data, is not clearly better overall than the unsupervised mode. Hence the advantage of creating human coded data for any new text source would be the ability to measure SentiStrength's performance rather than the ability to run it in supervised mode.

When evaluating supervised performance it is important to use a test set of texts that is different from the training set. The standard way of achieving this result is known as 10-fold cross validation and is used in Table 7.1. This is available as an option in SentiStrength. With this method a single set of human coded texts is used but is split into a training part (90 % of the tests) and a testing part (the remaining 10 % of the texts). This ensures that the training and test texts are different but does not give accurate results because only 10 % of the texts are used for testing. To circumvent this accuracy issue the process is repeated for each remaining set of 10 % of the texts and the ten results are averaged to give a more precise accuracy estimate.

7.7 Sarcasm, Irony and Politics

The SentiStrength results in Table 7.1 contain two correlations that are lower than the rest (BBC Forums and Digg, both for positive sentiment strength only). An examination of the data revealed that many incorrect matches were associated with discussions of political and other controversial issues. These often employ irony and sarcasm in the form of ostensibly positive statements with a negative meaning, such as "warmongers will be happy that another 10 soldiers were killed today". In response to this problem a number of options to detect sarcasm were tested for SentiStrength but none were adopted because all incorrectly identified sarcasm more often than not. The most promising rule was that a text was sarcastic if it contained both positive and negative sentiment and a politician's name and a winky emotion ;) but even this rule failed. Sarcasm is known to be difficult to automatically detect (Gonzalez-Ibanez et al. 2011) and is often also problematic for humans, perhaps because its power is partly due to the cleverness with which it is constructed. There have been some small successes with automated sarcasm detection, however. Book reviews are one example due to the repeated use of stock sarcastic types of phrase, such as "this book has a great cover" that can be learned from a training corpus (Tsur et al. 2010). Sarcasm in Portuguese political discussions can also be identified through a combination of features including the use of a politician's name in diminutive form (Carvalho et al. 2009). These successes do not seem to transfer well to general sarcasm detection in English and so this seems to be a major challenge.

A consequence of the difficulty in detecting sarcasm and the problems that it causes is that SentiStrength is likely to have lower accuracy than normal for positive sentiment strength in sets of texts in which sarcasm is common, including political discussions. Whilst the results in Table 7.1 are still significantly positive and hence may be useful in practice, the performance is below human levels of accuracy. This problem can be partly resolved by using supervised mode for SentiStrength.

7.8 Adaptations for Specific Topics

The typical sentiment of terms depends on the context in which they are used. For instance, in most contexts sentences containing any of the terms "horror", "frightened", "scream" or "scared" would be negative but if the context is a horror movie review then these terms might tend to indicate a positive review instead (good horror movies may be scary). Hence if SentiStrength is applied to texts from a relatively narrow context, such as a type of product review or discussions of a specific topic, then the lexicon may need to be modified to take into account the commonly used sentiment terms and strengths in the new context. This can be done manually with a human expert and a development corpus, perhaps also using common sense to reassess terms strengths as well as examining incorrect results on the development corpus. For instance if the development corpus suggests a new sentiment term then common sense may dictate that synonyms of that term should also be added.

SentiStrength also has a method to automatically suggest new terms based upon a development or training corpus. This *lexical extension method* proceeds as follows.

1. The user creates human coded training and testing corpora (sets of texts) and feeds the training corpus to SentiStrength (Java) with the command to identify extra words to add.
2. SentiStrength makes a list of all words in all texts in the training corpus and assigns each a score of zero.
3. All texts are classified by SentiStrength with its default lexicon.
4. For each text with a positive or negative score from SentiStrength different from the human coder score, a value is added to the score for each term in the text equal to the difference between the human coded and SentiStrength scores. This number is positive if SentiStrength is too high for the text and negative if it is too low.
5. SentiStrength prints out a sorted list of all terms with non-zero scores, sorted by the score value.
6. Either (automatic method) all terms with a sufficiently high or low score are added to the lexicon with a nominal weight (e.g., +3 or −3) or (manual method, recommended) a human expert scans the list and decides which terms to add and their strengths.
7. The performance of the new version of SentiStrength is assessed on the test corpus and compared with the original performance on the test corpus.

Table 7.2 Results of adding the topic-specific sentiment terms to the training and test corpora using two different methods

Corpus	Original correlation	Correlation with extra terms (1)	Correlation with extra terms (2)
Riots	0.4104	**0.4429**	0.4383
AV	0.4038	0.4124	**0.4126**
MySpace	0.5919(+)	**0.6134(+)**	0.6091(+)
	0.6023(−)	0.5963(−)	**0.6041(−)**
BBC	0.3357(+)	**0.3376(+)**	**0.3376(+)**
	0.6098(−)	0.6095(−)	**0.6104(−)**
Digg	**0.3566(+)**	0.3554(+)	0.3554(+)
	0.5709(−)	**0.5715(−)**	0.5709(−)
Runners World	**0.6318(+)**	0.6305(+)	**0.6318(+)**
	0.5632(−)	**0.5632(−)**	**0.5632(−)**
Twitter	**0.6027(+)**	0.6024(+)	0.6024(+)
	0.5160(−)	**0.5160(−)**	**0.5160(−)**
YouTube	**0.5933(+)**	0.5878(+)	0.5878(+)
	0.5498(−)	0.5461(−)	0.5462(−)

The highest values for each corpus are in bold (*source*: Thelwall and Buckley 2013). The first two corpora used the single scale sentiment output and the other six used the dual scale output

In summary, although the normal term optimisation method used in supervised mode SentiStrength can adjust lexicon weights for existing terms, the lexicon extension method is able to identify new terms to add to the lexicon. The method has been tested on two corpora and gave small accuracy improvements in both cases (Table 7.2). Although the manual term adding variant did not give more accurate results than the automatic method the latter added some irrelevant terms, such as *letters* which is undesirable. Hence the manual method is preferred.

7.9 Mood Adaption

SentiStrength can be assigned either a positive or a negative mood. This mood determines the polarity of sentences that do not contain explicit sentiment polarity indicators but have an indication of the presence of sentiment, either through excessive punctuation (e.g., "look here!!) or through deliberate misspellings (e.g., "loooook"). In cases where energy (the term arousal is used by psychologists) is perceived without an indication of sentiment polarity humans seem to use contextual information to fill the gap (Fox 2008). The original version of SentiStrength was developed for predominantly positive texts and had a fixed positive mood but the mood can be set to either positive or negative in the current version. To test which is better, both settings can be tried on a test or development set and the more accurate one retained. Setting the best mood can substantially improve performance (Thelwall and Buckley 2013).

7.10 Sentiment Anomaly Detection with Local Context

Any sentiment analysis program makes mistakes due to complex sentence constructions that it cannot decode (e.g., sarcasm). In some cases it may be possible to predict that a classification is likely to be incorrect and to automatically correct it without identifying the linguistic causes. This has been tested for SentiStrength with a simple rule to detect sudden jumps in either positive or negative sentiment. The rule is that if a sentiment score differs by over 1.5 from the average of the previous three contributions in the sequence then the new sentiment score is regarded as an anomaly and damped by adjusting it by bringing it 1 closer to the average of the previous three scores. This method has been shown to be capable of improving SentiStrength's accuracy in monologs (sequential lists of contributions by a single person, such as their tweet stream), dialogs (sequential lists of messages exchanged between two communication partners) and multi-participant discussions, but the improvement is only minor and depends upon the type and probably the topic of communication (Thelwall et al. 2013). The testing for this used human-coded data sets for which the coders were able to see the texts in the context of the texts preceding them and hence the human coders could also be influenced by communication patterns and may be subconsciously or consciously damping their recorded sentiment scores. Because the anomaly detection improvement probably depends on the type and topic of communication, more research is needed before this method is used outside of experimental settings.

7.11 Language Variants

SentiStrength can be customised for new languages by translating its sentiment lexicon and other resources, adjusting its optional settings to cope with language specific features, such as negating words occurring after sentiment terms in Germanic languages, and refining through testing on a human-coded development corpus. For some languages, additional processing will be needed to get good results, however. For example, the morphology of Turkish word formulation (Turkish is an agglutinative language) means that Turkish text must be parsed to separate out negating suffixes from sentiment terms and then the negation can be re-introduced by inserting an artificial negating word (e.g., _NOT_) prior to the sentiment word before being submitted to SentiStrength (Vural et al. 2013). Languages without word boundaries in standard form also need to be first processed with an algorithm to split the characters into words. This applies to Chinese and Japanese. The pre-processing approach has also been employed by some commercial SentiStrength users for languages like French, Spanish and Portuguese, where it is not strictly necessary, in order to fit in with existing systems that work exclusively with lemmatised text. In this case all the SentiStrength resource files should also contain equivalently lemmatised text (i.e., with words parsed into standard abstract units known as lemmas).

Converting SentiStrength to work with a new language requires no coding in most cases because all its language resources are stored externally in plain text files and because it has language customisation options built in, including a UTF-8 mode for non-ASCII characters. Nevertheless a good conversion will take at least a month to translate the resource files, human-code a development corpus of 1000 texts and refine the lexicon and options based upon an examination of incorrect classifications in the development corpus. The accuracy of the translated variant should also be assessed before use on a second human coded corpus, to determine its accuracy. This is likely to be lower than SentiStrength's accuracy for English due to the longer development time for this language.

7.12 Application: The Role of Sentiment in Major Media Events

To illustrate a simple application of SentiStrength to investigate sentiment patterns in social web texts, this section describes a case study of tweeting major events. Other chapters in this book illustrate applications of SentiStrength and other sentiment analysis programs to computing systems, modelling the spread or influence of sentiment online and psychological experiments to test it (Thelwall et al. 2011).

The Twitter study aimed to identify the typical patterns of sentiment changes during a major event and to determine whether changes in sentiment could be used to predict the amount of interest in an event during the early stages of its evolution. The raw data used was a collection of 35 million tweets from February 9, 2010 to March 9, 2010. Major events were detected with a time series analysis of relative word frequencies in tweets (Thelwall and Prabowo 2007), followed by human filtering. For each word a spike value was calculated—the biggest daily relative frequency increase in the proportion of tweets containing the word. For each day after the third the increase was calculated as the daily value minus the average of all previous daily values. The word list was then ranked in descending order of spike size and manually filtered to remove words describing the same event as a more highly-ranked word, words referring to purely online events (e.g., the Follow Friday hashtag #ff) and words associated with non-news events, such as Valentine's Day.

For each of the top 30 remaining events, the sentiment scores of all matching tweets were calculated using SentiStrength and the average levels of sentiment before, during and after the event were calculated. Across all 30 events, the spikes were typically associated with small (average 6 %) increases in negative sentiment

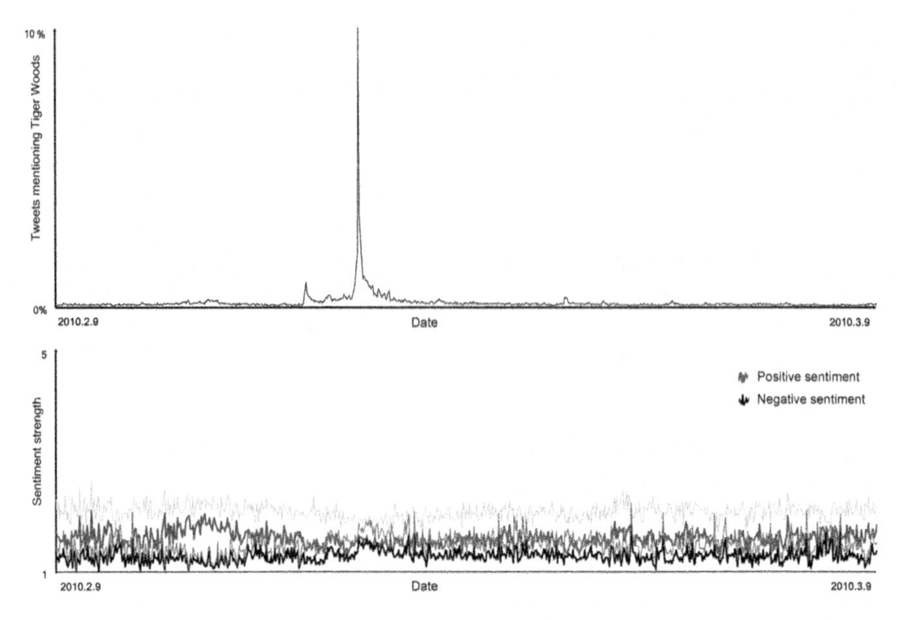

Fig. 7.1 Tweeting about Tiger Woods before, during and after his public announcement about having an affair. The *top graph* shows the total volume of tweeting mentioning Tiger Woods. *Black lines* in the *lower graph* reveal average negative sentiment strength for all Tiger Woods tweets (*thick line*) and for subjective tweets (*thin line*); *grey lines* reflect positive sentiment

but often no change in positive sentiment. Figure 7.1 shows a very negative event with only a small increase in average negative sentiment strength.

The main outcomes of the study were the result that sentiment changes were typically too small to be useful for predicting the importance of an event in its early stages, that negativity was the key sentiment for major media events in Twitter and that sentiment is rarely expressed explicitly in tweets about major events. This last fact is particularly surprising because major events presumably arouse strong emotions in order to trigger a sudden spike of tweeting so it seems that the sentiment is implicit in the sending of the tweet and does not need to be expressed.

7.13 Application: Sentiment in YouTube Comments

A second study examined the role of sentiment in YouTube comments; these are left by a small minority of viewers after or during a YouTube video and are interesting for the insights that they can give into viewer reactions to the video or its topic.

The investigation found that weak positivity was the most common sentiment in comments and this typically corresponded to mild praise for a video, its author or topic (Thelwall et al. 2012b). In addition, videos with stronger positive comments tended to have weaker negative comments and vice versa, suggesting a viewer consensus of opinions about the video. Some comments are replies to previous comments, however, and so any sentiments could be directed at other commenters rather than the video. Probably as a result of this, negativity significantly associated with the densest discussions within comment sections. In other words negativity was more successful than positively in fostering interactions, a phenomenon also found for other online contexts (Chmiel et al. 2011).

7.14 Conclusion

This chapter described the sentiment strength detection program SentiStrength that uses a dual positive/negative sentiment strength scoring system and is optimised for general social web texts. SentiStrength employs a lexicon of sentiment words and word stems together with average positive or negative sentiment strength scores for them. Texts are classified with the largest positive or negative scores of any constituent word unless these are modified by any of the additional classification rules, such as in the case of emotions, negations and booster words.

SentiStrength has near-human accuracy on general short social web texts but is less accurate when the texts often contain sarcasm, as in the case of political discussions. The accuracy of SentiStrength can be enhanced by extending its lexicon and altering its mood setting for sets of texts with a narrow topic focus. As the case studies illustrate, SentiStrength can be used to analyse large scale sentiment patterns in the social web in addition to its commercial uses.

Acknowledgements This work was supported by a European Union grant by the 7th Framework Programme, Theme 3: Science of complex systems for socially intelligent ICT. It is part of the CyberEmotions project (contract 231323).

References

Carvalho, P., Sarmento, L., Silva, M.J., de Oliveira, E.: Clues for detecting irony in user-generated contents: oh...!! it's "so easy" ;-). In: Jiang, M., Yu, B. (eds.) Proceedings of the 1st International CIKM Workshop on Topic-Sentiment Analysis for Mass Opinion, pp. 53–56. ACM, New York (2009). doi:10.1145/1651461.1651471

Chmiel, A., Sienkiewicz, J., Thelwall, M., Paltoglou, G., Buckley, K., Kappas, A., Hołyst, J.A.: Collective emotions online and their influence on community life. PLoS ONE **6**(7), e22207 (2011). doi:10.1371/journal.pone.0022207

Derks, D., Bos, A.E.R., von Grumbkow, J.: Emoticons and online message interpretation. Soc. Sci. Comput. Rev. **26**(3), 379–388 (2008). doi:10.1177/0894439307311611

Fox, E.: Emotion Science. Palgrave Macmillan, Basingstoke (2008)

Gonzalez-Ibanez, R., Muresan, S., Wacholder, N.: Identifying sarcasm in Twitter: a closer look. In: Lin, D., Matsumoto, Y., Mihalcea, R. (eds.) Proceedings of the 49th Annual Meeting of the Association for Computational Linguistics: Human Language Technologies: Short Papers, vol. 2, pp. 581–586. Association for Computational Linguistics, Portland (2011)

Krippendorff, K.: Content Analysis: An Introduction to Its Methodology. Sage, Thousand Oaks (2004)

Kucuktunc, O., Cambazoglu, B.B., Weber, I., Ferhatosmanoglu, H.: A large-scale sentiment analysis for Yahoo! Answers. In: Adar, E., Teevan, J., Agichten, E., Maarek, Y. (eds.) Proceedings of the 5th ACM International Conference on Web Search and Data Mining WSDM 2012, pp. 633–642. ACM, New York (2012). doi:10.1145/2124295.2124371

Norman, G.J., Norris, C.J., Gollan, J., Ito, T.A., Hawkley, L.C., Larsen, J.T., Cacioppo, J.T., Berntson, G.G.: The neurobiology of evaluative bivalence. Emot. Rev. **3**(3), 349–359 (2011). doi:10.1177/1754073911402403

Pennebaker, J., Mehl, M., Niederhoffer, K.: Psychological aspects of natural language use: our words, our selves. Annu. Rev. Psychol. **54**, 547–577 (2003). doi:10.1146/annurev.psych.54.101601.145041

Stone, P.J., Dunphy, D.C., Smith, M.S., Ogilvie, D.M.: The General Inquirer: A Computer Approach to Content Analysis. The MIT Press, Cambridge (1966)

Thelwall, M., Buckley, K.: Topic-based sentiment analysis for the social web: the role of mood and issue-related words. J. Am. Soc. Inf. Sci. Technol. **64**(8), 1608–1617 (2013). doi:10.1002/asi.22872

Thelwall, M., Prabowo, R.: Identifying and characterising public science-related concerns from RSS feeds. J. Am. Soc. Inf. Sci. Technol. **58**(3), 379–390 (2007). doi:10.1002/asi.20504

Thelwall, M., Buckley, K., Paltoglou, G.: Sentiment in Twitter events. J. Am. Soc. Inf. Sci. Technol. **62**(2), 406–418 (2011). doi:10.1002/asi.21462

Thelwall, M., Buckley, K., Paltoglou, G.: Sentiment strength detection for the social web. J. Am. Soc. Inf. Sci. Technol. **63**(1), 163–173 (2012a). doi:10.1002/asi.21662

Thelwall, M., Sud, P., Vis, F.: Commenting on YouTube videos: from Guatemalan rock to El Big Bang. J. Am. Soc. Inf. Sci. Technol. **63**(3), 616–629 (2012b). doi:10.1002/asi.21679

Thelwall, M., Buckley, K., Paltoglou, G., Skowron, M., García, D., Gobron, S., Ahn, J., Kappas, A., Küster, D., Hołyst, J.A.: Damping sentiment analysis in online communication: discussions, monologs and dialogs. In: Gelbukh, A. (ed.) Computational Linguistics and Intelligent Text Processing, 14th International Conference, CICLing 2013, Samos, Greece, 24–30 March 2013, Proceedings, Part II. Lecture Notes in Computer Science, vol. 7817, pp. 1–12. Springer, Berlin/Heidelberg (2013). doi:10.1007/978-3-642-37256-8_1

Tsur, O., Davidov, D., Rappoport, A.: ICWSM - A great catchy name: semi-supervised recognition of sarcastic sentences in online product reviews. In: Cohen, W.W., Gosling, S. (eds.) Proceedings of the 4th International AAAI Conference on Weblogs and Social Media, pp. 162–169. The AAAI Press, Washington, DC (2010)

Vural, G., Cambazoglu, B.B., Senkul, P., Tokgoz, O.: A framework for sentiment analysis in Turkish: application to polarity detection of movie reviews in Turkish. In: Gelenbe, E., Lent, R. (eds.) Computer and Information Sciences III: 27th International Symposium on Computer and Information Sciences, pp. 437–445. Springer, London (2013). doi:10.1007/978-1-4471-4594-3_45

Walther, J., Parks, M.: Cues filtered out, cues filtered in: computer-mediated communication and relationships. In: Knapp, M., Daly, J., Miller, G. (eds.) The Handbook of Interpersonal Communication, 3rd edn., pp. 529–563. Sage, Thousand Oaks (2002)

Weber, I., Ukkonen, A., Gionis, A.: Answers, not links: extracting tips from Yahoo! answers to address how-to web queries. In: Adar, E., Teevan, J., Agichten, E., Maarek, Y. (eds.) Proceedings of the 5th ACM International Conference on Web Search and Data Mining WSDM 2012, pp. 613–622. ACM, New York (2012). doi:10.1145/2124295.2124369

Part III
Modeling

Chapter 8
Detection and Modeling of Collective Emotions in Online Data

Janusz A. Hołyst, Anna Chmiel, and Julian Sienkiewicz

8.1 Introduction and Data Description

Collective phenomena appearing in physical systems, e.g., the emergence of regular crystal structures or ferromagnetic order are spectacular results of interactions between appropriate matter constituents, e.g., molecules of liquids or spins of magnetic materials. It is clear that emotions experienced, expressed and perceived by humans (Zajonc 1980; Feldman 1995) can influence human-human communication (Shimanoff 1984; Eviatar and Zaidel 1991; Fussell and Moss 1998) also when it is mediated by Internet (Sobkowicz and Sobkowicz 2010, 2012; Sobkowicz 2013). Can such affective interactions lead to collective emotional states (Stürner and Simon 2004; Sabucedo et al. 2011) shared by an online group (Szell et al. 2010)? How can one detect and quantify this order?

These questions are difficult to answer since it is not clear, if emotions in a given group are an outcome of individual group members' personalities, an outcome of external impact (e.g. news transmitted by mass-media) or if they result from *intra-group interactions*. In this chapter we show how to detect and to model the *collective* character of affective phenomena in online media (Chmiel et al. 2011a; Chmiel and Hołyst 2013). For the detection purposes we have taken over six million comments from various datasets and perform straightforward analysis by measuring lengths of clusters with similar emotional valences and comparing it to a random case. As an outcome we observe a general rule governing the ordered appearance of consecutive emotional comments—an affective preferential processes (Chmiel et al. 2011a). Then we propose a model equipped with the above observations and study its statistical behavior. Finally we investigate situations when the process of preferential cluster growth leads to the emergence of a critical cluster followed

J.A. Hołyst (✉) • A. Chmiel • J. Sienkiewicz
Faculty of Physics, Warsaw University of Technology, Koszykowa 75, 00-662 Warsaw, Poland
e-mail: jholyst@if.pw.edu.pl; anulachmiel@gmail.com; julas@if.pw.edu.pl

J.A. Hołyst (ed.), *Cyberemotions*, Understanding Complex Systems,
DOI 10.1007/978-3-319-43639-5_8

Table 8.1 Properties of investigated datasets: number of comments N, number of different users giving comments U, number of discussions/threads T, average valence in the dataset $\langle e \rangle$, probability of finding positive, negative or neutral emotion [respectively $p(+)$, $p(-)$ and $p(0)$]

Dataset	N	U	T	$\langle e \rangle$	$p(-)$	$p(0)$	$p(+)$
BBC	2,474,781	18,045	97,946	−0.44	0.65	0.16	0.19
Blogs	242,057	NA	1219	0.14	0.22	0.43	0.35
Digg	1,646,153	84,985	129,998	−0.16	0.48	0.21	0.31
IRC	1,889,120	NA	93,323	0.17	0.15	0.53	0.32

Each data set has a different overall average valence—BBC is strongly negative, Digg is mildly negative while Blogs and IRC are mildly positive

by posts always displaying the same valence (Chmiel and Hołyst 2013). The observation of collective character of emotions in online groups visible in the data is parallel to findings on persistence in IRC data presented in Chap. 10 (Garas et al. 2012), as well as to analysis of emotional avalanches shown in Chap. 11 (Mitrović et al. 2011) and observation of mirroring of exchanged sentiments in reciprocal messages (Chap. 12, Hillmann and Trier 2012).

Our analysis concerns over six million comments from four prominent interactive spaces: blogs (Ounis and Macdonald 2006; Weroński et al. 2012), BBC discussion forums (Chmiel et al. 2011a,b; Sobkowicz et al. 2013), the popular social news website Digg (Wu and Huberman 2007; Mitrović et al. 2011; Hogg and Lerman 2012; Pohorecki et al. 2013) and Internet Rely Chat (IRC) `#ubuntu` channel (Garas et al. 2012; Gligorijević et al. 2013; Sienkiewicz et al. 2013) (for key properties see Table 8.1). BBC Forum data included discussions posted on the Religion and Ethics[1] and World/UK News[2] message boards starting from the launch of the website (July 2005 and June 2005 respectively) until June 2009. The Blog06 dataset (later related to as "Blogs") is a subset of the collection of blog posts from 06/12/2005 to 21/02/2006. Only posts attracting more than 100 comments were extracted, as these seemed to initiate non-trivial discussions. The Digg[3] dataset comprises a full crawl of digg.com, one of the most popular social news websites. The data spans February to April 2009 and consists of all the stories, comments and users that contributed to the site during this period. Finally, the IRC channel,[4] obtained from the logs of the `#ubuntu` channel covers the period of 1st January 2007 and 31st December 2009. The texts were processed using sentiment analysis classifiers (see Chaps. 6 and 7) to predict their emotional valences. As an effect we worked with a ternary type of data—each comment is described with a single value $e \in \{-1, 0, 1\}$ reflecting its sentiment orientation (i.e. negative, positive or neutral).

[1]http://www.bbc.co.uk/dna/mbreligion.

[2]http://www.bbc.co.uk/dna/mbfivelive/F2148565, http://www.bbc.co.uk/dna/mbfivelive/F2148564.

[3]http://www.digg.com.

[4]http://help.ubuntu.com/community/InternetRelayChat.

8.2 Emotional Clusters as Effects of Affective Interactions

8.2.1 Why Are Emotional Discussions Not Well Described by i.i.d. or Markov Process?

To detect affective interactions between discussion participants we calculated statistics for groups of messages with similar emotion levels. Every discussion thread (identified by a unique URL) was analyzed separately and was converted into a chain corresponding to a temporal order of consecutive messages even if a tree structure had been present (messages had been posted as comments to comments). We define an emotional cluster of size n as a chain of n consecutive messages with the same sentiment orientation (i.e. negative, positive or neutral, see the upper row in Fig. 8.1). For comparison we present also shuffled data received from the same discussion (see the bottom row in Fig. 8.1). The clusters in the shuffled data are clearly shorter than the clusters in the original discussion. The reason for this could be emotional interactions between group members in the original data. To prove this hypothesis we checked the distribution of cluster lengths. Let us assume that at a given site a message with a sentiment $e = \{-1, 0, 1\}$ is observed. If affective interactions between group members were absent then a probability that this message starts a cluster of length n would be described by an independent and identically distributed (i.i.d.) random process with the cumulative distribution (for details see the text in the frame below):

$$P^{(e)}_{i.i.d.}(\geq n) = p(e)^{n-1}, \text{ where } n > 0 . \tag{8.1}$$

Here $p(e)$ is the probability of a message with sentiment e (for exact values see Table 8.1). Figure 8.2 shows BBC, Digg, Blogs and IRC data compared to predictions from an i.i.d. process. The agreement between data and Eq. (8.1)

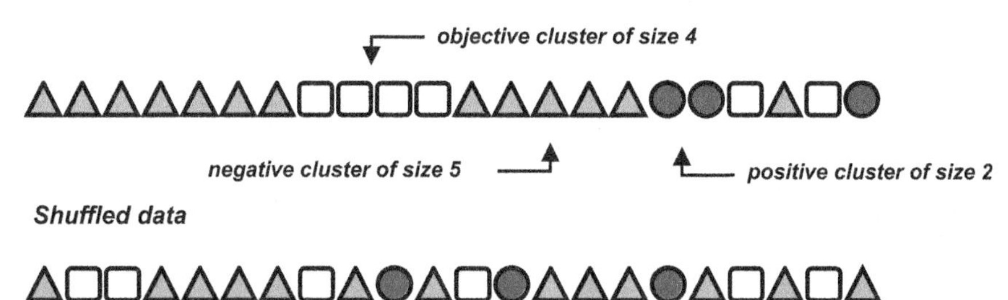

Fig. 8.1 An example of a discussion in the "Eastern religion" BBC Forum in September 2005 consisting of 22 posts (*upper row*). Each symbol represents one post: *red circles*, *blue triangles* or *white squares* indicate that the comment was classified as, respectively, positive, negative or neutral (objective). The *bottom row* presents shuffled data, i.e., the comments were arranged in a random order. Note the difference between the length of clusters in the original and in the shuffled data

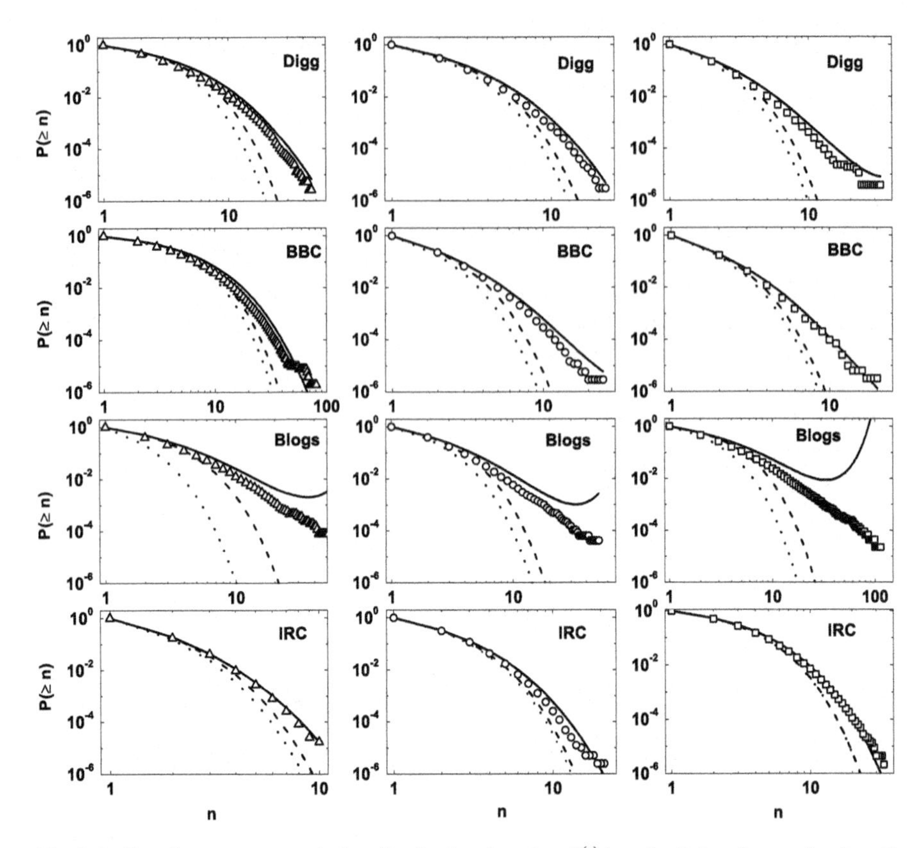

Fig. 8.2 Complementary cumulative distribution function $P^{(e)}(\geq n)$ of the cluster size for all data used in the study. Symbols are data (*triangles*, *circles* and *squares*, respectively for negative, positive and neutral clusters), *dotted lines* are i.i.d. processes given by Eq. (8.1), *dashed lines* are Markov processes given by Eq. (8.3) while *solid lines* come from Eq. (8.6) and represent distributions based on the preferential attraction rule. The spurious increase of $P^{(e)}(\geq n)$ for $n > 40$ for Blogs data is due to violation of the scaling $p(e|ne) = p(e|e)n^{\alpha}$

diverges for $n > 10$, and the frequency of long clusters of the same emotional valence is large compared to the frequency expected for mutually independent messages.

An independent and identically distributed (i.i.d.) random process (Feller 1968) corresponds to the simplest stochastic process where no statistical dependence exists between events at consecutive time-steps and at every time-step the event probability distribution is the same. Let the parameter $p(e)$ describe the probability of an event e at any time step (i.e., a probability of

(continued)

negative, positive or neutral emotion). If the event e has taken place then the conditional probability that this site is a beginning of cluster of n events e scales as

$$P^{(e)}_{i.i.d.}(n) \sim [1 - p(e)]^2 p(e)^{n-1}. \quad (8.2)$$

Here the pre-factor $[1 - p(e)]^2$ corresponds to the probability of events other than e that take place just before and just after the cluster, i.e., where the valence e is changed. Taking into account the normalization condition $\sum_{n=1}^{\infty} P^{(e)}_{i.i.d.}(n) = 1$ we can get the cluster distribution $P^{(e)}_{i.i.d.}(n) = [1 - p(e)]p(e)^{n-1}$. Note that because of the normalization this distribution is the same for any pre-factor standing in-front of the function $p(e)^{n-1}$ in (8.2). Thus the cumulative cluster distribution $P^{(e)}_{i.i.d.}(\geq n)$ is given by Eq. (8.1). The same result can be received by noting that when the event e is observed at a given site then it starts a cluster of size equal or larger than n when $n - 1$ events e take place in next time steps.

Since the i.i.d. process does not describe correctly the observed distribution of emotionally homogeneous clusters in online discussions we have looked for dynamics where messages are dependent one on another. A simple extension of i.i.d. process is a process with one memory step, i.e., the Markov chain where a probability $p(e_{t+1})$ of the state e_{t+1} at time $t + 1$ (in our case an emotional valence expressed at time $t+1$) is dependent on the state e_t observed at time t. If one neglects a longer memory and assumes that the process is stationary then a probability that two consecutive messages possess the same emotional valence e can be expressed as $p(e_t = e, e_{t+1} = e) \equiv p(ee) \equiv p(e)p(e|e)$ where $p(e|e)$ is the conditional probability that a valence e is observed at time $t+1$ if the same valence is observed at time t. Values of such conditional probabilities have been estimated from collected data and they are presented in Table 8.2. If affective interactions between group members were described by the Markov process then the probability that a message with the sentiment e starts a cluster of length n would be described by the cumulative distribution (for details see the text in the frame below)

$$P^{(e)}_{M}(\geq n) = p(e|e)^{n-1}, \text{ where } n > 0. \quad (8.3)$$

The distribution (8.3) is displayed in Fig. 8.2. The agreement between data from online discussions and the distribution (8.3) is a little bit better than for i.i.d. process described by the distribution (8.1) however we still observe a discrepancy for $n > 10$. It means the frequency of observed long clusters of the same emotional valence is larger than the frequency expected for messages described by the Markov process with one memory step.

Table 8.2 Values of model parameters fitted to real data sets Blog06, BBC Forum and Digg: α—exponent of preferential emotional interactions, $p(e|e)$—conditional probability of the appearance of two consecutive messages with the same emotional values, n_c—size of the critical cluster (needed for MET emergence) predicted by the model, n_{max}—the maximum cluster size found in datasets, m^d_{th}—the number of threads containing the critical cluster in datasets

Dataset	α	$p(e\|e)$	n_c	n_{max}	m^d_{th}	m^s_{th}
$Blog06_+$	0.23 ± 0.01	0.45	33	51	4	26
$Blog06_-$	0.19 ± 0.01	0.51	35	67	4	57
$Blog06_0$	0.16 ± 0.01	0.58	31	114	35	217
BBC_+	0.38 ± 0.02	0.27	32	25	0	2.08
BCC_-	0.051 ± 0.005	0.69	1672	81	0	0
BBC_0	0.45 ± 0.04	0.2	36	20	0	0.04
$Digg_+$	0.20 ± 0.01	0.37	115	22	0	0
$Digg_-$	0.11 ± 0.01	0.56	195	46	0	0
$Digg_0$	0.37 ± 0.04	0.27	35	33	0	0.001

The last column m^s_{th} is the average number of threads with the critical cluster received from numerical simulations on the real structure of the data i.e. the lengths of threads were acquired from the data with parameters α and $p(e|e)$. Simulations were repeated $R = 1000$ times for every thread

A Markov chain (Norris 1997) is a basic stochastic process with one memory step when the probability of the next time state depends only on the previous one by corresponding conditional probabilities. The probability of finding a cluster of size n scales as

$$P^{(e)}_M(n) \sim [1 - p(e)][1 - p(e|e)]p(e)p(e|e)^{n-1} \quad (8.4)$$

and similarly to the i.i.d. process [see Eq. (8.2)], the pre-factor $1 - p(e)$ corresponds to any event just before the cluster other than e and the pre-factor $1 - p(e|e)$ to the event just after the cluster. Note that similarly as for i.i.d. process both pre-factors are cancelled due to the normalization condition that leads to the cluster distribution $P^{(e)}_M(n) = [1 - p(e|e)]p(e|e)^{n-1}$. Finally the cumulative distribution is given by Eq. (8.3).

8.2.2 *Long Memory Model of Affective Interactions*

Now we will show that the observed frequency of emotional clusters can be well described by a model where expressed online emotions are governed by specific long-time affective interactions (Chmiel et al. 2011a). To quantify the strength of such interactions between authors of consecutive posts, consider the conditional

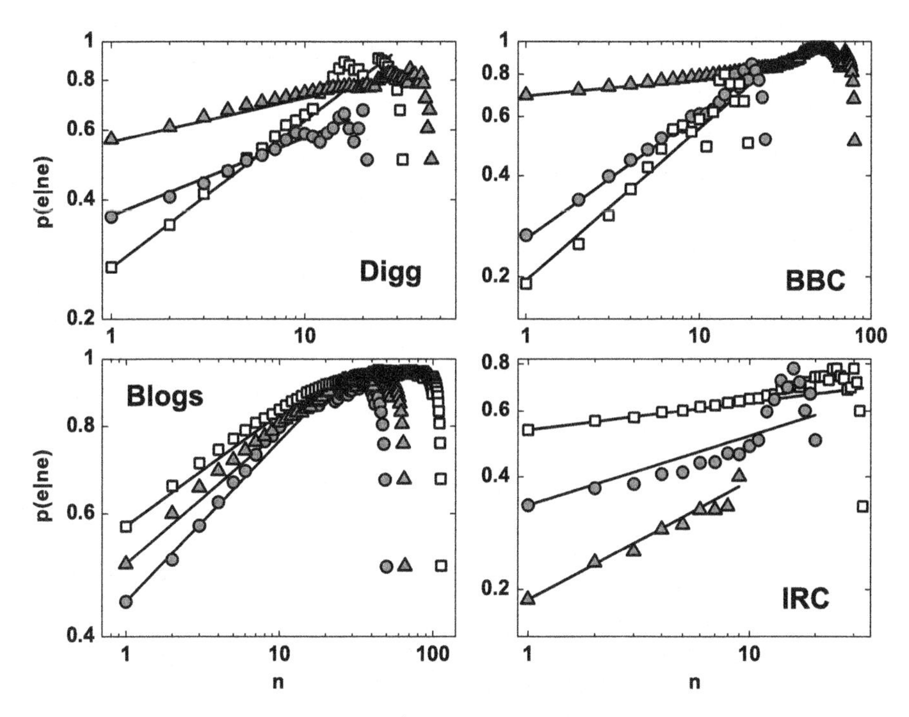

Fig. 8.3 The conditional probability $p(e|ne)$ of the next comment occurring having the same emotion for Digg, BBC, Blogs and IRC data. Symbols are data (*triangles*, *circles* and *squares*, respectively for negative, positive and neutral clusters) and lines reflect the fit to the preferential attraction relation $p(e|ne) = p(e|e)n^{\alpha}$

probability $p(e|ne)$ that after n comments with the same emotional valence the next comment will have the same valence. For the i.i.d. process such a conditional probability is independent from the parameter n since by definition $p(e|ne) = p(e)$ for the i.i.d. process. Similarly for the Markov process with one step memory we have $p(e|ne) = p(e|e)$. Figure 8.3 shows however that this probability for the original data is an increasing function of n for $n < 20$, i.e. the data reveals the relation $p(e|e) < p(e|ee) < \ldots < p(e|ne)$ thus finding a positive message after seven positive comments is more likely than after six. The observed *preferential growth* of mono-emotional clusters can be well described by a power law

$$p(e|ne) \approx p(e|e)n^{\alpha}. \tag{8.5}$$

The characteristic exponent $\alpha \geq 0$ represents the strength of the preferential process leading to the long-range attraction between posts of the same emotion. The case $\alpha = 0$ corresponds to the Markov process.

The phenomenon of preferential growth (8.5) is evident for all datasets presented in Fig. 8.3 however the range of this scaling varies for different e-communities and different emotional valences of clusters, e.g., for the BBC neutral clusters we find a good fit for the whole range of occurrence data. Note that the scaling (8.5) cannot be valid for very large n since $p(e|ne) \leq 1$.

Preferential processes are common in complex systems (Krapivsky et al. 2000) with positive feedback loop dynamics and they can be one of sources responsible for the emergence of fat-tailed distributions, including power-law scaling (Barabási and Albert 1999; Krapivsky and Redner 2001). Application of the relation (8.5) gives an analytical approximation (Chmiel et al. 2011a) to the cluster distribution (see text in the frame below)

$$P_{\alpha}^{(e)}(\geq n) \approx p(e|e)^{n-1}[(n-1)!]^{\alpha}, \text{ where } n > 0\,. \tag{8.6}$$

The solution (8.6) is presented in Fig. 8.2 with solid lines. The fit with the data is far better than for the cumulative distribution following from i.i.d. [Eq. (8.1)] or the Markov process (8.3), especially for large n.

A preferential growth of clusters with a similar emotional valence plays a crucial role for distribution of emotional clusters. An approximated analytical solution that well fits to real data can be obtained by extending the relation (8.5) up to the maximal cluster size n_{max} in the considered community. The resulting cluster probability distribution for clusters of the size n scales as

$$P_{\alpha}^{(e)}(n) \sim [1 - p(e|e)n^{\alpha}][p(e|e)]^{n-1}[(n-1)!]^{\alpha} \tag{8.7}$$

where the first factor corresponds to an event other than e just after the cluster of size n. This formula resembles the previous Markov formula (8.4) with additional factors reflecting the preferential effect. The analytical form of the missing normalization factor in the cumulative distribution (8.6) was received in Chmiel et al. (2011a) using the approximation $p(e|e)^{n_{max}}(n_{max})^{\alpha} \ll 1$. The condition holds for all of the datasets except for the neutral clusters in Blog06. As a result, the cumulative cluster distribution is given by Eq. (8.6).

Values of the exponent α for different communities and different cluster types are presented in Fig. 8.4 as a function of the probability $p(e)$ derived from the frequency of a given emotion. A good fit is $\alpha = 0.75 \exp[-4p(e)]$ although power-law and linear approximations also work well, comparing the values of R^2 (see Fig. 8.4 description). This behaviour means that for more frequent emotions the chance to attach a consecutive message with the same valence grows slower with n than for less frequent ones.

In short, one can conclude that the distribution of clusters with the same values of emotional valence observed in online communities can be explained by the process of preferential cluster growth that is a direct sign of affective interactions between

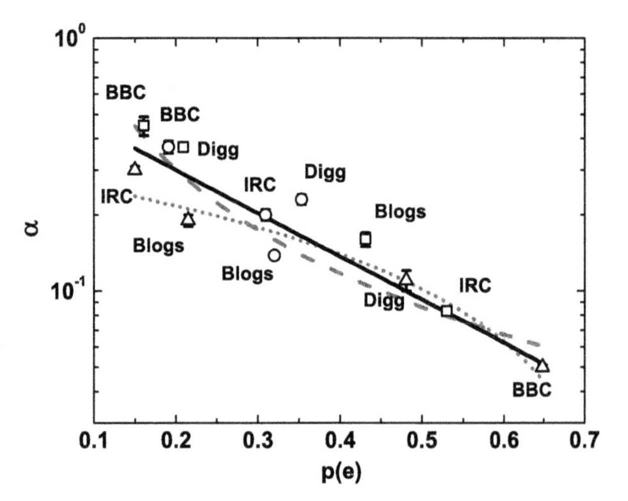

Fig. 8.4 The decay of the preferential exponent α with emotion frequency $p(e)$. Symbols are data (*triangles*, *circles* and *squares*, respectively for negative, positive and neutral clusters), the *solid line* follows the relation $\alpha = 0.75 \exp[-4p(e)]$ while *dotted* and *dashed curves* are, respectively, power-law and linear fits. The value of R^2 for exponential fit is 0.96, while for power-law and linear it is 0.94 and 0.90 respectively

community members. Emotional clusters are results of collective emotions present in the examined systems.

8.3 Critical Emotional Clusters

The relation (8.5) means that if $p(e|ne) = 1$, then the cluster reaches its *critical size* n_c^e,

$$n_c^e = [p(e/e)]^{-\frac{1}{\alpha_e}} \tag{8.8}$$

From the moment on when the critical cluster appears the discussion will be permanently ordered and all following messages in this thread will possess the same emotional valence e. Now we will study conditions for the emergence of such a critical cluster. We start (Sect. 8.3.1) by simulating the process of preferential cluster growth in an artificial environment (Chmiel and Hołyst 2013) and then present an analytical description of this phenomenon in Sect. 8.3.2. To make the problem simpler, we shall usually consider a two-state system where only positive $e = 1$ or negative $e = -1$ messages can appear in such an artificial discussion. In Sect. 8.3.3 we will show that the behavior of a three-state model is similar. Moreover we will assume that each simulated discussion thread possesses the same length L, unlike in real data, where the thread distribution was close to a power-law function (see Sect. 8.3.4 and Supporting Information in Chmiel et al. 2011a).

8.3.1 Numerical Simulations of Emergence of Critical Clusters

The evolution rules of our two-state system are as follows:

- The emotion in the first message is randomly chosen with even probabilities $p(e = 1) = p(e = -1) = 1/2$.
- The probability of emotion e in the next message is dependent on the discussion history. Information about this history is coded in size n of a recently observed emotional cluster. The cluster of size n is defined as a sub-chain of length n of consecutive states with the same values as the valences e (Chmiel et al. 2011a).
- The process of the cluster growth is based on the behavior observed in real data. The conditional probability that a cluster containing n consecutive messages with the same valence e increases its length to $n + 1$ is given by the equation:

$$p(e|ne) = x_e n^{\alpha_e} \tag{8.9}$$

  where $0 < \alpha_e < 1$ and x_e is a constant that can depend on the cluster valence e [compare (8.5)]. It amplifies the cluster growth rate and equals to a one-step memory parameter $p(e|e)$ that can be calculated from real data. Usually we will disregard the dependence x_e on the valence e and use a valence independent value x for simulations and analytical calculations. In the numerical simulations of emotional patterns in each time step we randomly choose a value between [0; 1]. If it is smaller than $p(e|ne)$, then the cluster of the emotion e is continued; otherwise, the cluster is terminated, and the opposite emotion $(-e)$ appears.
- If the cluster size reaches the critical value n_c^e given by Eq. (8.8) then all following messages in this thread are assigned the same emotional valence e corresponding to this critical cluster.

Since the critical cluster usually does not start in the beginning of a thread, a characteristic time T_c can be thus defined when the cluster reaches its critical length n_c. In numerical simulations we shall use the $\langle T_c \rangle$ as the average over R realizations (threads); in almost all cases we take $R = 10^4$. Unless otherwise stated we consider the simplest case $x = x_1 = x_{-1} = 0.5$ and $\alpha_{-1} = \alpha_1 = \alpha$. It follows we shall write n_c instead of n_c^e.

The probabilities of the appearance of both emotions when calculated in an unordered phase (before the critical cluster occurrence) are the same $p(-1) = p(1) = 0.5$, and the distribution of the observed cluster lengths is very similar to the one observed in real data. After transition time T_c, i.e., when the critical cluster appears, the discussion changes into a mono-emotional thread (MET). Starting here, the probabilities $p(-1)$ and $p(1)$ become 0 and 1 (or 1 and 0). This means that half of the threads is nearly all positive, and the other half nearly all negative (if the threads are long enough). It is obvious that the average critical time $\langle T_c \rangle$ should depend on the strength of emotional interactions, i.e., on the exponent α. It is also obvious that $\langle T_c \rangle$ has to be larger or equal to the critical size of the cluster $\langle T_c \rangle \geq n_c$

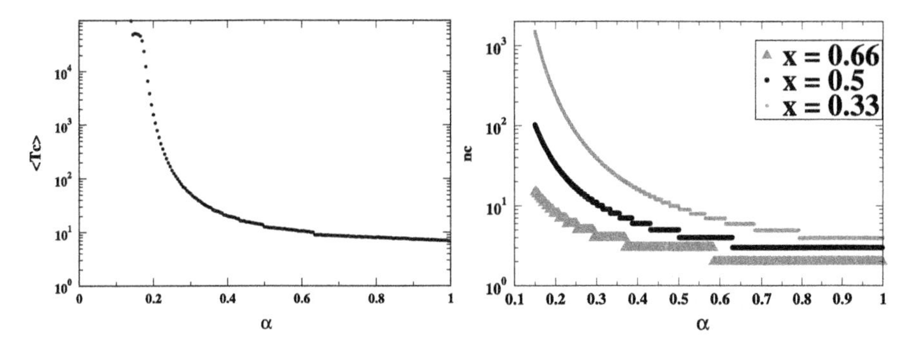

Fig. 8.5 (*Left*): Time T_c needed for the emergence of the critical cluster for $x = 0.5$, $L = 10^7$. (*Right*): Size of critical cluster as a function of α for different values of x

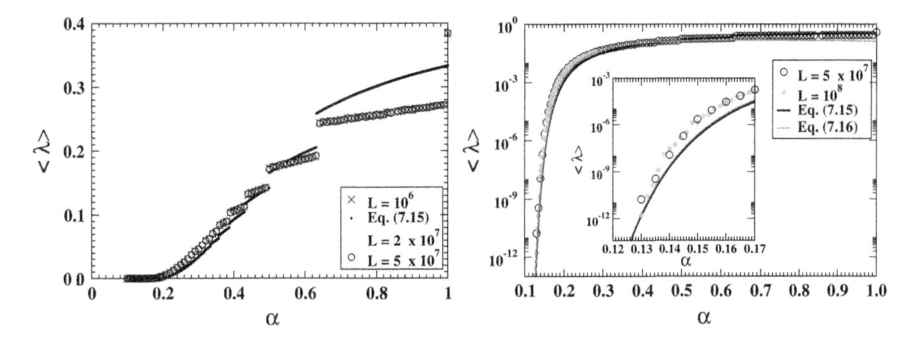

Fig. 8.6 Relation between the inverse of the critical time $\langle\lambda\rangle$ and the exponent of affective interactions α for $x = 0.5$ for different values of discussions lengths L. *Crosses*: $L = 10^6$, *squares*: $L = 2 \times 10^7$, *empty circles*: $L = 5 \times 10^7$, *diamonds*: $L = 10^8$ (only *left panel*). *Black circles* follow from Eq. (8.15) and are very close to the *solid line* from Eq. (8.16)

(see Fig. 8.5). Values of $\langle T_c\rangle$ are received from numerical simulations and n_c from Eq. (8.8).

Since in some threads the critical cluster is not observed at all, $\langle T_c\rangle$ is not an appropriate observable, and a more convenient measure is a mean inverse of the critical time

$$\langle\lambda(x,\alpha)\rangle = \frac{1}{\widetilde{R}}\sum_{i=1}^{i=\widetilde{R}}\frac{1}{T_c^i} \tag{8.10}$$

where $\widetilde{R}$ is the number of threads that were ordered during the simulation, which means that their critical times were smaller than the thread's length. In Fig. 8.6 we present a relation between $\langle\lambda\rangle$ and α. The left plot is presented in the linear scale and clearly displays the staircase shape of this dependence that follows from the integer values of T_c (compare Fig. 8.5). The right plot presents in the log-linear scale a rapid decrease in $\langle\lambda\rangle$ for $\alpha \approx 0.15$. The multi-steps shape for $\alpha > 0.3$ and a rapid decrease observed for $0.13 < \alpha < 0.2$ are only weakly dependent on the system size L. We tested this behavior for different values of L; for clarity, we show only representative simulations for $L = 10^6$, $L = 2 \times 10^7$ and $L = 5 \times 10^7$. Of course, the length of the thread L influences the value α when the order is observed for the first time in our ensemble of $R = 10.000$ samples. It is $\alpha = 0.13$ for a system of the size $L = 5 \times 10^7$ and $\alpha = 0.15$ when $L = 10^3$.

8.3.2 Analytical Estimation of Critical Times

Since the variable $\langle\lambda\rangle$ is the average $\langle 1/T_c\rangle$ it equals to a probability that a comment is at the end of the critical cluster. On the other hand, the probability of finding the critical cluster can be obtained from a distribution of cluster sizes:

$$P(n) = A(x, \alpha)x^n[(n-1)!]^\alpha, \tag{8.11}$$

that is analogous to Eq. (8.7). It follows

$$\langle\lambda(x, \alpha)\rangle = P(n_c) \tag{8.12}$$

where n_c is given by Eq. (8.8). The inverse of normalization constant in Eq. (8.11)

$$A(x, \alpha) = \left[\sum_{n=1}^{n=n_c} x^n[(n-1)!]^\alpha\right]^{-1} \tag{8.13}$$

can be calculated numerically (Chmiel and Hołyst 2013). A compact analytical approximation for this constant can be received when $\alpha = 0$ and $x < 1$

$$A(x, \alpha) \approx \frac{1-x}{x}. \tag{8.14}$$

After applying Eq. (8.8) to Eq. (8.11) we get from Eq. (8.12):

$$\langle\lambda(x, \alpha)\rangle = A(x, \alpha)\, x^{x^{-\frac{1}{\alpha}}} \left[\left(x^{-\frac{1}{\alpha}} - 1\right)!\right]^\alpha \tag{8.15}$$

that well fits the behavior of $\langle\lambda(\alpha)\rangle$ received from numerical simulations (see the right panel in Fig. 8.6). In the limit $\alpha \ll 1$ Eq. (8.15) reduces to

$$\langle\lambda(x, \alpha)\rangle \approx (1-x)\exp\left(-\alpha x^{-\frac{1}{\alpha}}\right) \tag{8.16}$$

and we get $\langle\lambda(x, 0)\rangle = 0$.

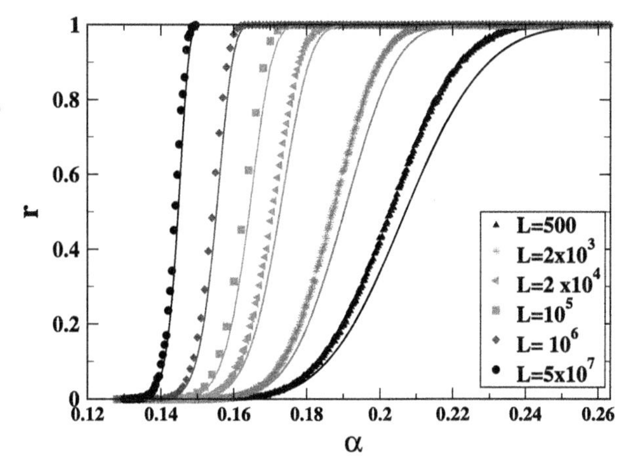

Fig. 8.7 Fraction of ordered threads as a function of the exponent α for various thread lengths L. *Lines* correspond to Eq. (8.18)

Let us consider discussions in an ensemble of threads of length L with affective interactions described by the characteristic exponent α and the parameter x and let us define a fraction of discussions that are mono-emotionally ordered from certain moments as $r(\alpha, x, L) = \frac{\widetilde{R}}{R}$. This value is also the probability of the MET occurrence before time $t = L$. It follows the value of r can be written as

$$r(\alpha, x, L) = 1 - [1 - \lambda(\alpha, x)]^L \tag{8.17}$$

where an explicit form can be received by inserting into Eq. (8.17) results (8.14) and (8.15)

$$r(\alpha, x, L) \approx 1 - \left[1 - (1 - x)\exp\left(-\alpha x^{-\frac{1}{\alpha}}\right)\right]^L \tag{8.18}$$

Results of numerical simulations and Eq. (8.18) are presented in Fig. 8.7. As one could expect, a fraction r of the MET phase in all threads increases with the increase of α exponent and thread length L. Moreover for longer threads the agreement between Eq. (8.18) and numerical simulations is better and the transition between the states $r \approx 0$ and $r \approx 1$ is steeper. In the thermodynamical limit $L \to \infty$ this transition becomes discontinuous since

$$\lim_{L\to\infty} \lim_{\alpha\to 0+} r(\alpha, x, L) = 0 \tag{8.19}$$

and

$$\lim_{L\to\infty} r(\alpha > 0, x > 0, L) = 1 \tag{8.20}$$

Until now we were focused on the relation of the fraction r of ordered realizations and the exponent α for a fixed value of the parameter x corresponding to the system

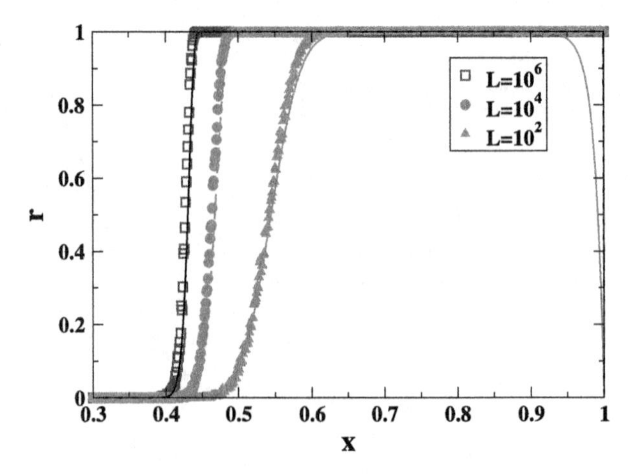

Fig. 8.8 Fraction of ordered threads as a function of the parameter x with fixed value $\alpha = 0.2$ for various thread lengths L. *Lines* correspond to Eq. (8.18). A decay of the function $r(\alpha, x)$ for $x \to 1^-$ is an artefact following from Eq. (8.14)

short memory $p(e|e)$ [see Eq. (8.9)]. Figure 8.8 shows the influence of x parameter on the fraction r where we used the solution Eq. (8.18). As it could be expected the increase of the short memory makes the MET occurrence more probable. In the thermodynamical limit $L \to \infty$ we have a discontinuous transition similar to that observed in Fig. 8.7. In fact we get from Eq. (8.18)

$$\lim_{L\to\infty} \lim_{x\to 0+} r(\alpha, x, L) = 0, \tag{8.21}$$

which should be compared to Eq. (8.19) and Eq. (8.20).

To show the combined influence of parameters α and x on the system dynamics we have presented the results of simulations for pairs (x, α) fulfilling the condition $r = 0.5$ and solutions of Eq. (8.18) for different values of L (see Fig. 8.9). There is a good agreement between numerical simulations and the approximate analytical solution (8.18) especially for $\alpha \ll 1$. As we can see in Fig. 8.9 a large fraction of MET phase can emerge only if both parameters α and x are high enough. For $x < 0.3$ the parameter α needed to observe a half of ordered realizations should be higher than 0.3 even for very long threads $L = 10^6$.

8.3.3 *Three-State System*

A natural extension of the two-state system is to add one more state, i.e., $e \in \{-1, 0, 1\}$. To compare properties of such systems with our previous results, we compare a symmetrical three-state model where $x_{-1} = x_0 = x_1 = 0.5$ and $\alpha_{-1} = \alpha_0 = \alpha_1$ with a symmetrical two-state model where $x_{-1} = x_1 = 0.5$ and

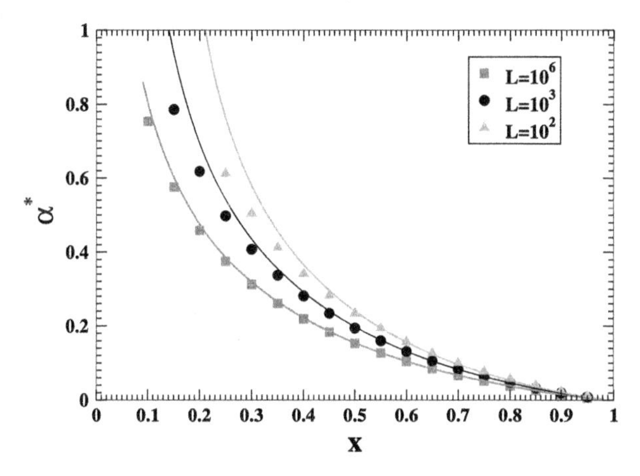

Fig. 8.9 For each value of x we find a corresponding value of α^* when half of realization are ordered, $r = 0.5$. *Points* represent numerical simulations for different values L and lines correspond to Eq. (8.18)

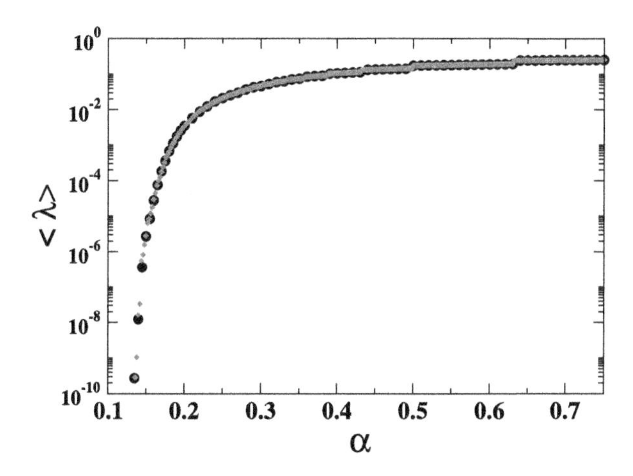

Fig. 8.10 Relation between the observable $\langle\lambda\rangle$ and the exponent α; *large circles*: two-state system with parameters $x_{-1} = x_1 = 0.5$, $\alpha_{-1} = \alpha_1$; *small circles*: three-states system with $x_{-1} = x_1 = x_0 = 0.5$ and $\alpha_{-1} = \alpha_1 = \alpha_0$

$\alpha_{-1} = \alpha_1$. Values of the inverse of critical time $\langle\lambda\rangle$ as a function of the exponent α are presented in Fig. 8.10. Since results for both systems follow the same line, we can state that the number of possible emotional states does not influence a critical time needed for the emergence of MET. This observation can be explained as follows: the occurrence of MET requires growth of a critical cluster of any emotion e. The growth process is dependent only on the conditional probability of cluster growth [Eq. (8.9)] that is insensitive to the number of possible emotional states. If initial probabilities $p(e)$ of a spontaneous occurrence of every emotional state e are equal and clusters of posts representing each emotion possess identical growth

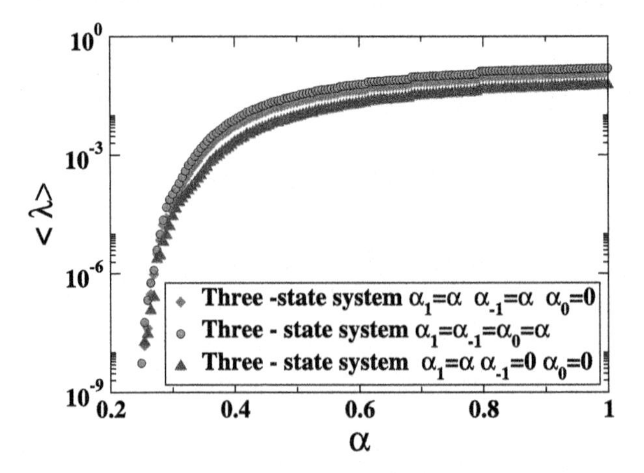

Fig. 8.11 Relation between the observable $\langle \lambda \rangle$ and the exponent α for symmetrical and asymmetrical three-state system for $x = 0.33$ and different values of α; $L = 2 \times 10^6$

parameters α_e and x_e then an average time needed for the emergence of *any* critical cluster should be independent from the number of possible emotional states.

Figure 8.11 shows the results for an asymmetrical three-state system with $x_{-1} = x_0 = x_1 = 0.33$. We considered models when one or two emotional states are random ($\alpha_{-1} = 0$ or/and $\alpha_0 = 0$) and the preferential process appears only for the remaining emotional state. We observe that for a small value of $\alpha < 0.25$ all three considered curves collapsed.

8.3.4 *Real-World Data*

Here we compare theoretical predictions of the MET phase emergence with data corresponding to various online communities: BBC Forum, Digg and blogs. One of the differences between our model and real data is a thread length distribution. In the simulation we analyzed systems with a fixed length L and various cases were tested for L between $L = 100$ and $L = 10^8$. In real datasets the thread length occurrence is described by specific distributions as presented in Fig. 8.12. The maximum observed value of L is around 10^3 for Digg and Blog06 and approximately 10^4 for BBC Forum, but those values appear only occasionally. Generally the character of these distributions is close to a power-law decay, thus, the majority of data comprises short discussions.

In Table 8.2, for three datasets we present values of exponents α (see Chmiel et al. 2011a), one-step memory parameters $p(e|e)$, values of critical clusters n_c following from Eq. (8.8) and n_{max}, i.e., maximum sizes of clusters observed in real data. The first step of our data analysis was to look for threads containing a cluster larger than a critical one. Assuming that only clusters appearing at the end of discussions can

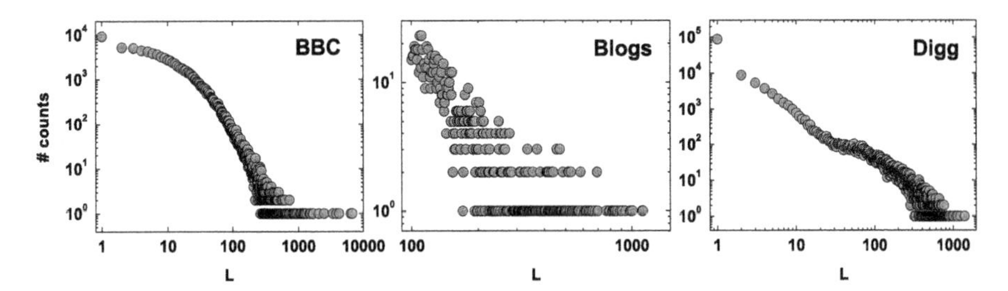

Fig. 8.12 Histograms of thread's lengths L for BBC Forum, Blog06 and Digg (see Supporting Material in Chmiel et al. 2011a)

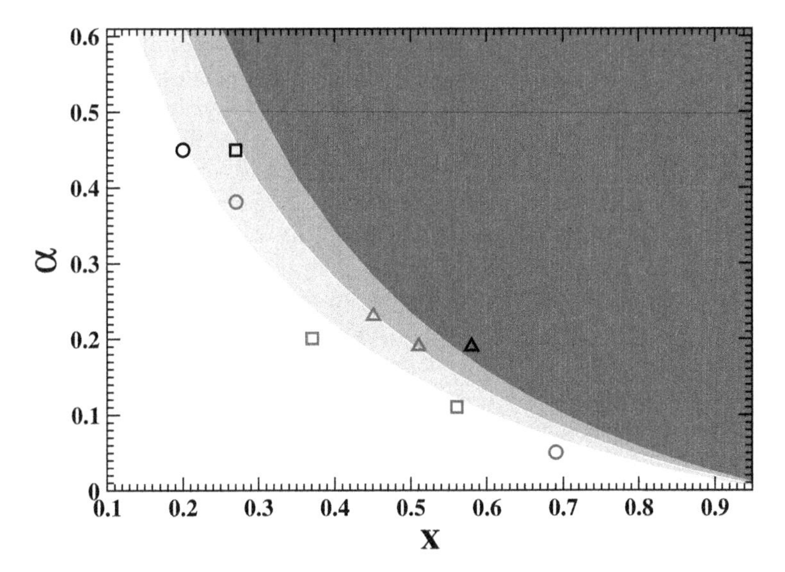

Fig. 8.13 Comparison of conditions for the MET phase emergence in the model and in real data. Results of numerical simulations show pairs of parameters (x, α) for which a half of simulated threads were ordered $(r = 0.5)$ and they are displayed by various colors for different threads lengths L. *White color* represents $L > 10^6$, *gray*—$10^6 < L < 10^3$, *orange*—$10^3 < L < 10^2$ and *red* is $L < 10^2$ (similar results can be found in Fig. 8.9). *Symbols* present values (x, α) calculated from real data (see Fig. 8.3 and Table 8.2). BBC Forum is marked with *circles*, Digg with *squares* and Blog06 with *triangles*; *red color* presents positive emotions, *blue*—negative and *black* is used for neutral ones. In the case of real data the parameter x was estimated as $p(e|e)$. The plot suggests that the best conditions for the MET phase occurrence are in the Blog06 dataset

be treated as examples of a MET phase in real data we have found threads with n_{max} larger than n_c only in Blog06 data (see Table 8.2).

The appearance of such clusters in Blog06 is in agreement with the analysis presented in Fig. 8.13, where results of simulations make up the colored background for real pairs (x, α) displayed by various symbols. In case of data corresponding to negative and neutral clusters in the BBC Forum, and negative and positive clusters in Digg the length of the system needed to statistically observe half of realizations with MET is larger than $L = 10^6$. In real Digg data we do not find discussions of such length thus the absence of MET phase is not puzzling for this community. A similar situation takes place in case of the neutral clusters in Digg and positive clusters in BBC. Here one needs threads less than $L = 10^6$ but more than 10^3. On the other hand for positive and negative clusters in Blog06 the data points lie very close to $L = 10^3$ line, while for neutral Blog06 clusters the system size necessary to find half of the ordered threads is only slightly larger than 10^2. These conditions are in a qualitative agreement with observations of the MET phase in this community. For Blog06 data we found 35 threads with neutral clusters larger than n_c, negative clusters larger than n_c were found in four threads and positive clusters appeared also in four threads.

In Blog06 dataset we find threads containing two or three clusters larger than n_c in the same thread. It means that even after the critical cluster emerged, hetero-emotional comments may afterwards be posted. Such a behavior is not predicted by our model, where after the mono-emotional cluster crosses its critical size no other emotional state can be observed. To explain this discrepancy let us point out that in case of real data there exist fluctuations that might result from mistakes in the classification algorithm[5] (Paltoglou et al. 2010), and also from a random emotional behavior of participants that is not provided for by model.

The last column of Table 8.2 follows from numerical simulations performed on the thread length distribution taken from real data. The average number m_{th}^s of threads with a critical cluster obtained in this way is always larger than the value observed in real data m_{th}^d. This discrepancy also follows from fluctuations that are absent in our model. In real data such fluctuations mean splitting of an emotionally homogeneous state into two or more parts by one or a few random messages with other emotional valence. If all separate parts were shorter than n_c our search algorithm did not detect MET in a given thread and thus the number of detected MET cases is lower. The effect of fluctuations is demonstrated in Fig. 8.14. The left panel is an example of a *clean* MET, where at the end of the discussion the users post only messages with a mono-emotional expression. The right panel is an example of a *noisy* MET where during the presumably mono-emotional phase messages with different emotions randomly appear—or at least are detected by the sentiment classifier. To quantify this behavior, for every thread with a critical cluster, we define a coefficient γ:

$$\gamma = \frac{l^e}{l} \tag{8.22}$$

Here l^e is a number of messages with a dominant emotional state e (corresponding to the critical cluster) observed in the part of the thread starting from the critical cluster, whereas l is a number of all messages in this part of a thread. In other words

[5]The accuracy of detection of the subjectivity amounts: 72 %. The accuracy of detection of polarity amounts: 67 %.

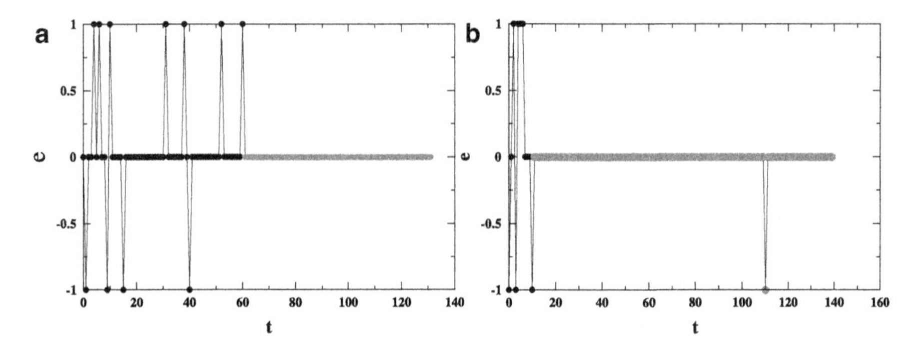

Fig. 8.14 Examples of the threads from Blog06 where l are marked in *gray*: (**a**) a thread with a cluster of the size $n_{max} = 72$ that closed the discussion ($\gamma = 1$); (**b**) a cluster with the size $n_{max} = 99$ that started a discussion with $l = 129$ comments, where $l^e = 128$ messages expressed the same emotional values and one message was different. The γ coefficient for this thread was 0.992

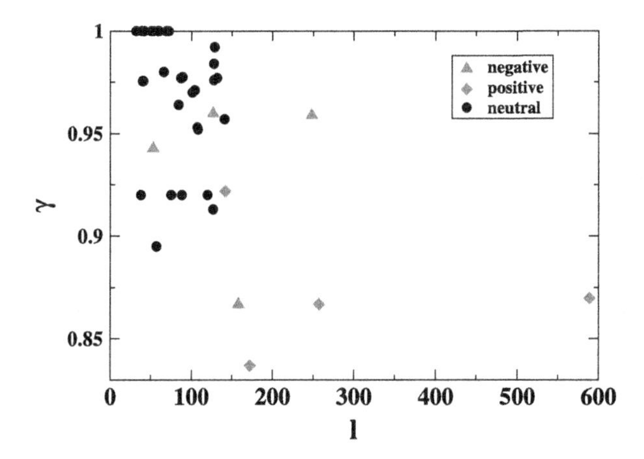

Fig. 8.15 Relation between γ coefficients and MET cluster sizes l for Blog06. *Diamonds* correspond to positive MET events, *triangles* to negative events and *circles* to neutral events

$l = l^e + l^d$ where l^d is a number of messages with emotional states different than e that were observed after the critical cluster. Let us note that $1 - \gamma$ is a measure of a noise level. Figure 8.15 presents values l and γ for all threads containing the MET phase. Mean values of this ratio for clusters of various emotions are $\langle\gamma\rangle_+ = 0.874$, $\langle\gamma\rangle_- = 0.932$ and $\langle\gamma\rangle_0 = 0.969$. It means that for a vast majority of threads the critical cluster was located close to the discussion end and not many posts expressing other emotions following critical clusters were observed.

Let us assume that the calculated ratio $\langle\gamma\rangle_e$ is the probability of a single post with emotion e being not affected by an additional independent and identically distributed random process. Such a process has been neglected in our model although it could lead to additional fluctuations disrupting the MET phase. One can then estimate a suppressing factor Γ that limits the observed number of MET events

$$\Gamma(\langle\gamma\rangle_e) = \langle\gamma\rangle_e^{n_c^e}. \tag{8.23}$$

Using the data from Table 8.2 and the $\langle\gamma\rangle_0$ ratio estimated above we get the value $\Gamma_0 = 0.39$ for neutral MET in Blog06. This suppressing factor can be compared to the ratio of a number of observed MET events to a number of predicted MET events following Table 8.2 $\Delta_0 = m_{th}^d / m_{th}^s = 0.16$. Although our approach to estimate the effects of fluctuations is elementary we receive a fairly good agreement between parameter values Γ_0 and Δ_0. The small statistics of positive and negative METs events ($m_{th}^d = 4$) was an obstacle for the noise level estimation in those cases. MET phase can be detected in real datasets if affective interactions described by parameters $p(e|e)$ and α are strong enough and the threads are long. In our case the necessary condition was met only for Blog06 data and here indeed we have observed a number of MET cases, however the effects of fluctuations suppress their emergence.

8.4 Conclusions

On the basis of automatic sentiment detection methods applied to huge datasets we have shown that Internet users' messages correlate at the simplest emotional level: positive, negative or neutral messages tend to provoke similar responses. This result agrees with observations of singular events corresponding to propagations of emotions in bipartite networks of blogs (Mitrović et al. 2010). Our simple approach demonstrates that the existence of many groups of consecutive messages (i.e., clusters) with the same emotional valence can be explained by preferential processes for cluster growth. The collective character of expressed emotions occurs in several different types of e-community. It was observed for BBC Forums and Digg, both communities where negative emotions dominate, and also for the Blog06 blogs where most comments are positive.

We hypothesize that the collective emotions are results of affective interactions in the observed groups. The strength of these interactions can be indirectly measured by the parameter α expressing the influence of the most recent emotional cluster on the probability that the next post has the same emotion [Eq. (8.5)]. Figure 8.4 shows that this strength depends not only on the kind of e-community but also on the value of the emotional valence e. Surprisingly, stronger collective behavior, corresponding to larger values of α, exists when a given emotion is less frequent in a given community. In other words, group members representing *minor emotions* display larger mutual interactions and tend to be much more clustered as compared to representatives of *major emotions*.

Motivated by those findings, in the second part of the chapter we studied a specific long-memory stochastic process that represents a data driven binary model of emotional online discussion threads. The dynamics is described by nonnegative parameters x and α corresponding respectively to the single step memory and to the characteristic exponent of the preferential process of mono-emotional cluster building. Analytical and numerical calculations show that in such a model persistent mono-emotional threads (MET) can emerge when a cluster reaches a critical size n_c. This phenomenon takes place in time T_c independent from the system size L. It follows that the longer the thread the more likely it is that it will be ordered. For finite threads fraction r of MET-containing threads increases continuously with parameters x and α. However, in the thermodynamical limit $L \to \infty$ there is a discontinuous transition between a phase without mono-emotional threads $r = 0$ and a completely ordered phase $r = 1$. The transition takes place when either $x \to 0_+$ or $\alpha \to 0_+$. The extension of the model to a three-state dynamics does not change its main properties, e.g. the critical time T_c depends in a similar way on the emotional interaction exponent α.

It is interesting that the emergence of MET phase was observed for 43 threads in Blog06 dataset and the presence of this phase could be explained by our model if corresponding parameter values were used. The number of MET events is much lower compared to theoretical estimations which can by explained be effects of fluctuations and sentiment classification errors. The absence of the MET events in BBC and Digg datasets is consistent with analytical and numerical calculations of MET density in our model.

Acknowledgements This work was supported by a European Union grant by the 7th Framework Programme, Theme 3: Science of complex systems for socially intelligent ICT. It is part of the CyberEmotions project (contract 231323). The work was also supported by Polish Ministry of Science Grant 1029/7.PR 631 UE/2009/7.

References

Barabási, A.L., Albert, R.: Emergence of scaling in random networks. Science **286**(5439), 509–512 (1999). doi:10.1126/science.286.5439.509

Chmiel, A., Hołyst, J.A.: Transition due to preferential cluster growth of collective emotions in online communities. Phys. Rev. E **87**(2), 022808 (2013). doi:10.1103/PhysRevE.87.022808

Chmiel, A., Sienkiewicz, J., Thelwall, M., Paltoglou, G., Buckley, K., Kappas, A., Hołyst, J.A.: Collective emotions online and their influence on community life. PLoS ONE **6**(7), e22207 (2011a). doi:10.1371/journal.pone.0022207

Chmiel, A., Sobkowicz, P., Sienkiewicz, J., Paltoglou, G., Buckley, K., Thelwall, M., Hołyst, J.A.: Negative emotions boost user activity at BBC forum. Physica A **390**(16), 2936–2944 (2011b). doi:10.1016/j.physa.2011.03.040

Eviatar, Z., Zaidel, E.: The effects of word-length and emotionality on hempispheric contribution to lexical decision. Neuropsychologia **29**(5), 415–428 (1991). doi:10.1016/0028-3932(91)90028-7

Feldman, L.A.: Valence focus and arousal focus - individual-differences in the structure of affective experience. J. Pers. Soc. Psychol. **69**(1), 153–166 (1995). doi:10.1037/0022-3514.69.1.153

Feller, W.: An Introduction to Probability: Theory and Its Applications. Wiley, New York (1968)

Fussell, S.R., Moss, M.M.: Figurative language in emotional communication. In: Fussell, S.R., Kreuz, R.J. (eds.) Social and Cognitive Approaches to Interpersonal Communication, pp. 113–142. Lawrence Erlbaum Associations, Mahwah (1998)

Garas, A., García, D., Skowron, M., Schweitzer, F.: Emotional persistence in online chatting communities. Sci. Rep. **2**, 402 (2012). doi:10.1038/srep00402

Gligorijević, V., Skowron, M., Tadić, B.: Structure and stability of online chat networks built on emotion-carrying links. Physica A **392**(3), 538–543 (2013). doi:10.1016/j.physa.2012.10.003
Hillmann, R., Trier, M.: Dissemination patterns and associated network effects of sentiment social networks. In: Proceedings of the 2012 IEEE/ACM International Conference on Advances in Social Networks Analysis and Mining, pp. 511–515. IEEE, Istanbul (2012). doi:10.1109/ASONAM.2012.88
Hogg, T., Lerman, K.: Social dynamics of Digg. EPJ Data Sci. **1**, 5 (2012). doi:10.1140/epjds5
Krapivsky, P.L., Redner, S.: Organization of growing random networks. Phys. Rev. E **63**(6), 066123 (2001). doi:10.1103/PhysRevE.63.066123
Krapivsky, P.L., Redner, S., Leyvraz, F.: Connectivity of growing random networks. Phys. Rev. Lett. **85**(21), 4629–4632 (2000). doi:10.1103/PhysRevLett.85.4629
Mitrović, M., Paltoglou, G., Tadić, B.: Networks and emotion-driven user communities at popular blogs. Eur. Phys. J. B **77**(4), 597–609 (2010). doi:10.1140/epjb/e2010-00279-x
Mitrović, M., Paltoglou, G., Tadić, B.: Quantitative analysis of bloggers' collective behavior powered by emotions. J. Stat. Mech. **2011**(2), P02005 (2011). doi:10.1088/1742-5468/2011/02/P02005
Norris, J.R.: Markov Chains. Cambridge University Press, Cambridge (1997)
Ounis, I., Macdonald, C.: The TREC Blogs06 collection: creating and analysing a blog test collection. Tech. Rep. TR-2006-24, Department of Computer Science, University of Glasgow (2006)
Paltoglou, G., Gobron, S., Skowron, M., Thelwall, M., Thalmann, D.: Sentiment analysis of informal textual communication in cyberspace. In: Proceedings of ENGAGE 2010. LNCS State-of-the-Art Survey, pp. 13–25. Springer, Heidelberg (2010)
Pohorecki, P., Sienkiewicz, J., Mitrović, M., Paltoglou, G., Hołyst, J.A.: Statistical analysis of emotions and opinions at Digg website. Acta Phys. Pol. A **123**(3), 604–614 (2013). doi:10.12693/APhysPolA.123.604
Sabucedo, J.M., Durán, M., Alzate, M., Barreto, I.: Emotions, ideology and collective political action. Univ. Psychol. **10**(1), 27–34 (2011)
Shimanoff, S.B.: Commonly named emotions in everyday conversations. Percept. Mot. Skills **58**(2), 514–514 (1984). doi:10.2466/pms.1984.58.2.514
Sienkiewicz, J., Skowron, M., Paltoglou, G., Hołyst, J.A.: Entropy-growth-based model of emotionally charged online dialogues. Adv. Complex Syst. **16**(4 and 5), 1350026 (2013). doi:10.1142/S0219525913500264
Sobkowicz, P.: Quantitative agent based model of user behavior in an internet discussion forum. PLoS ONE **8**(12), e80524 (2013). doi:10.1371/journal.pone.0080524
Sobkowicz, P., Sobkowicz, A.: Dynamics of hate based Internet user networks. Eur. Phys. J. B **73**(4), 633–643 (2010). doi:10.1140/epjb/e2010-00039-0
Sobkowicz, P., Sobkowicz, A.: Two-year study of emotion and communication patterns in a highly polarized political discussion forum. Soc. Sci. Comput. Rev. **30**(4), 448–469 (2012). doi:10.1177/0894439312436512
Sobkowicz, P., Thelwall, M., Buckley, K., Paltoglou, G., Sobkowicz, A.: Lognormal distributions of user post lengths in Internet discussions - a consequence of the Weber-Fechner law? EPJ Data Sci. **2**, 2 (2013). doi:10.1140/epjds14
Stürner, S., Simon, B.: Collective action: towards a dual pathway model. Eur. Rev. Soc. Psychol. **15**(1), 59–99 (2004). doi:10.1080/10463280340000117
Szell, M., Lambiotte, R., Thurner, S.: Multirelational organization of large-scale social networks in an online world. Proc. Natl. Acad. Sci. U. S. A. **107**(31), 13636–13641 (2010). doi:10.1073/pnas.1004008107
Weroński, P., Sienkiewicz, J., Paltoglou, G., Buckley, K., Thelwall, M., Hołyst, J.A.: Emotional analysis of blogs and forums data. Acta Phys. Pol. A **121**(2B), B128–B132 (2012)
Wu, F., Huberman, B.A.: Novelty and collective attention. Proc. Natl. Acad. Sci. U. S. A. **104**(45), 17599–17601 (2007). doi:10.1073/pnas.0704916104
Zajonc, R.B.: Feeling and thinking - preferences need no inferences. Am. Psychol. **35**(2), 151–175 (1980). doi:10.1037/0003-066X.35.2.151

Chapter 9
How Online Emotions Influence Community Life

Julian Sienkiewicz, Anna Chmiel, Paweł Sobkowicz, and Janusz A. Hołyst

9.1 Introduction

In this chapter we introduce data driven models that describe relations between expressed online emotions and individual activities as well as the entire community termination. We shall focus on a special aspect of the online emotional dynamics—its non-stationarity. Unlike in the previous case (Chap. 8, Chmiel et al. 2011a), where we focused on the dynamics *inside* the threads, revealing collective character of the emotional exchange, here we try to extract conditions necessary for discussion to be started and ceased. Our key idea is to demonstrate non-equilibrium stages of expressed emotions that are relaxed during the community life. We propose two different approaches—the first one is based on the observation of activity in online threads and on relations between a thread length and emotions included in it. The second approach uses a physical concept of entropy as an indicator of dialogue's forthcoming closing. Both studies express the same key condition: there has to be some kind of emotional imbalance for the discussion to last. The integral part of both approaches are models, designed to reproduce stylized facts observed in the datasets: power-law distributions of discussion lengths or activity as well as relations between emotions and duration of the thread.

J. Sienkiewicz (✉) • A. Chmiel • P. Sobkowicz • J.A. Hołyst
Faculty of Physics, Warsaw University of Technology, Koszykowa 75, 00-662 Warsaw, Poland
e-mail: julas@if.pw.edu.pl; anulachmiel@gmail.com; pawelsobko@gmail.com; jholyst@if.pw.edu.pl

J.A. Hołyst (ed.), *Cyberemotions*, Understanding Complex Systems,
DOI 10.1007/978-3-319-43639-5_9

Table 9.1 Fundamental properties of datasets studied in this chapter: number of comments C, number of dialogues (IRC)/discussions (BBC) N, shortest dialogue/discussion length L_{min}, longest dialogue/discussion length L_{max}, average valence $\langle e \rangle$, probability of finding negative, neutral or positive emotion (respectively $p(-)$, $p(0)$ and $p(+)$)

Dataset	C	N	L_{min}	L_{max}	$\langle e \rangle$	$p(-)$	$p(0)$	$p(+)$
IRC	1,889,120	93,323	11	339	0.17	0.15	0.53	0.32
BBC	2,474,781	97,946	1	6789	−0.44	0.65	0.16	0.19

9.2 Datasets Used

The sources on which we base our analysis comprise of two datasets: (1) BBC *Message Boards*, a public discussion Forum[1,2] (Chmiel et al. 2011b) and (2) automatically archived Internet Relay Chats (IRC) logs from the #ubuntu channel[3] (Sienkiewicz et al. 2013). In the case of the first source the data covers 4 years, from the launch of the website (June–July 2005) until the beginning of the data extraction process in June 2009. For the IRC channels we chose a 3-year period between 1st January 2007 and 31st December 2009. Data from the BBC Forum are used in both main parts of this chapter (Sects. 9.3 and 9.4)—the discussions cover a wide selection of topics, from politics through TV issues to religion. On the other hand, #ubuntu IRC channel is a typical place where you can seek advice and resolve operating system issues. This data will be used only in the second part of the chapter (Sect. 9.4). Fundamental properties of both datasets are gathered in Table 9.1.

The *emotional classifier* program (based on the hierarchical Language Model) used to analyze the emotional content of the discussions is described in detail in Chap. 6. As an outcome, each post (or comment) is annotated with a single value $e = \{-1, 0, 1\}$ to quantify its emotional content as negative, neutral or positive, respectively. The accuracy of the classifier checked for 950 humanly annotated comments in IRC data is 62.49 % for subjectivity detection and 70.25 % for polarity detection while in the case of BBC data the numbers are, respectively, 73.73 % and 80.92 % for 594 annotated documents. The average emotional value $\langle e \rangle$ as well as the frequencies of each valence $p(-)$, $p(0)$ and $p(+)$ for both datasets are given in Table 9.1.

[1]http://www.bbc.co.uk/dna/mbreligion.

[2]http://www.bbc.co.uk/dna/mbfivelive/F2148565, http://www.bbc.co.uk/dna/mbfivelive/F2148564.

[3]http://help.ubuntu.com/community/InternetRelayChat.

9.3 Activity and Emotions of BBC Forum Users

In this part of the chapter using the example of the BBC Forum we show that negative emotions can drive Internet discussions. We focus on statistical analysis of large record of user comments and emotions detected by automated tools. Our goal is to understand the relationship between user activities and the emotions that they express in individual discussion threads and in all of their posts. We compare the observations with agent-based simulation of a discussion board, which includes pairwise exchanges of comments as significant factor determining activity statistics as well as emotional content. The model reproduces many of the characteristic features of the forum discussions, suggesting their origin is akin to those derived from human analyses of smaller datasets (Rafaeli and Sudweeks 1997; Sobkowicz and Sobkowicz 2010). First we present our analysis methods, followed by detailed descriptions of user activity and emotions. Later a computer simulation model is described and its results compared with the observations. Lastly we discuss implications of the findings, especially correlation between user activity and negative emotions.

9.3.1 *User Activity*

In literature there are various observables introduced to characterize Internet user behavior. Very common is the analysis of inter-event time and waiting time distributions (Gonçalves and Ramasco 2008; Radicchi 2009; Barabási 2005; Vázquez et al. 2006; Masuda et al. 2009), which can be described by power-law relationships. Barabási (2005) suggested that the bursty nature of various human activities in cyberspace (e-mail, web-browsing) follows from decision-based queuing processes. Radicchi (2009) found that the distribution of inter-event times for a user is strongly dependent on the number of operations executed by that user.

Here we consider user activity a_i defined to be the total number of posts written by user i in all discussion threads during the observation period. For simplicity, this quantity will also be referred to as a. The maximum observed activity in the dataset is $a_{max} = 18{,}274$, i.e., one user authored more than eighteen thousand messages, while the average activity is $\langle a \rangle = 137$, and the median $m_a = 3$. The number of occurrences of a is presented in Fig. 9.1a and it is well fitted by the power-law relationship $h_a \sim a^{-\beta}$, where $\beta = 1.4$. The relatively small value of the exponent β suggests a high number of very active users of this Forum. All figures presenting the histogram of user activity posses the same layout: we show the real data and the numerical simulations of the model (see Sect. 9.3.4).

Since all discussions in the Forum are split into separate threads j, we define d_i^j (or d for short) to be the local activity of user i in thread j measured by the number of posts that this user submitted in the discussion. Whereas both its maximum and average value (respectively $d_{max} = 1582$ and $\langle d \rangle = 2.84$) are lower than in the

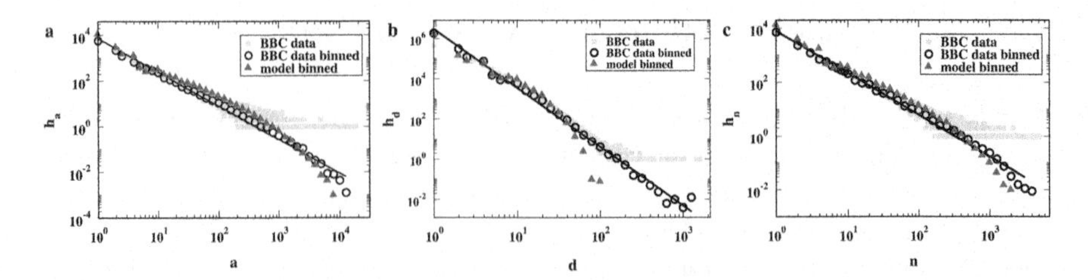

Fig. 9.1 Histograms of user activity and diversity of threads: (**a**) histograms of user total activity a (*squares*), binned data (*circles*), binned results of numerical simulations of the agent model (*triangles*), (**b**) histogram of user activity in thread d, (**c**) histogram of the diversity of threads n. Lines are fits to the data and they follow relations $h_a \sim a^{-\beta}$, $h_n \sim n^{-\tau}$, $h_d \sim d^{-\alpha}$ and with $\beta = 1.4, \tau = 1.5, \alpha = 2.9$

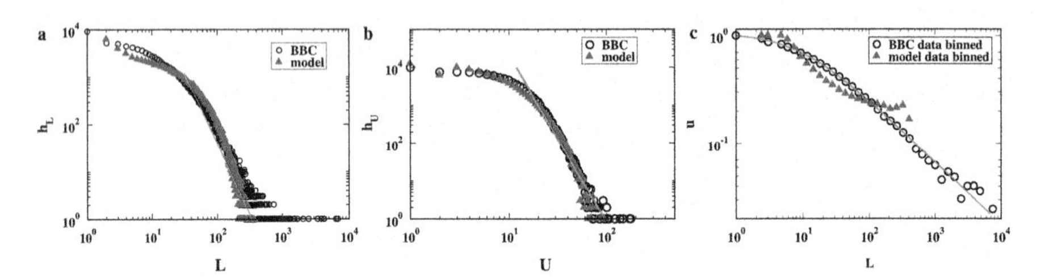

Fig. 9.2 (**a**) Histogram of threads with length L. The BBC data are represented by *empty circles*, the result of the numerical simulation of the agent model is shown by *red triangles*, the *black line* corresponds to $h_L \sim L^{-\eta}$ with $\eta = 2.5$, (**b**) histogram of the number of unique users U making a comment in the thread. *Lines* correspond to relation $h_U \sim U^{-\eta}$ with $\eta = 4.9$, (**c**) the normalized number of unique users u making a comment in a thread of length L, *circles*—data binned logarithmically, *red triangles*—model. *Green line* corresponds to relation $u = A(L + b)^{-\delta}$ with fitted parameters $\delta = 0.58$, $A = 3.72$ and $b = 8.6$

case of a, the number of occurrences of d, shown in Fig. 9.1b, still follows a similar relationship $h_d \sim d^{-\alpha}$ with exponent $\alpha = 2.9$, which is double that of a.

Taking into account the above described quantities: (1) How many threads are in the area of interest of a user? (2) How does the user diversify her activity among different discussions? To answer these issues, consider the number of different discussions n_i in which the user i takes part. The results shown in Fig. 9.1c present the number of occurrences of n_i. Again we find power-law scaling $h_n \sim n_i^{-\tau}$ with $\tau = 1.5$. The results reveal diversity in human habits: the overwhelming majority of users join just one discussion and usually post only one comment in it. However, there is also a significant number of those that write often and express themselves in several discussions.

While the user statistical behavior shows a strong tendency to be scale-invariant, this is not so clear for the thread statistics shown in Fig. 9.2a. Here, we consider the thread length L and the number of unique users U posting at least one comment in the thread. Histograms of both quantities h_L and h_U display power-law tails for

$U, L > 20$. This is most prominent in the case of h_U which is also characterized by a rather large exponent $\eta = 4.9$.

To understand the impact exerted by the most frequent users on the length of a thread consider the dependence between the normalized number of unique users in a single thread defined by $u = U/L$ and thread length (Fig. 9.2c). For short threads (L between 1 and 10) u is about $0.6 - 1$ while for threads larger than 400 comments it drops below 0.1. A good fit is $u(L) = A(L+b)^{-0.58}$ (green line in Fig. 9.2b) thus the number of unique users grows more slowly than linearly with thread lengths. This suggests that mutual discussions between specific users rather than a large number of independent comments submitted by many users sustain thread life.

9.3.2 User Emotions

As mentioned in the Introduction, the recent progress in automatic sentiment analysis gives the ability to quantify the emotional content of large scale textual data. This has already led to observations of emotionally-linked communities in blogs (Mitrović et al. 2010) and to tracing shifts in public opinion (González-Bailón et al. 2010). Other, indirect methods have also revealed the emergence of phenomena like the existence of the 'hate networks' in political discussions (Sobkowicz and Sobkowicz 2010) and emotional connections within communities in massive multi-player online games (Szell et al. 2010). Kappas et al. (2010) have demonstrated that BBC Forum posts elicited subjective experiences at emotions by participants of psychophysiological tests.

The following quantities describe the emotions of individual debaters and discussions threads. The average (global) emotion of a user $\langle e \rangle_a$ is the sum of all emotions e in posts written by the user i divided by her activity a_i. The average emotion of a thread $\langle e \rangle_L$ is the sum of all emotions in the thread j divided by its length L_j. The third value $\langle e \rangle_d$ is the average emotional expression of the user i in the thread j. The main features of the distribution $p(\langle e \rangle_a)$, presented in Fig. 9.3a, are peaks for $\langle e \rangle_a = -1, 0, 1$ which are a straightforward effect of the large number of users with $a = 1$ and threads with $L = 1$ (see Figs. 9.1a and 9.2a). The local maximum around $\langle e \rangle_a = -0.5$ is a specific attribute of the BBC Forum because it possesses a strong bias toward negative emotions, with an average value of $\langle e \rangle = -0.44$. Figure 9.3b presents distribution of $p(\langle e \rangle_L)$. Similar shape is observed for $p(\langle e \rangle_d)$.

So far we have treated user activities and emotions as mutually independent variables but consider now the relationship between them. Figure 9.4a (inset) presents users' global average emotions $\langle e \rangle_a$ versus their global activity a. Neglecting fluctuations for large values of a caused by small numbers of very active users, there is a constant mean emotion that is around the Forum's average value $\langle e \rangle$. Hence, on average, the user activity level a does not influence her emotions e. In the main panel of Fig. 9.4a the reversed relationship is plotted, i.e., the distribution of average global activity versus users' average emotions (black bars). For comparison we

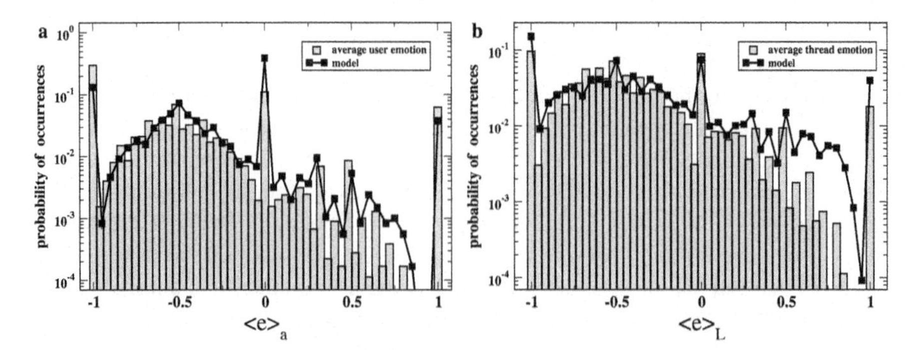

Fig. 9.3 Distribution of emotion. Probability distribution of (a) users global average emotions $\langle e\rangle_a$ and (b) threads global emotion $\langle e\rangle_L$. BBC data is presented by *gray bars*, results of numerical simulation of the agent model by *black squares*, *line* is only to guide the eyes

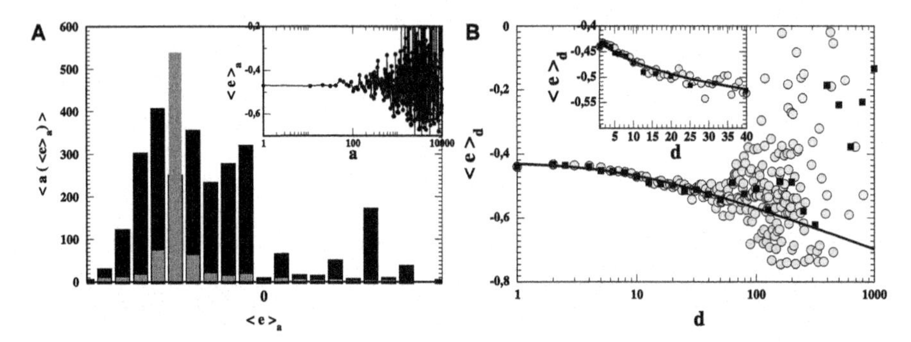

Fig. 9.4 (a) Average activity of user in the function of emotions of user. Users' global activity $\langle a(\langle e\rangle_a)\rangle$ versus their global average emotion $\langle e\rangle_a$: *black bars*—empirical data, *gray bars*—shuffled data. *Inset*: users' global average emotion $\langle e\rangle_a$ versus users' global activity a, (b) relationship between users' average emotion in a thread $\langle e\rangle_d$ and users' activity in the thread d. *Gray circles* are original data, *black squares* are binned data and the *black curve* corresponds to equation $\langle e\rangle_d = A_1 + B_1 \ln(d + b)$ with $A_1 = -0.31$, $B_1 = -0.054$ and $b = 8.6$. Inset is focused on the local activity less than 40

present shuffled data (gray bars) where the emotional values of posts were randomly interchanged between users. Whereas the second distribution follows a Gaussian-like function, the original set is characterized by a broad maximum stretching across almost all of the negative part of the plot and some minor fluctuations in the positive part. This implies that there is a dependence between user activity and emotions even at the global level.

Users can take part in many threads, thus their local and global activities as well as corresponding local emotions can be very different. But how are users' emotions $\langle e_i^j\rangle$ expressed in a thread connected to the activity level d_i^j in it? Figure 9.4b shows the average emotions of a user in a thread as a function of the user's local activity. In this case, an increase in activity in a particular thread leads to more negative average emotions in the thread. Recall that there was no relationship between a user's global

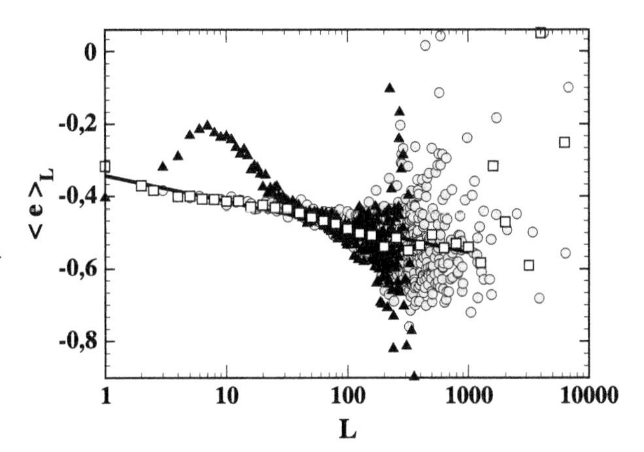

Fig. 9.5 Relationship between the average emotion in discussions with fixed thread length L (*circles*), *squares* are binned data, the *black line* corresponds to the relationship $\langle e\rangle_L = A' + B' \ln(L)$ with fitted parameters $A' = -0.34$ $B' = -0.03$, *triangles* represent the model results

activity and her emotions, as shown in the (inset) of Fig. 9.4a. For longer discussions there is a more homogeneous group of users (see Fig. 9.2b), thus on average one user writes a larger number of posts $\langle d\rangle(L) = \frac{1}{u(L)}$. As shown in Fig. 9.4b, the average emotions for users that are locally more active decreases. These two effects document that the longer threads possess, on average, more negative emotions. In fact in Fig. 9.5 there is a logarithmic decay in mean thread emotions $\langle e\rangle_L$ as a function of thread length L.

9.3.3 Life-Spans of Communities

Do community emotions evolve over time? This phenomenon was studied quantitatively as follows. Threads of the same size were grouped together and a moving average of the emotion type of the last ten comments was calculated for each point. As seen in Fig. 9.6a, shorter threads tend to start from a lower (i.e., less negative) emotional level than longer ones. On the other hand threads end with a similar mean emotional valence value regardless of their lengths: the last point of each data series in 9.6a (circles, squares, triangles and diamonds) is at almost the same level, about −0.42. This phenomenon is echoed in Fig. 9.6b where the average emotional valence of the first ten comments minus the average emotional valence of the last ten comments is plotted, showing that longer threads have larger changes in emotional valence. Figure 9.6c also suggests that the initial emotional content (whether positive or negative) may be used as an indicator of the expected length of a thread: low absolute average emotion valences lead to shorter discussions. A possible heuristic explanation is that the first few posts in a thread may give it the potential (emotional fuel) to propel further discussion. Once the emotions driving the discussion dry out, the thread is no longer of interest to its participants and it may die. For threads possessing higher initial levels of emotion it takes more comments

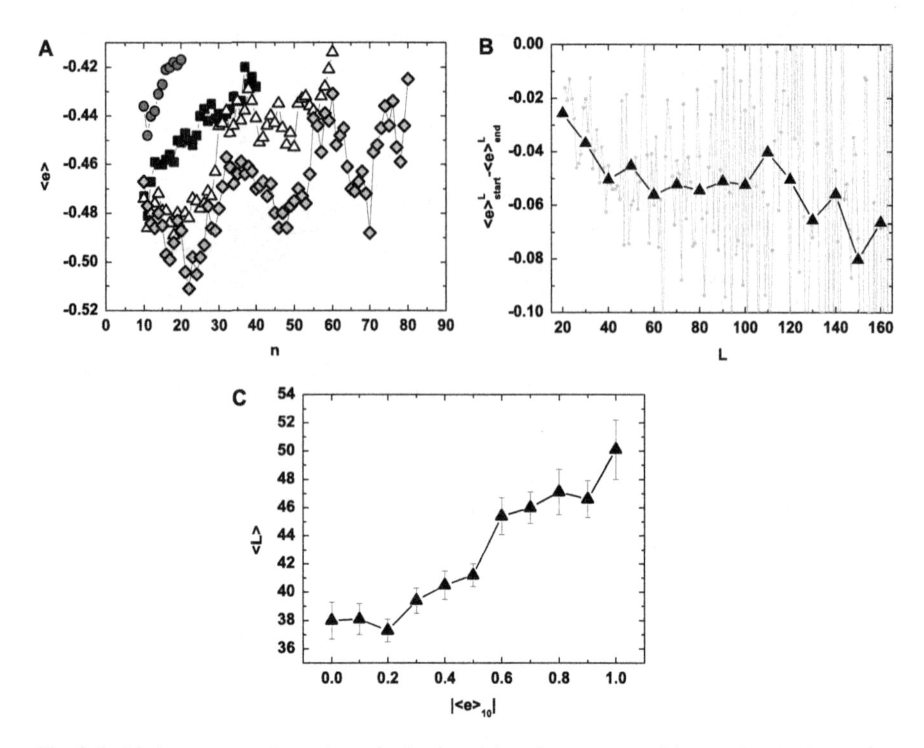

Fig. 9.6 (**a**) Average emotion valence in the thread (moving average of the previous ten messages in the thread). Four groups of threads of lengths 20, 40, 60 and 80 are represented by different *symbols* (respectively *circles*, *squares*, *triangles* and *diamonds*). Shorter threads start from emotional levels closer to zero, (**b**) emotional level (valence) at the beginning of a thread minus the emotional level at the end as a function of thread length (*grey symbols*). *Black triangles* display binned data. Longer threads use more emotional "fuel" over time, (**c**) average length of the thread as a function of the absolute value of the average emotion valence of the first ten comments. Emotional thread starts, whether positive or negative, usually lead to longer discussions. *Error bars* indicate standard deviations

to resolve the emotional issue, resulting in longer discussions [cf. results for the discussions at BBC forum in Chmiel et al. (2011b)]

9.3.4 Computer Model and Simulations

To analyze our data we have developed an extension of simple computer model of the community and discussion process. Such simulations, using agent-based computer models, have been previously proposed in Schweitzer and García (2010), Ding and Liu (2010), Sobkowicz and Sobkowicz (2010), Czaplicka et al. (2010). We use a modification of the model introduced in Sobkowicz and Sobkowicz (2010), which may be described as follows. The simulated users (agents) belong to three categories: those with deeply set, controversial opinions (denoted as A and B), and neutral agents, denoted as N. Comments always reflect opinion of the user, so we would use the same categories to describe users and posts. As the simulation makes no attempt to map real topics of BBC discussions, we have no knowledge regarding the relative numbers of disagreeing factions A and B. For this reason we have assumed that the size of both opinionated groups is symmetrical. Independent of the user community mix, the topics of discussion (called source messages) may also favor A, B or neutral viewpoint.

It should be noted that the comment opinion (A/B/N) is separate from the comment emotional content. The latter is determined by the comparison between categories of its author and the target of the post (which may be the source message of the thread or another post). Adding neutral agents allows more flexibility in treating emotions than in the original model (Sobkowicz and Sobkowicz 2010). Neutral agents' posts are always emotionally neutral. If the author of the comment and the target belong to the same non-neutral category (A-A, B-B) then the emotion expressed in the comment is positive. For A-B or B-A combinations emotions are negative. For A or B category authors commenting on neutral posts there is certain probability x_N that the comment would be emotionally negative, in remaining cases it would be neutral. In contrast to Sobkowicz and Sobkowicz (2010) we used only one class of agent activity.

We have assumed a population of 25,000 agents reading the forum. For each thread we randomly select agents who may participate in discussion. Each such agent 'reads' one of the messages within a thread (called target). The target may be the source of the thread (usually news item from BBC) or one of earlier posts by other users. The probability of the agents to read the source message, p_s, is one of the control parameters of the simulation. With probability $(1 - p_s)$ the target is another earlier post, where we assume preferential attachment rules to calculate probability of reader choosing specific comment as target. Specifically, the chance of reading a post is proportional to its total degree (outdegree of a post is always 1, indegree may be quite high). After reading the target post, the agent then decides whether or not to comment on it.

The probability of posting a comment is given by $p_c f(r, t)$, where a universal 'comment activity' ratio p_c, the same for all agents, is modified by a factor $f(r, t)$, depending on the reader r category and that of the target message t. This factor reflects greater probability of getting aroused by contrary views and motivated to post a comment. Thus for (r =A, t =B) and (r =B, t =A) pairs we have $f(r, t) = 1$, while for other combinations $f(r, t) = f^* < 1$, where f^* is an adjustable parameter.

After the agent has commented a post other than the source we enter into a 'quarrel' subroutine. Here, the author of the target post is 'given a chance' to respond, with probability determined by $p_r f(r, t)$, where p_r is independent parameter from p_c, but $f(r, t)$ is the same as in the main routine. If the response is placed, the roles of the two agents are reversed, and a chance for counter-response is evaluated. This subroutine is continued until one of the agents 'decides' to quit. Values of p_c and p_r determine the relative importance of quarrels within the thread.

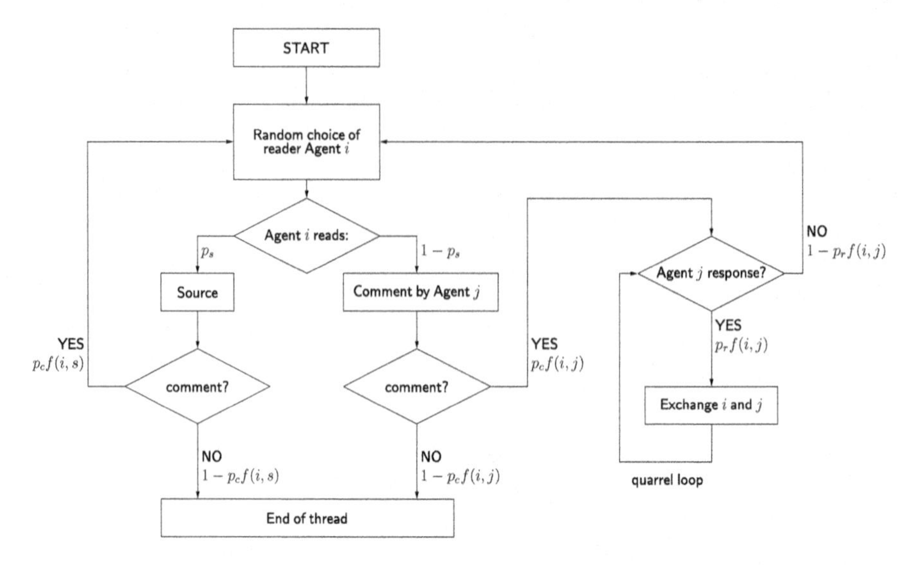

Fig. 9.7 Schematic diagram of the simulation process within a single thread. The process has been repeated 110,000 times. For (i =A, t =B) and (j, s =B, t =A) pairs $f(r, t) = 1$, while for other combinations $f(r, t) = f^* < 1$. Using values of parameters which: $p_s = 0.5, p_c = 0.93, p_r = 0.89$ and $f^* = 0.86$ resulted in $\sim$ 97,000 threads with nonzero number of comments

This simple simulation program returns then to the main routine of agents posting comments, until the currently selected agent decides not to post. The whole process is then repeated for a specified number of threads N_{th}. Figure 9.7 presents the diagram of the flow of the simulation program for a single thread.

The model does not include many features present in Internet discussions. The emotions are automatically determined by the category of agents/posts, rather than resulting from emotional content of previous messages. In real discussions it is commonly the use of offensive language that spurs replies in kind, not the opinion expressed in the post. Also, situations such as a single user posting more than one post within a thread (due to, for example, lack of space to express his/her viewpoint or repetition of the same message several times) are not considered. This leads to deviations of the model statistics from observations for threads with only a few comments. Despite the model simplicity it gives results quite similar to the ones derived from analysis of BBC forum.

Figures 9.1, 9.2, 9.3 and 9.5 compare results of simulations obtained for $N_{th} =$ 110,000, $p_s = 0.5$, $p_c = 0.93$, $p_r = 0.89$, $f^* = 0.86$ and $x_N = 0.91$ with statistics of the BBC forum. This choice of parameters resulted in simulations with about 2.5 million posts, ~970,000 active threads (i.e. thread topics to which at least one agent posted a comment) and $\sim$ 19,000 active agents (i.e. agents that posted at least one comment)—these values vary slightly between simulation runs, but correspond well to the observed data. The results of simulations are obtained using relative sizes of communities supporting A, B and neutral viewpoints of

A/B/N=32 %/32 %/36 %. For the topics of the threads (source messages) the fit resulted in ratio of $(A/B/N)_{source}$ =25 %/25 %/50 %.

About 70 % of the posts in simulation belonged to quarrels. The choice of f^* and x_N is determined by condition to obtain the same mean value of emotion $\langle e \rangle$ as observed in the BBC forum. The A/B/N ratios for users and source messages were chosen to reproduce the ratios of emotionally positive, neutral and negative messages. Simulation resulted in values of 20 %, 16 % and 64 %, in good agreement with data from BBC forum: 19 %, 16 % and 65 %, respectively. The distributions of average user and thread emotions (Fig. 9.3) follow closely the observed statistics. Other statistical characteristics of the discussions are such as user activity, thread lengths and unique authors as well as thread diversity are also rather close to the observed distributions.

The differences between simulation and reality are visible mainly for short threads. For example, the distribution of user activity within a thread, while being in the same range as the data from BBC forum, does not show power law behavior, with smaller counts for high activity values (Fig. 9.1b). A similar difference is seen in relationship between thread length and average emotion (Fig. 9.5), where simulation deviates from observations for low values of $L < 30$, remaining close to the observations for longer threads.

9.3.5 Discussion

In this part using sentiment analysis methods, we have found patterns in users' emotional behavior and observed scale-free distribution of users activity in the whole forum and in singular threads as well as power law tails for distribution of thread lengths and number of unique users in a thread. At the level of the entire Forum negative emotions boost users activity, i.e. participants with more negative emotions write more posts. At the level of individual threads users that are more active in a specific thread tend to express there more negative emotions and seem to be the key agents sustaining threads discussions. As result, longer threads possess more negative emotional content.

The simulation model which includes extended exchanges between pairs of users reproduces many of observed characteristics of the BBC forum reasonably well and is offered as possible explanation of the behavior. Similar model was used in Sobkowicz and Sobkowicz (2010), where thanks to combination of comment organization in the studied dataset and categorization analysis conducted by humans (as opposed to automated process) it has been possible to verify directly the 'quarrel' model of user activity. As quarrels are important part of the simulated system, we undertook to check if they are found in the original data. The nature of our dataset did not allow automatic recognition of all quarrels. The time ordering process flattened the dataset, hiding this information. Full attribution would require search through the text of the messages for tell-tales of pairwise exchanges of posts, for example references to other posts/users within a comment. A very simplified analysis based on temporal proximity of the posts, looking for series of consecutive posts such as ABAB..., where A and B are unique authors, gives an estimate of the quarrels at more than 40 %. A detailed analysis of a subset of threads shows that, especially for most active ones, the posts of pairs of agents are likely to be separated by other comments, so this statistics strongly underestimates the percentage of discussions.

We also give evidence that in BBC forums the initial emotional level of a discussion fuels its continuation: when this fuel is exhausted a discussion is likely to end (Fig. 9.6a, b). This is because higher levels of emotions in the first ten comments in a thread lead to longer discussions (Fig. 9.6c). This behavior agrees with observations concerning political discussions in Polish Internet forums (Sobkowicz and Sobkowicz 2010) where the growth of discussions was dependent on the degree of controversy of the subject and the intensity of personal conflict between the participants (Sobkowicz and Sobkowicz 2010)

9.4 Entropic Origin of Emotional Dialogues

Communication and its evolution is one of the key aspects of a modern life, which in an overwhelming part is governed by the circulation of information. In the most fundamental part, the communication is based on a *dialogue*—an exchange of information and ideas between two people (Sacks et al. 1974). Assuming an ideal situation, if the highest priority would be given to *acquiring* certain information, from a layman point of view the dialogue should be free from any additional components that could restrain conversation's participants from achieving the common goal. In reality, a holistic view on the communication should in fact treat it rather as a *discourse*, i.e., it needs to be defined by language use, communication of beliefs and social interactions (van Dijk 1997) or even social context (Littlejohn and Foss 2010). In this sense, the meaning emerges through a mutual relation between communicators and their social contexts. On the other hand, early models of communication focused on the generation of meaning by words themselves, creating a system of signs, governed by rules and used to signify objects (Saussure 1977). This is in fact, a very reductionist view, treating the language as made up of distinct units that can be studied in separation from their environment. Then again, using another approach, we can also treat the dialogue as an entity governed by conversational rules (Grice 1975). In this the concept of turn-taking is placed—apportioning of who is to speak next and when (Sacks et al. 1974). Recent studies in this area prove that although there are differences across the languages in the average gap between turns, all tested languages exhibit a universal behavior of avoidance to overlap and of minimizing the silence between discussion turns (Stivers et al. 2009). Clearly the approach that lays in the closest proximity to the area of complex systems is this given by Shannon and Weaver in the late 1940s (Shannon 1949). In their view, called the information theory, a message is transmitted by a channel from a source to a receiver that interprets it. The channel is characterized by its bandwidth, defining the capacity and resulting level of information. Thus a channel with high quality transmits the message itself while a poor quality channel may convey a contaminated content. Such an approach is deliberately free from taking into account the content of the message.

As compared to the off-line communication, the exchange of information in the Internet is claimed to be more biased toward the emotional aspect (Sobkowicz and Sobkowicz 2010). It can be explained by a online disinhibition effect (Suler 2004)—the sense of anonymity that almost all Internet users possess while submitting their opinions on various fora or blogs. In this part we argue that a simple physical approach based on the observation of entropy of emotional probability distribution during the conversation can serve as an indicator of a discussion about to finish. We give arguments supporting the observation of the maximum entropy rule in the emotional dialogues regardless of the type of the medium in question (i.e., negative, neutral), which results in creation of a tool that can be used to distinguish between the initial and final stage of the dialogue. The process of entropy maximization is claimed to be responsible for a power-law distribution of the discussion length and serves as a key idea for the numerical simulations of the dialogues which confirm that such assumed rules lead to good agreement between the observed and simulated discussion lengths.

9.4.1 Data Preparation

As mentioned in Sect. 9.2, we used the IRC `#ubuntu` channel logs—in total, there were 994 daily files with 4600–18,000 utterances in each file that share a following format: *post_number* [*timestamp*] $\langle$*user_id*$\rangle$ *sentiment_class* with the *sentimentclass* $e = \{-1; 0, 1\}$ used as marker for the emotional valence through this study. Moreover, we could also obtain the information that specifies which user communicates, i.e., directly addresses, another user shown as $\langle$*addressing_user_id*$\rangle \rightarrow$ $\langle$*addressed_user_id*$\rangle$). The discovery of the direct communication links between two users in the IRC channel was based on the discovery of another userID at the beginning of an utterance, followed by a comma or semicolon signs; a scheme commonly used in various multiple users communication channels. However, one has to bear in mind that this kind of information can be sometimes incomplete, i.e., in many cases users do not explicitly specify the receiver of his/her post. Another issue that arises is that the data consist of several overlapping dialogues held simultaneously on one channel. It is also sometimes difficult to indicate the receiver of the message as only part of them are annotated with a user id they are dedicated to. We created an algorithm that addresses this issue. It consists of two different approaches:

(a) if user A addresses user B in some moment in time and later A writes consecutive messages without addressing anybody specific we assume that he/she is still having a conversation with B

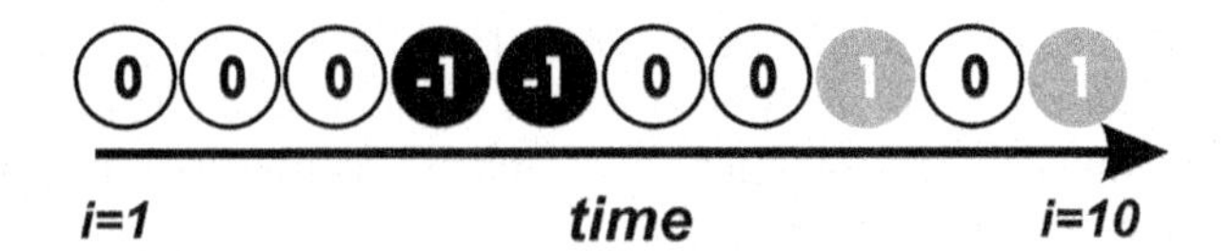

Fig. 9.8 An exemplary dialogue of $L = 10$ comments. Each *bullet* corresponds to a comment with a negative (marked as -1), neutral (marked as 0) or positive (marked as 1) content

(b) if user A addresses user B and then B writes a message without addressing anybody specific we assume that he/she is answering to A.

The main parameter of such algorithm is the time t in which the searching is being done; in our study we use $t = 5$ min as the threshold value. In this way we are able to extract a set of dialogues from each of the daily files. After processing the file according to above described rules another issue emerges: it often happens that a user gives a set of consecutive messages directed to one receiver. To create a standardized version of the dialogue (A to B, B to A, A to B and so on), we decided to accumulate the consecutive emotional messages of the same user, calculate the average value $\bar{e}$ in such series and then transform it back into a three-state value according to the formula

$$\begin{cases} e^i = -1 & \bar{e} \in [-1; -\frac{1}{3}] \\ e^i = 0 & \bar{e} \in (-\frac{1}{3}; \frac{1}{3}) \\ e^i = 1 & \bar{e} \in [\frac{1}{3}; 1] \end{cases} \tag{9.1}$$

The choice of the transformation form is selected in such a way that a continuous range $[-1; 1]$ is separated into an equal-range division in order to recover the original set of values $\{-1, 0, 1\}$. One could also use other ways to transform consecutive emotional messages into one value—we have also tried taking only the last valence, however it did not have any impact on the further analysis and results. The final step of the data preparation is to divide it into separate dialogues; in total, the algorithm produces $N = 93{,}329$ dialogues with the length between $L = 11$ and $L = 339$ (all the dialogues with $L \leq 10$ were omitted) that can be represented as a chain of messages (see Fig. 9.8) where all odd posts are submitted by one user and all even by another one.

9.4.2 Common Features

The obtained dialogues have been divided into groups of constant dialogue length L. For such data we follow the evolution of mean emotional value $\langle e \rangle_i^L$ and average emotional probabilities $\langle p(e) \rangle_i^L$. In both cases the $\langle \ldots \rangle_i^L$ symbol indicates taking all dialogues with a specific length L and averaging over all comments with number i, thus, for example, $\langle p(-) \rangle_i^L$ is the probability that at the position i in all dialogues of

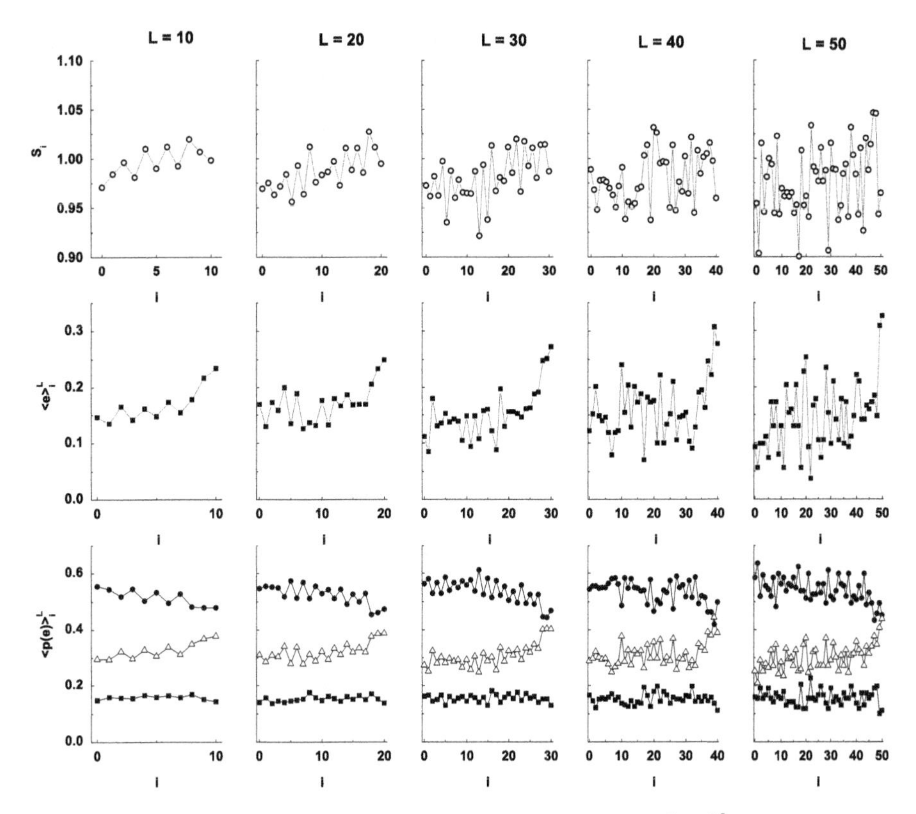

Fig. 9.9 IRC data: entropy S_i of the emotional probabilities distribution $\langle p(e)\rangle_i^L$ (*top row*), average emotional value $\langle e\rangle_i^L$ (*middle row*) and average emotional probabilities $\langle p(-)\rangle_i^L$ (*squares*), $\langle p(0)\rangle_i^L$ (*circles*), $\langle p(+)\rangle_i^L$ (*triangles*) in the i-th timestep for dialogues of specific $L = 10$ (*first column*), $L = 20$ (*second column*), $L = 30$ (*third column*), $L = 40$ (*fourth column*) and $L = 50$ (*fifth column*)

length L there is a negative statement. The characteristic feature observed regardless of the dialogue length is that the $\langle e\rangle_i^L$ at the end of the dialogue is higher than at the beginning (upper row in Fig. 9.9). In fact, there is especially a rapid growth close the very end of the dialogue, which is probably caused by participants who acknowledge others' support issuing comments like "thank you", "you were most helpful", etc.

The direct reason for such behavior is shown in the bottom row of Fig. 9.9, which presents the evolution of the average emotional probabilities $\langle p(-)\rangle_i^L$, $\langle p(0)\rangle_i^L$ and $\langle p(+)\rangle_i^L$. The observations can be summarized in the following way:

- the negative emotional probability $\langle p(-)\rangle_i^L$ remains almost constant,
- $\langle p(+)\rangle_i^L$ increases and $\langle p(0)\rangle_i^L$ has an opposite tendency,
- $\langle p(+)\rangle_i^L$ and $\langle p(0)\rangle_i^L$ tend to equalize in the vicinity of dialogue end.

Other manifestation of the system's features can be spotted by examining the level of the entropy S of the emotional probabilities $\langle p(e)\rangle_i^L$. Entropy or other information theoretic quantities as mutual information (Cover and Thomas 1991), Kullback-Leiber divergence (Lin 1991) or Jensen-Shannon divergence (Lin 1991) have been already used to quantify certain aspects of human mobility (Song et al. 2010), semantic resemblance or flow between Wikipedia pages (Masucci et al. 2011a,b) or correlations between consecutive emotional posts (Weroński et al. 2012). Moreover, basing on entropy, it has also been shown how the coherent structures in the e-mail dialogues arise (Eckmann et al. 2004) or how to predict conversation patterns in face-to-face meetings (Takaguchi et al. 2011). The concept of entropy is often used in such non-physical areas as ecology, for example as a tool for tracing the biodiversity (Shipley et al. 2006). However, as Bailey (1983)—the initiator of the social entropy theory (SET)—states, in the case of social sciences the term "entropy" had hardly been used until 1980s, spare the works of Miller (1953), Rothstein (1958) and Buckley (1967) who employed it for the examination of sociological organization structure. In this chapter, the entropy is used after Shannon's definition (Shannon 1948), i.e.,

$$S_i^{sh} = - \sum_{e=-1,0,1} \langle p(e) \rangle_i^L \ln \langle p(e) \rangle_i^L. \tag{9.2}$$

In Fig. 9.10 we show a schematic plot illustrating the direct consequences of Eq. (9.2). If the distribution of some feature is equiprobable (e.g., each of three political parties get exactly 1/3 of the total number of votes, 9.10a), the resulting value of entropy is maximal. In the opposite situation (one party gets the majority of votes, 9.10b), the entropy is very low—in an extreme situation, when all the votes are gathered by one party, the entropy is minimal and equals 0. Thus, entropy can

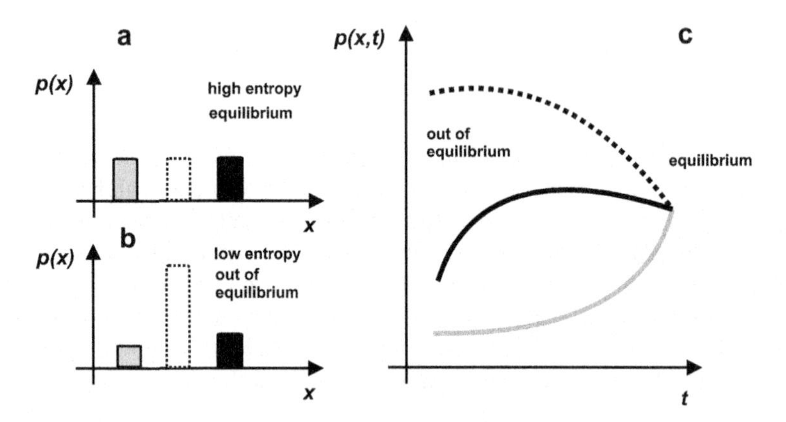

Fig. 9.10 Schematic plot illustrating the consequences of Eq. (9.2). (**a**) Equiprobable distribution of some feature, resulting value of entropy is maximal, (**b**) a dominant feature present in the probability distribution—resulting value of entropy is very low, (**c**) system is initially out of equilibrium equilibrates in the course of time, acquiring the state of maximal entropy

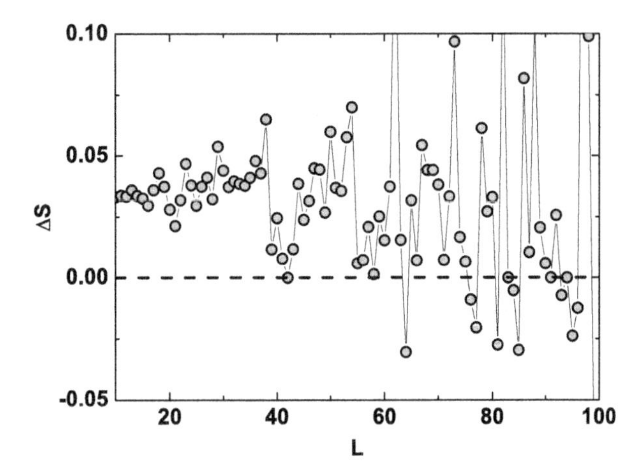

Fig. 9.11 Difference between terminal and initial entropy value ΔS versus the dialogue length L. The *dashed line* marks $\Delta S = 0$ level

serve as an indicator of the state of the system, showing if it is ordered (low S) or disordered (high S).

Here, taking into account the fact that $\langle p(-)\rangle_i^L$ is constant in the course of dialogue, we paid attention only to $\langle p(+)\rangle_i^L$ and $\langle p(0)\rangle_i^L$, thus the observed entropy had a form of

$$S_i = -\left[\langle p(0)\rangle_i^L \ln\langle p(0)\rangle_i^L + \langle p(+)\rangle_i^L \ln\langle p(+)\rangle_i^L\right]. \tag{9.3}$$

Plotting the difference between terminal and initial entropy ΔS versus the length of the dialogue L it is possible to see that for the dialogues up to $L \approx 50$ this difference is always above zero (see Fig. 9.11). It implies a following likely scenario for the dialogue: it evolves in the direction of growing entropy. In the beginning of the dialogue, the probabilities $\langle p(0)\rangle_i^L$ and $\langle p(+)\rangle_i^L$ are separated from each other, contributing to low value of initial entropy S_p. However, then the entropy grows, the probabilities $\langle p(0)\rangle_i^L$ and $\langle p(+)\rangle_i^L$ equalize leading to high value entropy (i.e., higher then the initial one) at the end of the dialogue.

However, it is essential to notice that the observed behavior in the IRC data is only one of the possible scenarios of the more general phenomenon of the principle of maximum entropy (Jaynes 1957), governing also certain aspects of biological (Williams 2011) or social systems (Johnson et al. 2010) (at the level of social networks). The tendency for the isolated system to increase its entropy and to evolve to reach the state characterized by the maximum entropy (MaxEnt) is a well-know physical phenomenon previously observed in many real-world systems (Harte 2011). It is a sign of the situation when the system is initially out of equilibrium and in the course of time it equilibrates (Fig 9.10c), acquiring the state of maximal entropy. One needs to stress that the division of the dataset into dialogues gives opportunity to claim the isolation of the system in question. Social sciences had incorporated the idea of equilibrium long before entropy (Spencer 1864), although it has then been used rather as a synonym of system integration and stability (Pareto

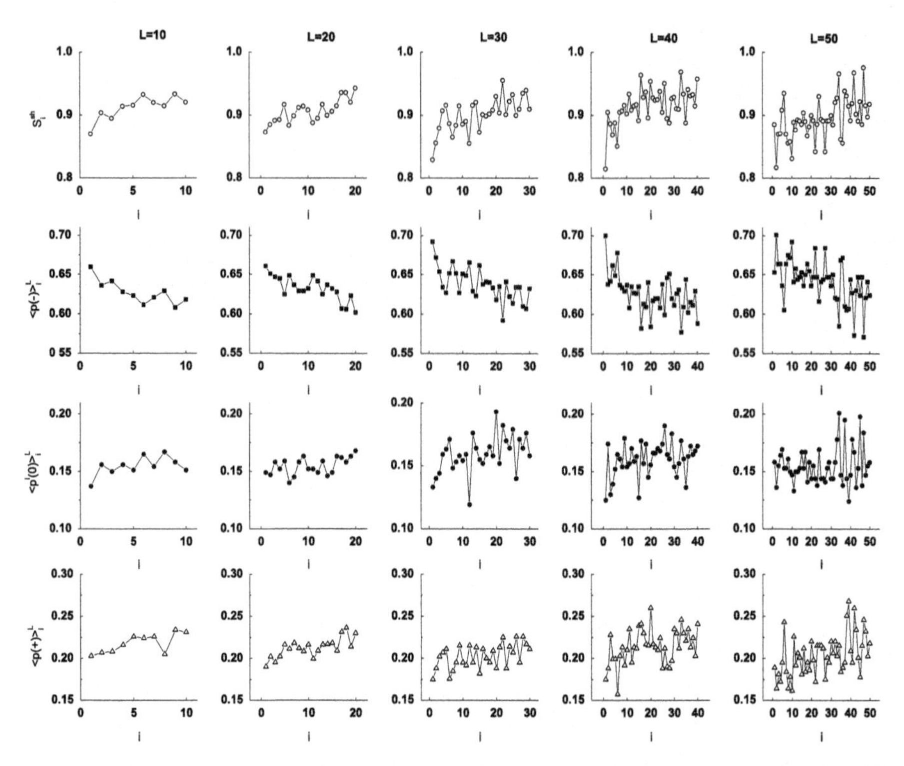

Fig. 9.12 BBC data: entropy S_i^{sh} of the average emotional probabilities distribution $\langle p(e)\rangle_i^L$ (*topmost row*) and average emotional probabilities $\langle p(-)\rangle_i^L$ (*squares*) $\langle p(0)\rangle_i^L$ (*circles*) and $\langle p(+)\rangle_i^L$ (*triangles*) in the i-th timestep discussions of specific length $L = 10$ (*first column*), $L = 20$ (*second column*), $L = 30$ (*third column*), $L = 40$ (*fourth column*) and $L = 50$ (*fifth column*)

1935; Parsons 1951). In the physical case (and also in this study) it is essential that growing entropy indicates the direction of time. Thus this behavior should be irrelevant of the type of the system in question.

In order to test this assumption, we performed an analysis analogous to this for the IRC data with respect to emotionally annotated dataset from the BBC Forum. In this case each discussion was treated as a natural "dialogue", although it usually consisted of more than two users communicating to each other. Following the line of thought presented for IRC data we grouped all discussion of constant length and calculated the quantities $\langle p(-)\rangle_i^L$, $\langle p(0)\rangle_i^L$, $\langle p(+)\rangle_i^L$ and S_i^{sh}. The results, shown in Fig. 9.12, bear close resemblance to those obtained for IRC data: one can clearly see that while the negative component decreases, the positive and objective (partially) ones increase. In has an instant effect on the value of entropy which grows during the evolution of the discussion (topmost row in Fig. 9.12). Less prominent entropy growth tendency (as compared to IRC) could be connected to the fact that in the case of BBC threads we can have users that appear in the discussion and then leave it, which violets the assumption of the system isolation. The main difference between IRC and BBC Forum results concerns the component whose value decreases during the discussion evolution: for IRC it is the $\langle p(0)\rangle_i^L$ while for BBC Forum - $\langle p(-)\rangle_i^L$. It is directly connected to the fact that the above mentioned components play the role of "discussion fuel" (Chmiel et al. 2011a) propelling thread's evolution. BBC Forum data come from such categories as `World News` and `UK News` and as such may lead the discussion participants to place comments of very negative valence. On the other hand `#ubuntu` IRC channel servers rather as a source of professional help which is normally expressed in terms of neutral dialogue. As the discussion lasts, the topic dilutes (BBC Forum) or the problem is being solved (IRC) and the dominating component dies out leading to maximization of entropy. Here entropy can serve as a kind of indicator measuring the way emotional states are changing. It can be directly applied as a tool to distinguish between the initial state that is later subject to a sort of thermalization and the final phase where all the emotions get mixed up. Thus, one may regard it as an index of the dialogue phase - regardless of the overall emotional character of the medium (i.e., neutral, negative).

There is also another process taking place in the system in question that displays a non-trivial behavior. As shown in Chap. 8, we can talk about grouping of similarly emotional messages. To quantify the persistence of a specific emotion one can consider the conditional probability $p(e|ne)$ that after n comments with the same emotional valence the next comment has the same sign. As it easy to prove, if e would be treated as an identical and independently distributed (i.i.d.) variable the conditional probability $p(e|ne)$ should be independent of n and equal to $p(e)$, i.e., the probability of a specific emotion in the whole dataset (see Table 9.1) In the case of the IRC data, the analysis shows (see Fig. 8.3 in Chap. 8) that $p(e|ne)$ is well approximated by

$$p(e|ne) = p(e|e)n^{\alpha}. \tag{9.4}$$

where $p(e|e)$ is the conditional probability that two consecutive messages have the same emotion. The exponents α and the conditional probabilities $p(e|e)$ are gathered in Table 9.2.

Table 9.2 Conditional probabilities $p(e|e)$ and scaling exponents for the power-law cluster growth α_e with errors

Emotion sign	$p(e\|e)$	α_e
Positive ($e = 1$)	0.34	0.138 ± 0.004
Neutral ($e = 0$)	0.53	0.083 ± 0.001
Negative ($e = -1$)	0.19	0.30 ± 0.01

9.4.3 Model and Computer Simulations

The methodology described above proves to be successful in finding the prominent characteristic of the data in question, however it is rather useless if one would like to perform the simulations of the dialogues. It is crucial to choose other way for calculating the average emotional probabilities "on the fly" and, using the results, decide on the further dialogue evolution. Thus, we decided to work with moving time window, i.e., the probability of the specific valences in the i-th timestep are

$$\begin{cases} \bar{p}_i^M(+) = \frac{1}{M}\sum_{j=1}^{j=M} \delta_{e(i-j),+1}, \\ \bar{p}_i^M(0) = \frac{1}{M}\sum_{j=1}^{j=M} \delta_{e(i-j),0}, \\ \bar{p}_i^M(-) = \frac{1}{M}\sum_{j=1}^{j=M} \delta_{e(i-j),-1}, \end{cases} \tag{9.5}$$

for $i \geq M$, where δ is the Kronecker delta symbol and M is the size of the window. Consequently, entropy S_i is also calculated using the probabilities $\bar{p}_i^M(+)$ and $\bar{p}_i^M(0)$ as

$$\bar{S}_i^M = -\left[\bar{p}_i^M(0)\ln\bar{p}_i^M(0) + \bar{p}_i^M(+)\ln\bar{p}_i^M(+)\right]. \tag{9.6}$$

expressing in fact the entropy in the i-th time window. The practical way of application is shown in Fig. 9.13 for a dialogue of $L = 30$ comments. In this case the size of the time window is set to $M = 10$.

The data-driven facts presented in the previous section lie at the basis of the simulation of dialogues in IRC channels data. The key point treated as an input

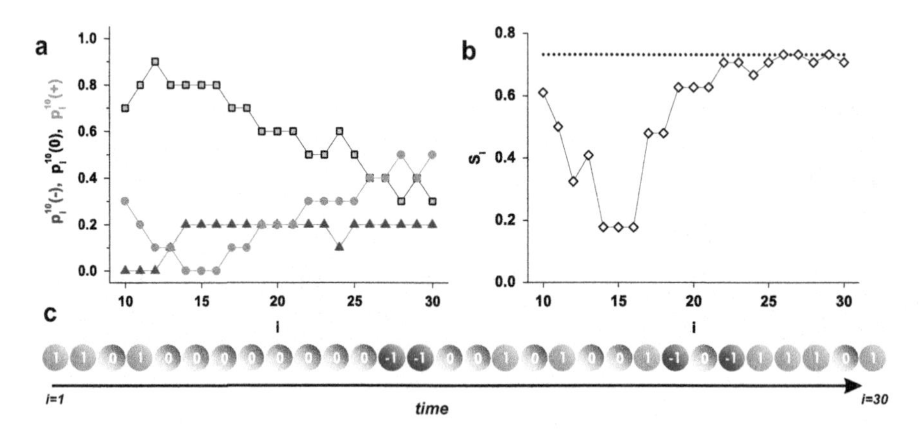

Fig. 9.13 (a) Probabilities of specific valence $\bar{p}_i^M(-)$ (*triangles*), $\bar{p}_i^M(0)$ (*squares*) and $\bar{p}_i^M(+)$ (*circles*) in the i-th time window given by Eq. (9.5) for the exemplary dialogue shown in panel c, (b) entropy $\bar{S}_i$ in the i-th time window defined by Eq. (9.6) for the exemplary dialog shown in panel c. The *dotted line* marks the maximal value of entropy in Eq. (9.6) i.e., $\bar{S}_i^{max} = \frac{2}{5}\ln\frac{5}{2} \approx 0.73$. The dialogue is real-world example from IRC data

parameter for this model is the observation of the preferential attraction of consecutive emotional messages. This idea "runs" the dialogue, whereas the discussion is terminated once the difference between the entropy in the given moment and its initial value exceeds certain threshold. Those features are implemented in the following algorithm:

1. start the dialogue by drawing the first emotional comment with probability $p(e)$ (values taken from Table 9.1),
2. set the next comment to have emotional valence e of the previous comment with probability $p(e|ne) = p(e|e)n^{\alpha_e}$
3. if the drawn probability is higher than $p(e|ne)$, set the next comment one of two other emotional values (i.e., if the original $e = 1$, then the next comment valence is 0 with probability $p(0)/[p(0) + p(-)]$ or -1 with probability $p(-)/[p(0) + p(-)])$
4. if the difference between entropy in this time-step and the initial entropy is higher than threshold level ΔS terminate the simulation, otherwise go to point (2).

The observed valence probabilities in this simulation are always calculated using quantities in a moving time window given by Eqs. (9.5) and (9.6) with $M = 10$.

There is another crucial parameter connected to the simulation process, i.e., the initial entropy threshold S_T. When time-step $i = M$ is reached, the entropy $\bar{S}_i^M$ is calculated for the first time and then decision is taken: if $\bar{S}_M^M < S_T$ the simulation runs further, otherwise it is canceled and repeated. The total number of successfully simulated dialogues is equal to this observed in the real data.

Figure 9.14 shows a comparison of the average emotional value $\langle e\rangle_i^L$ and average emotional probabilities $\langle p(e)\rangle_i^L$ for the real data and simulations performed according to the algorithm described in the previous section for dialogues of length

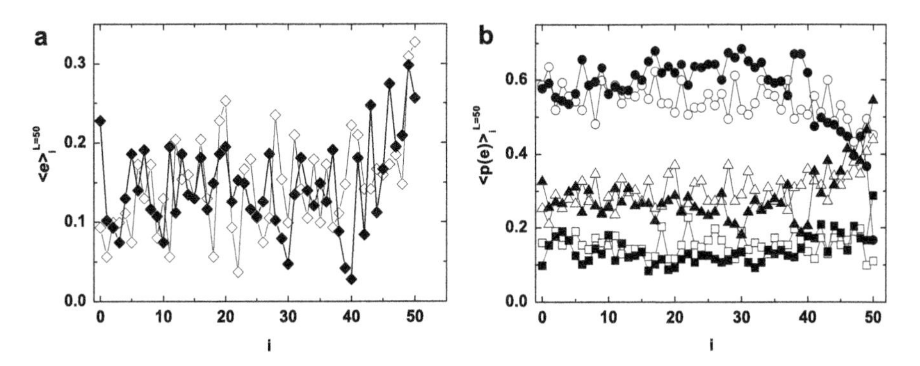

Fig. 9.14 Comparison of average emotional value $\langle e\rangle$ (**a**) and probability of specific emotion (**b** $\langle p(-)\rangle_i^{L=50}$—*squares*, $\langle p(0)\rangle_i^{L=50}$—*circles*, $\langle p(+)\rangle_i^{L=50}$—*triangles*) for simulations performed according to the procedure presented in Sect. 9.4.3 (*full symbols*) and for real data (*empty symbols*) for dialogue length $L = 50$. The real data shown are identical with those shown in the fifth column of Fig. 9.9

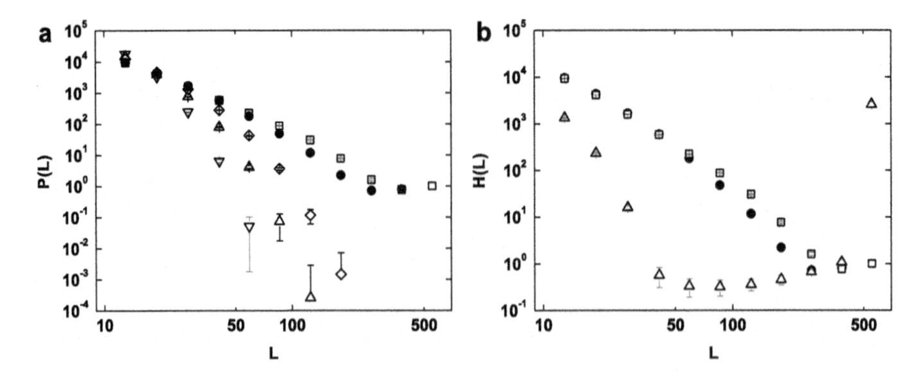

Fig. 9.15 (**a**) Dialogue length distribution $P(L)$ for real data (*circles*) and simulations for different values of the initial entropy threshold S_T parameter: $S_T = 0.1$ (*downward triangles*), $S_T = 0.5$ (*upward triangles*), $S_T = 0.6$ (*diamonds*) and $S_T = 0.63$ (*squares*), (**b**) dialogue length distribution $P(L)$ for: real data (*circles*), simulations with $S_T = 0.63$ (*squares*) and simulations with $S_T = 0.63$ and insertion of the additional neutral comments (*triangles*). Each simulation data point is an average over 100 realizations, error bars correspond to standard deviations. Data are logarithmically binned with the power of 1.45

$L = 50$. As one can see the plots bear close resemblance apart from only one detail, i.e., the rising value for the $\langle p(-)\rangle_i^L$ close to the end of the dialogue.

Moreover, the simulation strongly depends on the exact value of the initial entropy threshold S_T which can be clearly seen in Fig. 9.15a, where the dialogue length distribution is presented. If the entropy threshold S_T is restricted to values between 0.1 and 0.5 (downward and upward triangles) the distribution of dialogue lengths is exponential and does not follow the one observed in the real data (circles). Higher values of S_T ($S_T = 0.6$, diamonds) shift the curve closer to the data points, nevertheless the character is still exponential. Its only after tuning the S_T parameter to 0.63 that the results obtained from the simulations (squares) are qualitatively comparable with the real data.

9.4.4 Application

It is possible to consider a direct application of the above described model for changing the "trajectory" of the dialogue. For example let us assume that a dialogue system (Bohus and Rudnicky 2003; Williams et al. 2005; Skowron 2010) is included as part of the conversation and that its task is to prolong the discussion. In such situation, the system that could rely on the above presented properties would attempt to detect any signs indicating that the dialogue might come to an end and react against it. According to observations presented in Sect. 9.4.2 a marker for such event should be the growth of the entropy. In other words the dialogue system should prevent an increase of the entropy in the consecutive time-steps.

In the described case, such action would be an equivalent to an insertion of an objective comment. In this way, an equalization between $\bar{p}_i^M(+)$ and $\bar{p}_i^M(0)$ is prevented and dialogue can last further. An implementation of this rule is presented in Fig. 9.15b, where one can compare the real data (again empty circles), a simulation including the entropy-growth rule (again full circles) and a simulation following the insertion of objective comments (empty triangles). While there is a drop-down in the numbers for the small dialogue lengths, the vast majority of the dialogues has the maximal length (a point in the top-right corner). In this way the insertion of the objective comments is in line with the expected idea of dialogue prolonging.

It is essential to stress that this kind of a theoretical application could be presumably useful and suited only in certain situations and only for particular interactive environments. The key feature observed in the IRC channels data, i.e., the equalization of the emotional probabilities and entropy growth during the time of dialogue does not need to be present in other situations (see e.g., Chmiel et al. 2011a,b).

On the other hand one could argue that prolonging the discussion on an IRC channel that serves for resolving problems is of little use. We would like to stress that this analysis aims at showing the outline of a more general problem. In fact, this idea could be applied to such media as BBC Forums as well as have a therapeutic usage. To some extent, introduction of deliberately biased emotional comments and scenarios in a human-bot discussion has already taken place (Skowron et al. 2011b) resulting in congruent responses issued by participants.

9.4.5 Discussion

Analysis performed on the emotionally annotated dialogues extracted from IRC data demonstrate that following such simple metrics as probability of specific emotion can be useful to predict the future evolution of the discussion. Moreover, all the analyzed dialogues share the same property, i.e., the tendency to evolve in the direction of a growing entropy. Those features, combined together with the observations regarding the preferential growth of clusters, are sufficient to reproduce the real data by a rather straightforward simulation model. We have also proposed a procedure to directly apply the observed rules in order to modify the way the dialogue evolves. It appears, for example that insertion of objective comments prolongs the discussion by lowering the entropy value. Those observations may be of help for designing the next generation of interactive software tools (Skowron et al. 2011a,b; Gobron et al. 2010) intended to support e-communities by measuring various features of their interactions patterns, including their emotional state at the individual, group and collective levels.

9.5 Conclusions

The chapter presented two different approaches to tackle the problem of emotional non-stationarity in communication of online media participants. In the first one, we underlined a crucial connection between the emotional content of the discussion and its length (duration measured in number of comments). Our research directly shows that only the threads where a certain emotional potential is present can evolve into a vivid full discussion. In the case of the BBC Forum this necessary condition has to do with the amount of *negative* emotions. This fact was then confirmed in the second study, where we calculated the entropy of the emotional probability distribution and showed that the BBC discussions and IRC dialogues evolve in the direction of increasing entropy. While in the case of BBC Forum it means that the initial negative *fuel* dries out, the situation in IRC is different—it is connected to biddings of gratitude for resolving a computer-related problem. Nonetheless, in both cases it possible to find clear signs of the discussion coming to an end. We proposed two different modeling approaches (a model of quarrels and a model of increasing entropy basing on the preferential attachment of emotional comments) that, fed with the real-data parameters, are capable of mimicking certain stylized facts. Finally, we believe that using the proposed approaches it is possible to track the evolution of online discussions and that they might of use while designing bot machines for online portals.

Acknowledgements We would like to thank to Mike Thelwall, Georgios Paltoglou, Kevan Buckley and Marcin Skowron for providing us an access to data used in this study, to Agnieszka Czaplicka, Piotr Pohorecki and Paweł Weroński for their help with data cleansing and to Arvid Kappas for illuminating discussions on emotion modeling. This work was supported by a European Union grant by the 7th Framework Programme, Theme 3: Science of complex systems for socially intelligent ICT. It is part of the CyberEmotions (Collective Emotions in Cyberspace) project (contract 231323). The work was also supported by Polish Ministry of Science Grant 1029/7.PR UE/2009/7.

References

Bailey, K.D.: Sociological entropy theory: toward a statistical and verbal congruence. Qual. Quant. **18**(1), 113–133 (1983). doi:10.1007/BF00221453

Barabási, A.L.: The origin of bursts and heavy tails in human dynamics. Nature **435**, 207–211 (2005). doi:10.1038/nature03459)

Bohus, D., Rudnicky, A.: Ravenclaw: dialog management using hierarchical task decomposition and an expectation agenda. In: Proceedings of the 8th European Conference on Speech Communication and Technology, EUROSPEECH 2003 - INTERSPEECH 2003, ISCA, Geneva, pp. 597–600 (2003)

Buckley, W.: Sociology and Modern Systems Theory. Prentice-Hall, Englewood Cliffs, NJ (1967)

Chmiel, A., Sienkiewicz, J., Thelwall, M., Paltoglou, G., Buckley, K., Kappas, A., Hołyst, J.A.: Collective emotions online and their influence on community life. PLoS ONE **6**(7), e22207 (2011a). doi:10.1371/journal.pone.0022207

Chmiel, A., Sobkowicz, P., Sienkiewicz, J., Paltoglou, G., Buckley, K., Thelwall, M., Hołyst, J.A.: Negative emotions boost user activity at BBC forum. Physica A **390**(16), 2936–2944 (2011b). doi:10.1016/j.physa.2011.03.040
Cover, T.M., Thomas, J.A.: Elements of Information Theory, pp 18–26. Wiley, New York (1991)
Czaplicka, A., Chmiel, A., Hołyst, J.A.: Emotional Agents at Square Lattice. Acta Phys. Pol. A **117**(4), 688–694 (2010)
Ding, F., Liu, Y.: Modeling opinion interactions in a BBS community. Eur. Phys. J. B **78**(2), 245–252 (2010). doi:10.1140/epjb/e2010-10453-9
Eckmann, J.P., Moses, E., Sergi, D.: Entropy of dialogues creates coherent structures in e-mail traffic. Proc. Natl. Acad. Sci. U. S. A. **101**(40), 14333–14337 (2004). doi:10.1073/pnas.0405728101
Gobron, S., Ahn, J., Paltoglou, G., Thelwall, M., Thalmann, D.: From sentence to emotion: a real-time three-dimensional graphics metaphor of emotions extracted from text. Vis. Comput. **26**(6–8), 505–519 (2010). doi:10.1007/s00371-010-0446-x
Gonçalves, B., Ramasco, J.J.: Human dynamics revealed through web analytics. Phys. Rev. E **78**(2), 026123 (2008). doi:10.1103/PhysRevE.78.026123
González-Bailón, S., Banchs, R.E., Kaltenbrunner, A.: Emotional reactions and the pulse of public opinion: measuring the impact of political events on the Sentiment of Online Discussions (2010). e-print: arXiv:1009.4019
Grice, H.P.: Logic and Conversation. In: Cole, P., Morgan, J. (eds.) Syntax and Semantics, vol. 3. Academic Press, New York (1975)
Harte, J.: Maximum Entropy and Ecology. Oxford University Press, Oxford (2010)
Jaynes, E.T.: Information theory and statistical mechanics. Phys. Rev. **106**, 620–630 (1957). doi:10.1103/PhysRev.106.620
Johnson, S., Torres, J.J., Marro, J., Muñoz, M.A.: Entropic origin of disassortativity in complex networks. Phys. Rev. Lett. **104**(10), 108702 (2010). doi:10.1103/PhysRevLett.104.108702
Kappas, A., Küster, D., Theunis, M., Tsankova, E.: Cyberemotions: subjective and physiological responses to reading online discussion forums. Poster presented at the 50th annual meeting of the society for psychophysiological research, Portland, OR, September 2010
Lin, J.: Divergence measures based on the Shannon entropy. IEEE Trans. Inf. Theory **37**(1), 145–151 (1991). doi:10.1109/18.61115
Littlejohn, S.W., Foss, K.A.: Theories of Human Communication. Waveland Press, Long Grove (2010)
Masucci, A.P., Kalampokis, A., Eguíluz, V.M., Hernández-García, E.: Extracting directed information flow networks: an application to genetics and semantics. Phys. Rev. E **83**(2), 026103 (2011a). doi:10.1103/PhysRevE.83.026103
Masucci, A.P., Kalampokis, A., Eguíluz, V.M., Hernández-García, E.: Wikipedia information flow analysis reveals the scale-free architecture of the semantic space. PLoS ONE **6**(2), e17333 (2011b). doi:10.1371/journal.pone.0017333
Masuda, N., Kim, J.S., Kahng, B.: Priority queues with bursty arrivals of incoming tasks. Phys. Rev. E **79**(3), 036106 (2009). doi:10.1103/PhysRevE.79.036106
Miller, G.A.: What is information measurement? Am. Psychol. **8**(1), 3–11 (1953). doi:10.1037/h0057808
Mitrović, M., Paltoglou, G., Tadić, B.: Networks and emotion-driven user communities at popular blogs. Eur. Phys. J. B **77**(4), 597–609 (2010). doi:10.1140/epjb/e2010-00279-x
Pareto, V.: The Mind and Society. Harcourt, Brace, New York (1935)
Parsons, T.: The Social System. Free Press, Glencoe (1951)
Radicchi, F.: Human activity in the web. Phys. Rev. E **80**(2), 026118 (2009). doi:10.1103/PhysRevE.80.026118
Rafaeli, S., Sudweeks, F.: Networked interactivity. J. Comput.-Mediat. Commun. **2**(4) (1974). doi:10.1111/j.1083-6101.1997.tb00201.x
Rothstein, J.: Communication, Organization, and Science. Wing Press, Indian Hills (1958)
Sacks, H., Schegloff, E.A., Jefferson, G.: A simplest systematics for the organisation of turn-taking for conversation. Language **50**(4), 696–735 (1974)

Saussure, F.: Course in General Linguistics. Fontana/Collins, Glasgow (1977)
Schweitzer, F., García, D.: An agent-based model of collective emotions in online communities. Eur. Phys. J. B **77**(4), 533–545 (2010). doi:10.1140/epjb/e2010-00292-1
Shannon, C.E.: A mathematical theory of communication. Bell Syst. Tech. J. **27**(3), 379–423 (1948). doi:10.1002/j.1538-7305.1948.tb01338.x
Shannon, C.E., Weaver, W.: The Mathematical Theory of Communication. The University of Illinois Press, Urbana (1949)
Shipley, B., Vile, D., Garner, É.: From plant traits to plant communities: a statistical mechanistic approach to biodiversity. Science **314**(5800), 812–814 (2006). doi:10.1126/science.1131344
Sienkiewicz, J., Skowron, M., Paltoglou, G., Hołyst, J.A.: Entropy-growth-based model of emotionally charged online dialogues. Adv. Comput. Syst. **16**(04n05), 1350026 (2013). doi:10.1142/S0219525913500264
Skowron, M.: Affect listeners: acquisition of affective states by means of conversational systems. In: Esposito, A., Cambell, N., Vogel, C., Hussain, A., Nijholt, A. (eds.) Development of Multimodal Interfaces - Active Listening and Synchrony: Development of Multimodal Interfaces: Active Listening and Synchrony. Lecture Notes in Computer Science, vol. 5967, pp. 169–181. Springer, Berlin/Heidelberg (2010). doi:10.1007/978-3-642-12397-9_14
Skowron, M., Pirker, H., Rank, S., Paltoglou, G., Ahn, J., Gobron, S.: No peanuts! affective cues for the virtual bartender. In: Murray, R.C., McCarthy, P.M. (eds.) Proceedings of the 24th International Florida Artificial Intelligence Research Society Conference, pp. 117–122. AAAI Press, Menlo Park, CA (2011a)
Skowron, M., Rank, S., Theunis, M., Sienkiewicz, J. The good, the bad and the neutral: affective profile in dialog system-user communication. In: D'Mello, S., Graesser, A., Schuller, B., Martin, J.-C. (eds.) Affective Computing and Intelligent Interaction: 4th International Conference, ACII 2011, Memphis, TN, October 9–12, 2011, Proceedings, Part I. Lecture Notes in Computer Science, vol. 6974. Springer, Berlin, Heidelberg, pp. 337–346 (2011b). doi:10.1007/978-3-642-24600-5_37
Sobkowicz, P., Sobkowicz, A.: Dynamics of hate based Internet user networks. Eur. Phys. J. B **73**(4), 633–643 (2010). doi:10.1140/epjb/e2010-00039-0
Song, Ch., Qu, Z., Blumm, N., Barabási, A.L.: Limits of predictability in human mobility. Science **327**(5968), 1018–1021 (2010). doi: 10.1126/science.1177170
Spencer, H.: First Principles. Appleton, New York (1864)
Stivers, T., Enfield, N.J., Brown, P., Englert, Ch., Hayashi, M., Heinemann, T., Hoymann, G., Rossano, F., de Ruiter, J.P., Yoon, K.E., Levinson, S.C.: Universals and cultural variation in turn-taking in conversation. Proc. Natl. Acad. Sci. U. S. A. **106**(26), 10587–10592 (2009). doi:10.1073/pnas.0903616106
Suler, J.: The online disinhibition effect. CyberPsychol Behav **7**(3), 321–326 (2004). doi:10.1089/1094931041291295
Szell, M., Lambiotte, R., Thurner, S.: Multirelational organization of large-scale social networks in an online world. Proc. Natl. Acad. Sci. U. S. A. **107**(31), 13636–13641 (2010). doi:10.1073/pnas.1004008107
Takaguchi, T., Nakamura, M., Sato, N., Yano, K., Masuda, N.: Predictability of conversation partners. Phys. Rev. X **1**, 011008 (2011). doi:10.1103/PhysRevX.1.011008
van Dijk, T.A.: Cognitive context models and discourse. In: Stamenow, M. (ed.) Language Structure, Discourse and the Access to Consciousness, pp. 189–226. Benjamins, Amsterdam (1997)
Vázquez, A., Oliveira, J., Dezső, Z., Goh, K., Kondor, I., Barabási, A.L.: Modeling bursts and heavy tails in human dynamics. Phys. Rev. E **73**(3), 036127 (2006). doi:10.1103/PhysRevE.73.036127
Weroński, P., Sienkiewicz, J., Paltoglou, G., Buckley, K., Thelwall, M., Hołyst, J.A.: Emotional analysis of blogs and forums data. Acta Phys. Pol. A **121**(2B), B128–B132 (2012)

Williams, R.J.: Biology, methodology or chance? The degree distributions of bipartite ecological networks. PLoS ONE **6**(3), e17645 (2011). doi:10.1371/journal.pone.0017645

Williams, J.D., Poupart, P., Young, S.: Partially observable Markov decision processes with continuous observations for dialogue management. In: Dybkjaer, L., Minker, W. (eds.) Recent Trends in Discourse and Dialogue. Text, Speech and Language Technology, vol. 39, pp 25–34. Springer, Dordrecht (2005). doi:10.1007/978-1-4020-6821-8_8

Chapter 10
An Agent-Based Modeling Framework for Online Collective Emotions

David Garcia, Antonios Garas, and Frank Schweitzer

10.1 Introduction

A special feature of online communities is the frequent occurrence of collective emotions, which are not so easily observable in offline interaction. Spontaneously, large amounts of users share similar emotional states, due to their ability to reach many other users in a quick, and often anonymous way. Such collective emotions can result from exogenous as well as from endogenous causes. For example, external events, such as a catastrophe or a large marketing campaign, are able to trigger the online expression of emotions of millions of users. But collective emotions can be also created within online communities, in various forms such as *memes* in social networks (Leskovec et al. 2009), heated discussions in forums (Chmiel et al. 2011), and cascades of emotions in microblogs (Alvarez et al. 2015).

The increasing importance of online communication not only changes the way people interact everyday, but also offers a great chance to retrieve and analyze large amounts of data on human behavior. Everyday, millions of Internet users leave online traces that are publicly accessible, in the form of comments, video downloads, or product reviews. The unprecedented size of these datasets allows the quantitative testing of previous theories and hypotheses formulated in the social sciences, for example about social influence (Onnela and Reed-Tsochas 2010; Lorenz 2009), cooperation, and trust (Walter et al. 2009).

Furthermore, sentiment analysis techniques (Thelwall et al. 2013) allow the analysis of the emotions expressed through the Internet. For example, the emotional content of millions of `Twitter` messages has been used to study the daily patterns of mood (Golder and Macy 2011), and the assortativity of happiness in social networks (Bollen et al. 2011). Our aim is to study emergent collective emotions,

D. Garcia (✉) • A. Garas • F. Schweitzer
ETH Zürich, Zürich, Switzerland
e-mail: dgarcia@ethz.ch; agaras@ethz.ch; fschweitzer@ethz.ch

J.A. Hołyst (ed.), *Cyberemotions*, Understanding Complex Systems,
DOI 10.1007/978-3-319-43639-5_10

as their dynamics and preconditions can be studied based on the textual expressions of users in online communities. Interestingly, one finds a large degree of regularity in such online phenomena. For example, the lifetime of a forum discussion can be related to the level of negative emotions expressed in it (Chmiel et al. 2011). This opens the question about the mechanisms that lead to such collective emotional states.

Online data allows us to measure how and when collective emotional states emerge, but the analysis of this spontaneous behavior cannot be simply reduced to the activity of single users. Instead, these collective states should be treated as emergent phenomena resulting from the interaction of a large number of individuals. In our approach, we relate the statistical regularities observed in online communities to the interactions between users. The distinction between the micro level of individual users and the macro level at which their collective behavior can be observed is one of the specific features of the theory of complex systems. Over the last 40 years methods and tools from computer science, statistical physics, and applied mathematics have been utilized to address this micro-macro link and to predict the collective dynamics of a system from individual interactions of many system elements, or agents.

To study collective emotions, we need an appropriate description of the agents and their interactions, but we also need an appropriate framework to predict the collective dynamics of the system from its basic ingredients. Without such a framework, we are only left with extensive computer simulations of multi-agent systems, in which, for given assumptions of the interactions, we have to probe the entire parameter space, to find out the conditions for certain collective phenomena. Furthermore, collective emotions appear in different online communities, which often have different interaction mechanisms. Models of collective emotions in each of these communities, if designed and analyzed separately, might shed light on the particular properties of collective emotions in each one of them. Such approach, on the other hand, would not allow to draw conclusions on universal properties of collective emotions across communities. If designed within a unifying framework, models of collective emotions in different communities can be compared between different scenarios of online interaction.

In this chapter, we present a framework to describe collective emotions in online communities through agent-based models. In an agent-based model, we first need to describe the emotional states of individual agents, which should be based on insights obtained in psychology. We follow Russell's representation of *core affect* (Russell 1980) , modeling emotions as short-lived psychological states of the individual. This established theoretical perspective is based on two dimensions: *valence*, indicating whether the emotion is pleasant or unpleasant, and *arousal*, indicating the degree of activity or inactivity induced by the emotion. Therefore, the internal states of our agents will be composed of two independent variables of valence and arousal. Previous research has already analyzed the dynamics of individual valence and arousal (Kuppens et al. 2010), proving that agent-based models are useful for research in psychology (Smith and Conrey 2007).

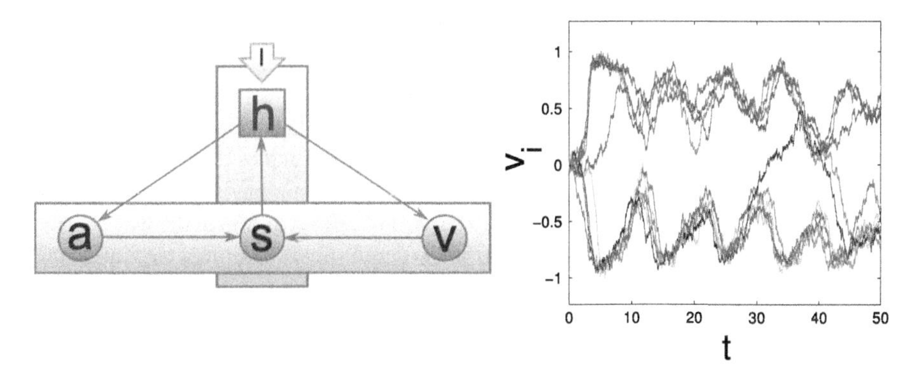

Fig. 10.1 *Left*: Schema of the components of the modeling framework. *Right*: Sample trajectories of valence of ten agents in a simulation

10.2 The Cyberemotions Modeling Framework

Our framework is specific enough to allow analytical results to predict simulation outcomes (Schweitzer and Garcia 2010), but general enough to cover a wide range of online emotional interactions. The main feedback loops of this framework, as sketched in the left panel of Fig. 10.1, are comprised of two orthogonal layers: an internal layer describing the agent (shown horizontally) and an external layer describing the communication process (shown vertically). In the internal layer, the arousal a and the valence v of an agent determine its emotional expression s, which reaches the external layer by contributing to the communication field h. The latter one has its independent dynamics and can, in addition to contributions from other agents, also consider input from external sources, I. The causality loop is closed by considering that both valence and arousal of an agent are affected by the communication field.

Since we are interested in modeling the emotional dynamics of Internet communities, this general framework can be easily adjusted to consider the particularities of various online platforms such as user expression limitations, external influence on users, communication in networks as opposed to broadcast, etc. In the following, we describe the framework and provide different examples of how to specify our modeling framework to cope with different online communities.

10.2.1 Brownian Agents

Our modeling framework is based on the principle of Brownian Agents, where each agent is a person interacting in the online medium. This modeling principle was successfully applied other contexts, describing the dynamics of opinions (Schweitzer and Hołyst 2000) as well as a large variety of other systems, from urban growth and economic agglomeration, to chemical pattern formation and swarming in biological systems (Schweitzer 2003). Brownian agents are described by a set of K state variables u_i^k, where the sub-index $i = 1, \ldots, N$ refers to each individual agent i, and the super-index $k = 1, \ldots, K$ refers to each variable. These variables could be *external* if they can be observed in empirical data, or *internal* if they can only be indirectly concluded from the observable data. Each of these state variables can be time dependent due to interaction with the agent's environment, or due to internal dynamics that do not require external influence. In a general way, we can formalize the dynamics of each state variable u_i^k as a superposition of two influences of different nature:

$$\frac{d\,u_i^k}{dt} = f_i^k + \mathscr{F}_i^{\text{stoch}} \tag{10.1}$$

This formulation is based on the principle of causality: the change in time of any variable, noted as $\frac{d\,u_i^k}{dt}$, is produced by some causes which are listed on the right hand side of the equation. In the case of Brownian agents, these causes are assumed to be described by a superposition of deterministic (f_i^k) and stochastic influences ($\mathscr{F}_i^{\text{stoch}}$).

The stochastic term models all the influences on the variables that are not observable on the time and length scale of the available data. This stochastic term does not direct the dynamics of the agent state in any particular direction, and it is commonly, but not necessarily, modeled by white noise. Furthermore, the strength of the stochastic influences might be different among agents, depending on local parameters of the agents, as in Schweitzer (2003).

The deterministic term f_i^k represents all the specified influences that change the corresponding state variable u_i^k. For example, nonlinear interactions with other agents can be modeled as a function that depends on the state variables of any set of agents, which can also include agent i itself. f_i^k can also describe the agent's response to the available information, which is the case for our modeling framework. Additionally, f_i^k can reflect the *eigendynamics* of the agent, which are the changes in the variables u_i^k not caused by any influence external to the agent. Examples of eigendynamics are saturation or exhaustion, common in the modeling of human behavior (Lorenz 2009; Kuppens et al. 2010). In order to design a multiagent system, we have to define the agent's state variables, u_i^k, and the dynamics of their change, f_i^k, specifying the interaction among agents. These dynamics are defined at the level of the individual agent and not at the collective level, in a way that the macroscopic dynamics emerge from the interaction of many agents, just as collective emotions emerge in online communities from the interaction of many users.

10.2.2 Emotional States and Their Internal Dynamics

Following the bidimensional representation of core affect (Russell 1980), we quantify the emotional state of an agent through the variables of valence $v_i(t)$, and arousal $a_i(t)$. As explained in Chap. 5, these two variables are known to capture most of the information of emotional experience, and represent the level of pleasure and activity associated with an emotion. In our model, we define the state of the agent as $e_i(t) = \{v_i(t), a_i(t)\}$. Note that valence and arousal are internal variables, i.e. cannot be directly observed on the agent. They can only be indirectly observed, for example through physiological measurements or individual reports.

In the absence of interaction emotions relax to an equilibrium state. This is supported by empirical studies that show how emotional states exponentially decay (Kuppens et al. 2010). This relaxation, $e_i(t) \rightarrow 0$, implies $v_i(t) \rightarrow 0$, $a_i(t) \rightarrow 0$. Thus, following Eq. (10.1), we define the dynamics of the Brownian agent as follows:

$$\frac{d\,v_i}{dt} = -\gamma_{vi}\, v_i(t) + \mathscr{F}_v + A_{vi}\,\xi_v(t)$$
$$\frac{d\,a_i}{dt} = -\gamma_{ai}\, a_i(t) + \mathscr{F}_a + A_{ai}\,\xi_a(t) \tag{10.2}$$

The first terms on the right-hand side of the equations describe the exponential relaxation of valence and arousal towards the equilibrium state. The parameters γ_{vi} and γ_{ai} define the time scales of this relaxation, which can be different for valence and arousal and across agents. The second terms describe the deterministic influences as explained below, and the third terms model the stochastic influences. $\xi_v(t)$, $\xi_a(t)$ are random numbers drawn from a given distribution of white noise, i.e., they have zero mean and no temporal correlations. The strengths of the stochastic components are quantified by A_{vi} and A_{ai}, which can also vary across agents.

The deterministic influences on the emotional state of the agent are described by the functions $\mathscr{F}_v$, $\mathscr{F}_a$. They depend on specific assumptions applicable to online collective emotions, in particular the agents' interaction, access to information, or their response to the media. These functions should also reflect possible dependencies on the emotional state of the agent itself, as emotional states could be more affected by certain emotions and less by other emotions. In the following sections, we extend the description of the agent by defining the actions an agent can take, to then follow in specifying the forms of these functions.

10.2.3 Emotional Communication in Online Communities

If information with emotional content becomes available to the agent, there should be excited emotional states, which are not externally observable unless the agent decides to communicate, creating a message or posting a comment in a discussion. Consequently, our assumption for the expression of emotions is that the agent expresses its valence through the externally observable variable $s_i(t)$ if its arousal exceeds certain individual threshold, $\mathscr{T}_i$

$$s_i(t) = r(v_i(t))\,\Theta[a_i(t) - \mathscr{T}_i] \tag{10.3}$$

where $\Theta[x]$ is the Heaviside's function which is one only if $x \geq 0$ and zero otherwise. If $\Theta[x] = 1$, we assume that the agent is not able to perfectly communicate its valence state, i.e. the exact value of $v_i(t)$, and its expression is simplified through a function $r(v)$. Thus, it is essential to specify this function depending on a coarse-grained representation of the valence of individual agents, which can be adjusted to the accuracy of the data analysis techniques available. In the following, we assume that empirical data only allows us to know the polarity of a message, choosing $r(v) = \text{sign}(v)$. Additionally, the agent might not be able to immediately express its emotions if the arousal hits the threshold at a particular time t. This expression might be delayed with certain delay Δt, as the agent might not have immediate access to communication media.

After describing the dynamics of emotional states and emotional expression, we need to specify how this emotional expression is communicated to the other agents. In line with previous models of social interaction (Schweitzer and Hołyst 2000), we assume that every positive and negative expression is stored in a communication field $h_{\pm}(t)$ with a component for positive communication $h_+(t)$, and another component for negative information $h_-(t)$. This variable essentially stores the "amount" of available comments of a certain emotional content at a given moment in time. We propose the following equation for the dynamics of the field:

$$\frac{dh_{\pm}}{dt} = -\gamma_{\pm} h_{\pm}(t) + c n_{\pm}(t) + I_{\pm}(t) \tag{10.4}$$

where each agent contribution $s_i(t)$ increases the corresponding field component by a fixed amount c at the exact time the expression occurred. This parameter c represents the impact of the information created by the agent to the information field, defining a time scale.

The variable $n_{\pm}(t)$ shows the total number of agents contributing positive or negative emotional expression at time t. These expressions are in general time dependent, i.e. they lose importance as they become older, usually due to the creation of new information in the community. This is represented by the exponential decay present in the first term of the right-hand side of Eq. (10.4), which is parametrized through $\gamma_{\pm}$. In addition, externally produced positive or negative emotional content might change the communication field, as for example news can have a great impact in the overall emotional state of an online community. We model this mechanism through the agent-independent term $I_{\pm}(t)$, which can be modeled as a stochastic input, or used to analyze the reactions of the model to external stimuli.

To finish the description of our framework, we need to specify how the available information influences the state of the individual agents, which is covered by the functions $\mathscr{F}_v$ and $\mathscr{F}_a$ of Eq. (10.2).

10.2.4 Feedback of Communication into Emotional States

The target of our model is to reproduce the emergence of a collective emotion, assuming that it cannot be understood as a simple superposition of individual emotional states. Our assumption is that the emotional expression of an agent may change the emotional state of a number of other agents, either directly or indirectly. For this influence we can define its form and investigate the various possible scenarios through computer simulations and mathematical analysis. Additionally, these can also be empirically tested when individual users are exposed to different emotional content, as discussed in Gianotti et al. (2008) and explored in ongoing experiments (Kappas 2011) like the ones described in Chap. 5.

In the communication field of our model, there are two components for positive, $h_+(t)$, and negative, $h_-(t)$, emotional information. Depending on the state of an agent, it might be affected by these different kinds of information in different ways. A general assumption for this function is that the valence increases with the respective information perceived by the agent. The strength of this influence should depend on the emotional state of the agent, often in a nonlinear manner. A general formulation for this kind of dynamics has the form:

$$\mathscr{F}_v(h_\pm(t), v_i(t)) = h_\pm(t) \sum_{k=0}^{n} b_k v^k(t) \tag{10.5}$$

where the key assumption is that the coefficients b_k are constants that does not depend on the value of the valence.

Arousal measures the degree in which the emotion encourages or discourages activity. It becomes important when it reaches a threshold $\mathscr{T}_i$, which is assumed to be the precondition for emotional expression (Rimé et al. 1998; Rimé 2009). Emotional expression should have some impact on the arousal, and we assume that the arousal is lowered after producing a message, or set back to the ground state in the most simple case. This means that the dynamics of arousal should be divided into two parts: one applying before the arousal reaches the threshold, and one at the exact moment when it is reached. Hence, we define the dynamics of the arousal $a_i(t)$ as:

$$\frac{d\,a_i}{dt} = \frac{d\,\bar{a}_i}{dt}\,\Theta[\mathscr{T}_i - a_i(t)] - a_i(t)\,\Theta[a_i(t) - \mathscr{T}_i] \tag{10.6}$$

As long as $x = \mathscr{T}_i - a_i(t) > 1$, $\Theta[x] = 1$ and the arousal dynamics are defined by $\frac{d\bar{a}_i}{dt}$ as in Eq. (10.2). Once the threshold is reached, $x \geq 0$, $\Theta[x] = 0$ and $\Theta[-x] = 1$, deterministically resetting the arousal back to zero.

To conclude the dynamics of arousal, we must specify the function $\mathscr{F}_a$, which applies when the arousal is below the threshold. The arousal was designed to be an orthogonal variable to valence, measuring the activity level of an emotion. It is reasonable to assume that agents respond to all the emotional content available in the community, i.e. the sum of both field components, in a way that depends on their own arousal in a nonlinear manner, regardless of the valence dimension. Following the same general point of view as for the case of valence, we may propose the following nonlinear dependence:

$$\mathscr{F}_a \propto [h_+(t) + h_-(t)] \sum_{k=0}^{n} d_k a^k(t) \tag{10.7}$$

The above description defines a complete framework to design agent-based models of collective emotions in online communities. Simulation and statistical analysis of the properties of these models can explain the reasons for the emergence of collective emotional states from the online interaction of large amounts of users.

10.2.5 Simulation of Collective Emotions

Models within this framework have the advantage of being tractable, allowing researchers to find analytical solutions that explain the emergence of collective emotions. We illustrate how a simulation of our model creates collective emotional states of polarized emotions in the right panel of Fig. 10.1, where we show the trajectory of valence for ten agents in a simulation. One can notice a quite synchronized change of the emotions, which is not surprising as the dynamics mainly depends on the value of h, which is the same for all agents and all other parameters are kept constant. More details about the mathematical analysis of this model can be found elsewhere (Schweitzer and Garcia 2010), allowing us to understand under which conditions we can observe such collective emotions. In particular, two important aspects can be highlighted: (1) when collective emotions are unipolar (positive or negative), or bipolar, and (2) when collective emotions are triggered by external influences, and when they are endogenously emerging from user interactions, leading to the appearance and disappearance of collective behavior like the one shown in Fig. 10.1.

10.3 A Model for Emotions in Product Reviews Communities

10.3.1 Applying the Framework to Review Emotions

The structure of this model is the same as the one shown in Fig. 10.1, where the emotional state of the agents is composed of valence and arousal, and is influenced by a collective information field. To model emotions in product reviews, we use specific assumptions about this kind of communication, which are explained in detail in Garcia and Schweitzer (2011). In our model, we focus on the discussion at the product level, ignoring relations between products. This means that the communication between agents always refers to the reviewed product. It is a particular property of a product that every user is allowed to review it only once. We introduce this constraint in the arousal dynamics. Specifically, after an agent's arousal reaches its threshold $\mathcal{T}_i$, the threshold is reset to a value of ∞, preventing the agent from making a second review on the same product. We assume that the initial values of these thresholds are heterogeneous among agents, sampled from a normal distribution with mean μ and standard deviation σ.

For this application, we assume that the arousal dynamics depends on the sum of both components of the field (h_+ and h_-), as formalized in Eq. (10.7). For this case, the polynomial function of Eq. (10.7) goes up to the second degree, modeling a quadratic dependence on the agent's own arousal. Our simulation results (Schweitzer and Garcia 2010) show that this form of arousal dynamics is able to produce the spontaneous emergence and disappearance of collective emotional states. For the valence dynamics, we assume that the influence of the information field in the agent's valence $\mathscr{F}_v$ depends on the previous value of the agents valence. This means that previous negative experiences of the product lead to a tendency to pay less attention to the positive expression of other agents. On the other hand, agents with positive experiences will be more influenced by positive emotional information than by negative one. We can formalize this asymmetry of agent perception through an exponential function with a cubic decay, as explained in Garcia and Schweitzer (2011).

Writing reviews is heavily influenced by preferences of the users and their relation to the properties of the product. In our model, user preferences are included as an agent internal variable u_i, constant in time. The heterogeneity on these preferences is captured by sampling u_i from a uniform distribution in the interval $[0, 1]$. This way we do not assume any kind of general preference towards a particular value, as preferences simply determine what is subjectively preferred and not what is better or worse. Product properties are represented in the same scale as user preferences, as described by a parameter $q \in [0, 1]$.

It is a common assumption in product review communities that a reviewer has previously purchased or experienced the reviewed product. In our model, this experience determines the initial value of the valence, calculated as the difference between the agent's preference u_i and the product property q. If a product is at perfect match with a user's preference $|u_i - q| = 0$, the agent starts with a maximum initial valence ($v_i(0) = 1$). If the product happens to be the complete opposite to the agent's expectations, the value of the difference between both would be maximum and the agent's valence $v_i(0) = -1$.

According to our framework, the value of an agent's expression s_i is determined by its valence v_i. We assume that agent expressions influence the field more the more emotional they are. As product reviews are fairly long texts compared to other kinds of online communication, sentiment analysis techniques are able to provide values for different degrees of emotionality. A review might contain only factual information and not influence the emotions of a reader, but it could also contain mild or extreme emotional content. Following the scheme of SentiStrength, explained in Chap. 6, the value of an agent's expression s_i ranges from -5 to 5, according to the value of its valence when creating the review.

10.3.2 Reproducing Emotions in Reviews Data

Our model for emotions in product reviews aims at reproducing collective properties of emotional expression towards products. Our dataset of reviews from `Amazon.com` contains more than 1.7 million reviews for more than 16.000

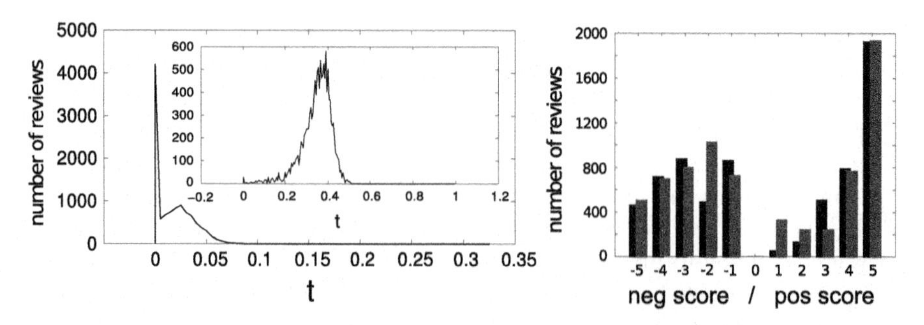

Fig. 10.2 *Left*: Amount of reviews for two simulations of the product reviews model. Rate of reviews and emotions for a strong media impulse and when the emotions spread through the community (inset). *Right*: Comparison between the emotional distribution of the reviews for "Harry Potter" (*black*) and the simulation results (*gray*) (Garcia and Schweitzer 2011)

products. Each review has been processed with SentiStrength (Thelwall et al. 2013), a sentiment analysis tool that gives values of positive and negative emotions in a text in a scale from 1 to 5. Statistical analysis of this dataset (Garcia and Schweitzer 2011) showed the existence of two patterns of the reaction of the community to the release of a product. Furthermore, emotional expression regarding products followed distributions of a characteristic shape, which our model should reproduce.

Given a particular set of values for the parameters of our model, the initial value of the communication field determines the type of collective dynamics of a simulation. This way the model is able to reproduce the different scenarios we found in the real data, which correspond to reviews resulting from mass media versus word of mouth influence. The left panel of Fig. 10.2 shows the time series of emotional expression in two simulations of the model. The outer plot shows the case when there is a strong input to the field at the beginning of the simulation. This initial impulse, simulating marketing campaigns, forces the dynamics of the community into a vastly decaying single spike. The inset on the left of Fig. 10.2 shows the alternative case of a slower increase of the activity in the community. The simulated time series shows that, in the absence of initial information, the model can build up endogenous cascades of reviews. This kind of dynamics requires a variance of the threshold distribution large enough to trigger some agents that lead the activity in early stages of the simulation.

The valence dynamics of this model were designed to reproduce different patterns of positive and negative emotional expression in product reviews. The black histogram in the right panel of Fig. 10.2 shows a typical histogram of emotional expression in our `Amazon.com` dataset. In general, the distribution of negative emotions is more uniformly distributed than the expression of positive emotions, which usually have a large bias towards the maximum value. Gray bars in Fig. 10.2 show the histogram of emotional expression from our simulations. The similarity between both histograms shows how we are able to reproduce the distribution of emotional expression in product reviews, given certain parameter values.

To conclude, Fig. 10.2 shows that the outcome of our model has macroscopic properties similar to real world data on product reviews. Our model provides a phenomenological explanation based on psychological principles, linking the microscopic interaction between agents with the macroscopic behavior we observed in our `Amazon.com` dataset. In particular, the different time responses and distributions of emotions expressed in the community have the same qualitative properties in model simulations and real data. Within our framework, further explorations of the relation between model and data are possible. For example, each product can be mapped to a set of parameter values that reproduce the collective properties of the community reaction. This would provide a measure of the impact of product properties and marketing in the psychometric space of the customers.

10.4 Modeling Real-Time Online Emotional Interaction

Another application of our modeling framework provides insights on the nature of human communication in real-time online discussions, i.e. chatrooms. Online communication like the one in chatrooms received recently much attention from the scientific community (Sienkiewicz et al. 2013; Garas et al. 2012). Relevant questions have been identified, such as the role of influential users (Borge-Holthoefer and Moreno 2012), or the time patterns between user actions (Radicchi 2009). The analysis of the times between message creations is a useful tool to detect communication bursts (Wu et al. 2010), as well as periods of inactivity (Garcia et al. 2013). As a result, many statistical regularities of our communication patterns are revealed, like the power-law nature of the waiting time distribution $P(\tau)$, where τ is the elapsed time between two consecutive actions of the same user. Such regularities should be, and are, considered in the design of our model. I.e., instead of being driven by the arousal dynamics the level of activity is sampled from the real inter-activity time distribution $P(\tau) \sim \tau^{-1.54}$, as reported in Garas et al. (2012). The causal relationships between the elements of this model is summarized in the left panel of Fig. 10.3.

Using our framework, the valence dynamics should follow Eq. (10.2) and is composed by a superposition of stochastic and deterministic influences:

$$\frac{d v_i}{dt} = -\gamma_v v_i + b(h_+ - h_-)v + A_v \xi_i \tag{10.8}$$

The exponential decay of the valence is determined by γ_v and the influence of the information fields is modeled through $b(h_+ - h_-)v$. The parameter b quantifies the valence change per time unit due to the discussion of emotional content. This change depends on the balance between positive, h_+, and negative, h_-, components of the field. This differs from the previous assumption used for the modeling of product review communities, but is more appropriate to capture communication in chatrooms. Chat discussions are usually very fast, real-time interactions that display

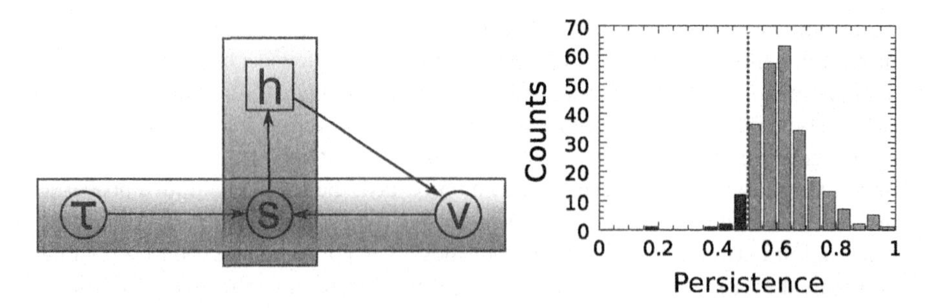

Fig. 10.3 *Left*: Schema of the model for emotional persistence in online chatroom communities. *Left*: distribution of persistence for simulations of the model, taking similar values as the empirical data from IRC channel (Garas et al. 2012)

a limited amount of messages to the users. Unlike in collective discussions where large amounts of messages can be accessed at any time, the emotional information in chatrooms cannot grow up to large value. In this model, the aim is to reproduce plausible chatroom interaction, in which users are just able to read a smaller amount of messages created in a short time.

As mentioned before, agents create messages with time intervals sampled from the empirical inter-event distribution. When posting a message, the variable s_i of the agent is set to a value that depends on its valence v_i. As chat messages are usually very short, we cannot assume the existence of very rich emotional content like in the case of product reviews, but just some emotional orientation as positive, negative, or neutral messages. We formalize the expression of valence polarity as:

$$s_i = \begin{cases} -1 & \text{if} \quad v_i < V_- \\ +1 & \text{if} \quad v_i > V_+ \\ 0 & \text{otherwise} \end{cases} \tag{10.9}$$

where the thresholds V_- and V_+ represent the limit values that determine the emotional content of the agent's expression. These thresholds do not need to be symmetric around zero, as human expression is systematically positively biased (Garcia et al. 2012). If humans communicate in the presence of social norms that encourage positive expression, thresholds should satisfy $|V_+| < |V_-|$.

In this application of our framework, the communication field is formulated exactly as in Eq. (10.4), i.e. it increases by a fixed amount c when an agent expresses its emotions. By analyzing the parameter space of the model, we are able to identify parameter values that reproduce observable patterns of real human communication. In Garas et al. (2012), it was shown that there is emotional persistence in online human communication, which reveals that there are collective emotions shared by the participants of the discussion. This emotional persistence can be reproduced by simulated conversations between agents chatting. The distribution of the emergence of this simulated persistence is shown in the right panel of Fig. 10.3.

The insights provided by agent-based models within our framework are of special use for certain ICT applications. Dialog systems, more commonly known because of the use of *chatbots*, benefit from this framework, as agent-based models can be formulated as computational entities that can simulate human behavior, and interact with users of a dialog system. Our agent-based approach is used for the next generation of emotionally reactive dialog systems (Rank et al. 2013).

10.5 Models of Collective Emotions in Social Networks

The general modeling framework is also flexible enough to capture models of collective emotions in online social networks. The first application to online social networks is introduced in Chap. 11, and here we outline the relation of that model with our modeling framework. This model of emotional influence between `MySpace` users builds on the empirical findings about (1) their interaction network, (2) their temporal activity patterns, (3) the entry rate of new users, and (4) the emotionality of their messages. However, we cannot identify using data analysis how messages influence the activity and the emotional state of other users. Thus, in our model we provide hypotheses about this feedback which are tested against the aggregated outcome.

Different from the previous examples where stochasticity was modeled simply by a additive stochastic force, we assume here that stochasticity results from three sources: (1) sampling from the empirical inter-activity time distribution $P(\Delta t)$, (2) sampling from the empirical rate $p(t)$ at which new users enter the network, (3) a spontaneous reset of both valence and arousal to a predefined value $(\tilde{v}, \tilde{a})$ with a rate r. The latter captures our uncertainty in determining the external influences on an agent's state and is treated as a tunable parameter as explained below.

To model how agents are affected by the messages they perceive, we designed three levels of aggregated information in our model: (1) aggregation of messages on the agent's wall which shall be captured by an information field h_i, (2) aggregation of messages perceived on the friends' walls, captured by an information field $\overline{h}_i$, (3) aggregation of messages on all walls, i.e. a mean-field information h_{mf} that that captures a kind of "atmosphere" of the whole community. Because each of the messages in the empirical data has a valence value and an arousal value, the information field h also has a valence and arousal component h^v, h^a which results from the respective aggregation. Specifically, different from previous modeling assumptions, we assume here that the agent's arousal and activity (e.g. in choosing conversation partners) is only affected by the arousal information, whereas the agent's valence is only affected by the valence information. This way we explore the role of richer emotional communication in an advanced model within the cyberemotions modeling framework (Schweitzer and Garcia 2010), which allows us to compare the results to previous models for different online communities.

The right panel of Fig. 10.4 shows the application of our framework to this model: agents A_i can post messages M_{ij} on the wall h_j of agent A_j, which would

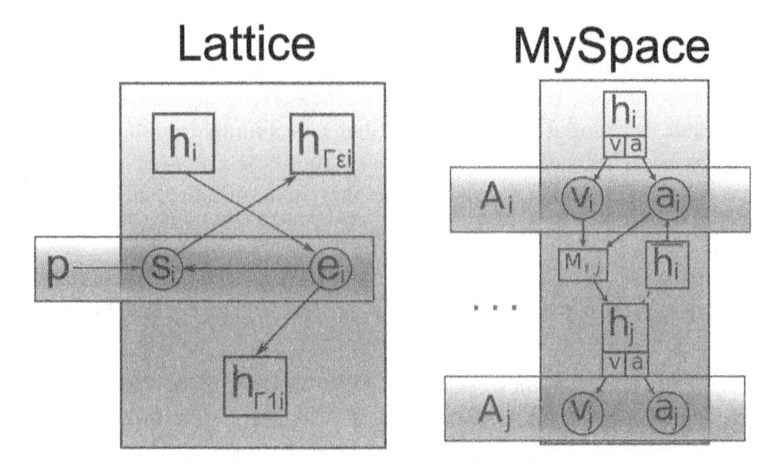

Fig. 10.4 Schema of the lattice model (Czaplicka and Hołyst 2012) and of the MySpace network model of Chap. 11 (*right*)

in turn influence the valence and arousal of A_j. The aggregation of the walls in the neighborhood of A_i is represented as the field $\overline{h}_i$, for which the wall h_j contributes to increase the arousal of A_i. Simulations of this model reproduce certain aspects of cascades of collective emotions, as shown in Chap. 11.

The second model variant refers to agents interacting on a square lattice (Czaplicka and Hołyst 2012). The left panel of Fig. 10.4 illustrates the feedback processes involved with reference to the general modeling framework.

Agents express their emotions through the externally observable variable s_i, determined by the agent's internal emotional valence e_i. This valence is assumed to be a discrete variable $e_i \in \{1-, 0, +1\}$ which can change to any state with a given probability p_s as a representation of a spontaneous emotional arousal. The agent expression influences the field of agents around a neighborhood within ϵ distance, $h_{\Gamma \epsilon i}$, and it takes place at events sampled with constant probability p. In addition, the internal state of the agent can influence the field of its neighbors at distance 1 $h_{\Gamma 1 i}$, and be influenced by the own agent's field h_i. Simulations of this model, explained in Czaplicka and Hołyst (2012), produce fluctuations of collective emotions that emerge from this local dynamics.

10.6 A Data-Driven Model of Emotions of Virtual Humans

In this section we describe a model to capture the mechanisms of emotional communication in a virtual society. The main purpose of the model is to provide means of integration between the available information provided by machine learning tools, and the avatar system that represents the emotional state of the people in a conversation, to be applied to the system shown in Chap. 13. This model relies

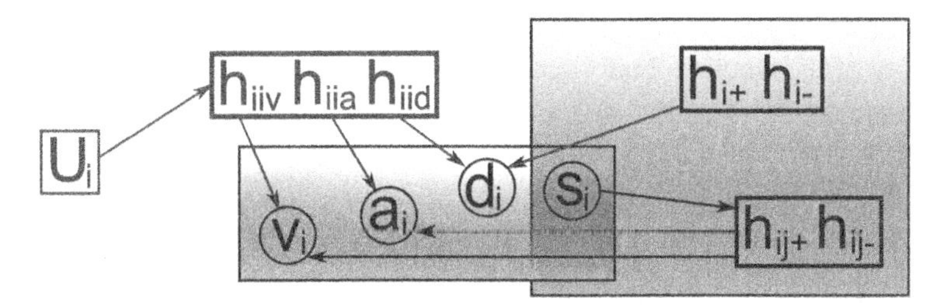

Fig. 10.5 Schema of the individual emotions model for virtual humans. Agents are described by their emotional state (VAD). They communicate through a conversation field $h_{ij\pm}$, perceive references to them through a personal field $h_{ii\pm}$, and update their state given utterances about themselves and user input $h_{iiv}, h_{iia}, h_{iid}$. U_i is the explicit user input about their emotional state

on a particular set of parameters and influence functions that can be tested from physiological data and from Internet communication, in experiments like the ones explained in Chap. 5. In addition, future user tests with this model will provide feedback from the participants. This way we will have means of testing the quality of particular assumptions or ranges of the parameters from a believable behavior observed by laymen in a controlled setup.

The agents have an emotional state defined by valence, arousal and dominance (VAD), as shown in Fig. 10.5. These are continuous variables that can take positive and negative values and that are not explicitly bounded, but that will be finite given their dynamics. As explained in Sect. 10.2.2, they will have an internal relaxation factor that does not need to be the same for all of them (γ_v,γ_a,γ_d). We assume that agent expressions are given (chat system), so the process that determines the creation of s_i (user expression) is not defined. The expression (an utterance) is composed of:

1. Sentiment (positive, negative, neutral)
2. Valence, arousal and dominance
3. Target person (I, you, them)

There are three types of fields that represent the communication of the system:

- h_{ij+} and h_{ij-} (conversation fields), that store the emotional communication between the agents A_i and A_j. This assumes that the system simulated is a one-on-one chat. The utterances will create a constant increase in this fields depending on their sentiment sign.
- h_{iiv}, h_{iia}, and h_{iid} that are the self-influence fields for the refinement of the representation of the emotions of the user. The values of valence, arousal and dominance of the utterances will create a change in these fields that will influence directly the state of the agent towards those values. This change will be stronger when the utterance created has the target *I* than another one. These fields receive an additional, overriding input from the user that shifts the state of the agent towards what the user decided. This input will come from explicit assessments of valence, arousal and dominance from a visual interface available in some experiments.

- h_{i+} and h_{i-} are the identity fields of the agent A_i. This fields represent the history of the emotional information targeted to this agent, in the sense that the information stored in this fields is specially relevant for the individual A_i. The changes in this fields are created by utterances with the class *You* detected from the target detector.

The valence is generally affected by the conversation fields with a shift parameter that reweights them to give more importance to one than another given the sign of the valence. The second change comes from the self-influence fields, forcing the attractors to particular values stated by the user or inferred from the expression.

$$\mathscr{F}_v = (\alpha h_{ij+} + (1-\alpha)h_{ij-})(b_1 v - b_3 v^3) + \beta_v(h_{iiv} - v) \tag{10.10}$$

When $v > 0$ and switching h_{ij+} and h_{ij-} when $v < 0$. In this function, the balance between the attention towards content of the same or different valence is parametrized through α. b_1 models the direct influence that the field has on the valence, and b_3 is a saturation parameter that ensures that the valence cannot go to infinity. β_v is the strength factor of the update to known values of the state from utterances or user input.

The arousal is supposed to increase with information in general, regardless of its valence but depending on how relevant is this information for the individual. This way, the arousal will be increase with all the fields and decrease only based in internal assessments of low arousal, coming from the user input or the arousal of the expression.

$$\mathscr{F}_a = ((1-\eta)h_{ij} + \eta h_i)(d_2 a^2 - d_3 a^3) + \beta_a(h_{iia} - a) \tag{10.11}$$

In this arousal dynamics, η balances how stronger is the identity field compared to the conversation field. d_2 and d_3 work in a similar way that the valence counterparts. The quadratic term makes sure that the influence of that term is always positive. Similarly, β_a refines the knowledge of the arousal like in the valence.

Our first approximation to the dynamics of the dominance can be based on the identity fields, and how the information directed to the individual changes its power regarding emotions.

$$F_d = g_+ h_{i+} - g_- h_{i-} + \beta_d(h_{iid} - d) \tag{10.12}$$

This way, the influence on the dominance would be independent of its own value and just induced by the social identity of the agent. The decay term γ_d will ensure that the dominance does not go to infinite values. The parameters g_+ and g_- represent the asymmetric effect on the dominance, if the fear reaction is supposed to be fast, it should satisfy $g_- > g_+$.

In this model there are two types of fields, signed fields like the conversation field, and VAD field that store a particular point in the emotion space rather than an amount of information. Signed fields have two components: a positive one (h_+) and a negative one (h_-). The input to these fields is multiplexed to the positive or the negative part given the polarity of the message relevant to them.

$$\frac{dh_\pm}{dt} = -\gamma_h h_\pm + sM_\pm(t) \tag{10.13}$$

As for previously defined fields, γ_h is the decay factor and s measures the impact that a user has on the field with each message. $M_\pm(t)$ is the amount of messages directed to the field component in a particular moment. Each message creates an increase only once, as an impulse of size s in the field.

In the model implementation, there are two size of changes s for the two different signed fields: s_d if the impact is produced by the utterance of one of the participants, and s_t when the utterance is directed to a particular individual. This way, we can adjust the balance between the quality of the information received from targeted utterances versus the aggregated amount of information in the general conversation.

A VAD field is different to a signed field in the sense that it stored values of valence, arousal and dominance rather than generalized positive or negative emotional information. The dynamics of this type of fields are different as they have bounded values and the input has a different nature. It has the same eigendynamics:

$$\frac{dh_v}{dt} = -\gamma_v h_v \tag{10.14}$$

An input to the field changes instantly its value to

$$h_v = (1 - s_v)h_v + s_v S_v(t) \tag{10.15}$$

where $S_v(t)$ is the valence of the utterance that changes the field at time t, and s_v is the importance of this utterance, according to its origin, differing between texts and explicit user inputs.

This way, the changes in the VAD field associated to an agent will be sharp, but the changes in the internal variables of the agent, which are the ones to be used to calculate the facial expression, will evolve smoothly but at different speeds. The equations apply the same way to arousal and dominance. In the model, there are three kinds of influence to the self-influence field, according to their origin they will cause different changes: (1) influence due to generalized expression, s_s, which correspond to the inherent individual emotions expressed in any text, (2) influence due to self-reference expression ("I" from target detector) s_i, coming from utterances classified as first person, and (3) influence due to user input s_u, triggered when the users make an assessment about their emotional state. This should be the most important one and close to one.

This model defines a data-driven system which, if run during the interaction of two or more people, provides the time evolution of their emotions in the three dimensions of valence, arousal, and dominance. Thanks to this, virtual human platforms can display rich facial expressions that cover a large variety of states, and these states evolve smoothly in time according to two principles: sentiment analysis from the utterances of a user, and emotion reactions according to the dynamics explained above.

10.7 Discussion

The applications of our modeling framework are not limited to the ones presented here. For example, a recent article (Mitrović and Tadić 2012) proposes an agent-based model based on our framework, to model emotional interaction in blog sites. The collective behavior of this model was empirically tested versus data from blogs and `Digg.com`. Additionally, our model has been used to define an agent-based model for emotional behavior in social networking sites, as presented in Chap. 11.

The insights provided by agent-based models within our framework are of special use for certain ICT applications. Dialog systems, more commonly known because of the use of *chatbots*, benefit from this framework, as agent-based models can be formulated as computational entities that can simulate human behavior, and interact with users of a dialog system. This connection between our agent-based approach and its applications for affective computing are explained in Rank (2010). Furthermore, data-driven simulations of our model have already been implemented in virtual human platforms, in which three-dimensional avatars show facial emotional expression (Ahn et al. 2012). Those platforms run simulations of individual agents to estimate the emotional state of the user, visualizing its emotions through the facial expression of the avatar.

To summarize, our modeling framework provides the means to understand and predict the emergence of collective emotional states, based on the interaction between individual agents. Its analytical tractability allows to find conditions when these states appear and disappear, leading us to the formulation of testable hypothesis of emotion dynamics. We tested some of these hypotheses against datasets of online origin, providing support to the existence of asymmetries in emotional expression. Instances of our models have been proven successful in reproducing collective behavior in product review communities and chatrooms. Future applications aim at applying our framework to other types of online communication, such as forum discussions, and open source communities.

Acknowledgements This research has received funding from the European Community's Seventh Framework Programme FP7-ICT-2008-3 under grant agreement no 231323 (CYBEREMOTIONS).

References

Ahn, J., Gobron, S., Garcia, D., Silvestre, Q., Thalmann, D., Boulic, R.: An NVC emotional model for conversational virtual humans in a 3D chatting environment. In: Perales, F.J., Fisher, R.B., Moeslund, T.B. (eds.) Articulated Motion and Deformable Objects: 7th International Conference, AMDO 2012, Port d'Andratx, Mallorca, July 11–13, 2012, Proceedings. Lecture Notes in Computer Science (Image Processing, Computer Vision, Pattern Recognition, and Graphics), vol. 7378, pp. 47–57 (2012). doi:10.1007/978-3-642-31567-1_5

Alvarez, R., Garcia, D., Moreno, Y., Schweitzer, F.: Sentiment cascades in the 15M movement. EPJ Data Sci. **4**(1), 6 (2015). doi:10.1140/epjds/s13688-015-0042-4

Bollen, J., Gonçalves, B., Ruan, G., Mao, H.: Happiness is assortative in online social networks. Artif. Life **17**(3), 237–251 (2011). doi:10.1162/artl_a_00034

Borge-Holthoefer, J., Moreno, Y.: Absence of influential spreaders in rumor dynamics. Phys. Rev. E **85**(2), 026116 (2012). doi:10.1103/PhysRevE.85.026116

Chmiel, A., Sienkiewicz, J., Thelwall, M., Paltoglou, G., Buckley, K., Kappas, A., Hołyst, J.A.: Collective emotions online and their influence on community life. PLoS ONE **6**(7), e22207 (2011). doi:10.1371/journal.pone.0022207

Czaplicka, A., Hołyst, J.A.: Modeling of Internet influence on group emotion. Int. J. Mod. Phys. C **23**(3), 1250020 (2012). doi:10.1142/S0129183112500209

Garas, A., Garcia, D., Skowron, M., Schweitzer, F.: Emotional persistence in online chatting communities. Sci. Rep. **2**, 402 (2012). doi:10.1038/srep00402

Garcia, D., Schweitzer, F.: Emotions in Product Reviews – Empirics and Models. In: Proceedings of 2011 IEEE International Conference on Privacy, Security, Risk, and Trust, and IEEE International Conference on Social Computing, PASSAT/SocialCom, pp. 483–488 (2011). doi:10.1109/PASSAT/SocialCom.2011.219

Garcia, D., Garas, A., Schweitzer, F.: Positive words carry less information than negative words. EPJ Data Sci. **1**(1), 3 (2012). doi:0.1140/epjds3

Garcia, D., Zanetti, M.S., Schweitzer, F.: The role of emotions in contributors activity: a case study on the GENTOO community. In: 2013 Third International Conference on Cloud and Green Computing (CGC), pp. 410–417 (2013). doi:10.1109/CGC.2013.71

Gianotti, L.R.R., Faber, P.L., Schuler, M., Pascual-Marqui, R.D., Kochi, K., Lehmann, D.: First valence, then arousal: the temporal dynamics of brain electric activity evoked by emotional stimuli. Brain Topogr. **20**(3), 143–156 (2008). doi:10.1007/s10548-007-0041-2

Golder, S.A., Macy, M.W. Diurnal and seasonal mood vary with work, sleep, and daylength across diverse cultures. Science **333**(6051), 1878–1881 (2011). doi:10.1126/science.1202775

Kappas, A.: Emotion and regulation are one! Emotion Rev. **3**(1), 17–25 (2011). doi:10.1177/1754073910380971

Kuppens, P., Oravecz, Z., Tuerlinckx, F.: Feelings change: accounting for individual differences in the temporal dynamics of affect. J. Pers. Soc. Psychol. **99**(6), 1042–60 (2010). doi:10.1037/a0020962

Leskovec, J., Backstrom, L., Kleinberg, J.: Meme-tracking and the dynamics of the news cycle. In: Proceedings of the 15th ACM SIGKDD International Conference on Knowledge Discovery and Data Mining - KDD '09, pp. 497–506. ACM, New York, NY (2009). doi:10.1145/1557019.1557077

Lorenz, J.: Universality in movie rating distributions. Eur. Phys. J. B **71**(2), 251–258 (2009). doi:10.1140/epjb/e2009-00283-3

Mitrović, M., Tadić, B.: Dynamics of bloggers' communities: bipartite networks from empirical data and agent-based modeling. Physica A **391**(21), 5264–5278 (2012). doi:10.1016/j.physa.2012.06.004

Onnela, J.P., Reed-Tsochas, F.: Spontaneous emergence of social influence in online systems. Proc. Natl. Acad. Sci. U. S. A. **107**(43), 18375–18380 (2010). doi:10.1073/pnas.0914572107

Radicchi, F.: Human activity in the web. Phys. Rev. E **80**(2), 026118 (2009). doi:10.1103/PhysRevE.80.026118

Rank, S.: Docking agent-based simulation of collective emotion to equation-based models and interactive agents. In: McGraw, R., Imsand, E., Chinni, M.J. (eds.) Proceedings of the 2010 Spring Simulation Multiconference on - SpringSim'10, p. 6. Society for Computer Simulation International, San Diego, CA (2010). doi:10.1145/1878537.1878544

Rank, S., Skowron, M., Garcia, D.: Dyads to groups : modeling interactions with affective dialog systems. Int. J. Comput. Linguistic Res. **4**(1), 22–37 (2013)

Rimé, B.: Emotion elicits the social sharing of emotion: theory and empirical review. Emot. Rev. **1**(1), 60–85 (2009). doi:10.1177/1754073908097189

Rimé, B., Finkenauer, C., Luminet, O., Zech, E., Philippot, P.: Social sharing of emotion: new evidence and new questions. Eur. Rev. Soc. Psychol. **9**(1), 145–189 (1998). doi:10.1080/14792779843000072

Russell, J.A.: A circumplex model of affect. J. Pers. Soc. Psychol. **39**(6), 1161–1178 (1980). doi:10.1037/h0077714

Schweitzer, F.: Brownian Agents and Active Particles. Collective Dynamics in the Natural and Social Sciences. Springer Series in Synergetics, 1st edn. Springer, Berlin, Heidelberg (2003). doi:10.1007/978-3-540-73845-9

Schweitzer, F., Garcia, D.: An agent-based model of collective emotions in online communities. Eur. Phys. J. B **77**(4), 533–545 (2010). doi:10.1140/epjb/e2010-00292-1

Schweitzer, F., Hołyst, J.A. (2000). Modelling collective opinion formation by means of active Brownian particles. Eur. Phys. J. B **15**(4), 723–732. doi:10.1007/s100510051177

Sienkiewicz, J., Skowron, M., Paltoglou, G., Hołyst, J.: Entropy-growth-based model of emotionally charged online dialogues. Adv. Comput. Syst **16**(04n05), 1350026 (2013). doi:10.1142/S0219525913500264

Smith, E.R., Conrey, F.R.: Agent-based modeling: a new approach for theory building in social psychology. Pers. Soc. Psychol. Rev. **11**(1), 87–104 (2007). doi:10.1177/1088868306294789

Thelwall, M., Buckley, K., Paltoglou, G., Skowron, M., Garcia, D., Gobron, S., Ahn, L., Kappas, A., Küster, D., Hołyst, J.: Damping Sentiment Analysis in Online Communication: Discussions, Monologs and Dialogs. Computational Linguistics and Intelligent Text Processing, Lecture Notes in Computer Science, vol. 7817, pp. 1–12 (2013). doi:10.1007/978-3-642-37256-8_1

Walter, F.E., Battiston, S., Schweitzer, F.: Personalised and dynamic trust in social networks. In: Bergman, L., Tuzhilin, A., Burke, R., Felfering, A., Schmidt-Thieme, L. (eds.) Proceedings of the 3rd ACM conference on Recommender systems - RecSys '09. ACM Press, New York, NY, pp. 197–204 (2009). doi:10.1145/1639714.1639747

Wu, Y., Zhou, C., Xiao, J., Kurths, J., Schellnhuber, H.J.: Evidence for a bimodal distribution in human communication. Proc. Natl. Acad. Sci. U. S. A. **107**(44), 18803–18808 (2010). doi:10.1073/pnas.1013140107

Chapter 11
Agent-Based Simulations of Emotional Dialogs in the Online Social Network MySpace

Bosiljka Tadić, Milovan Šuvakov, David Garcia, and Frank Schweitzer

11.1 Introduction

The Internet is increasingly recognized not only as a tool that people use, but as an environment where they function and live. The amount of time, energy and emotions spent on using online social networks (Šuvakov et al. 2012a; Giles 2010; Amichai-Hamburger and Vinitzky 2010; Cheung et al. 2011; Ryan and Xsenos 2011), online games (Szell et al. 2010; Szell and Thurner 2010), emails (Guimerà et al. 2003), blogs (Mitrović et al. 2010) or chats (Garas et al. 2012; Gligorijević et al. 2013) are getting unprecedented scores. Hence the social implications of the Internet (DiMaggio et al. 2001; Amichai-Hamburger 2002), the intricate relationships between the offline and the online worlds (Szell et al. 2010; Szell and Thurner 2010; Johnson et al. 2009), and mechanisms governing new techno-social phenomena (Kleinberg 2008; Mitrović and Tadić 2010, 2012a; Chmiel et al. 2011) pose new challenges for interdisciplinary scientific research. Researchers are faced with the problem of transferring concepts of offline human behaviors into the online world of social networks (Giles 2011). The question whether humans behave completely differently as "users" is tackled in various empirical investigations (DiMaggio et al. 2001; Johnson et al. 2009; Garcia et al. 2012; Mitrović et al. 2011). A particular question regards the emotional interaction between users in online social networks: How is emotional influence exerted if only written text is exchanged? What kind of

B. Tadić (✉) • M. Šuvakov
Department of Theoretical Physics, Jožef Stefan Institute, Ljubljana, Slovenia

Institute of Physics Belgrade, University of Belgrade, Pregrevica 118, 11080 Belgrade, Serbia
e-mail: Bosiljka.Tadic@ijs.si; milovan.suvakov@ipb.ac.rs

D. Garcia • F. Schweitzer
ETH Zurich, Zurich, Switzerland
e-mail: dgarcia@ethz.ch; fschweitzer@ethz.ch

J.A. Hołyst (ed.), *Cyberemotions*, Understanding Complex Systems,
DOI 10.1007/978-3-319-43639-5_11

emotions are actually involved? What is the role of the underlying network structure in spreading the emotions? Therefore, study of the stochastic processes related with stepping of the users into the online world and spreading of their behaviors and emotions through the online social network, are of key importance.

In our recent work (Šuvakov et al. 2012a), we have compiled and analysed a large dataset which contains the dialogs between the users in the MySpace social network, currently ranked as the fourth largest social networking site after Facebook, Google+, and Linkedin. By combining the methods of statistical physics with a machine learning approach of text analysis, by which the emotional content of the messages was extracted, we have found a strong evidence of the user's collective behavior in which emotions are involved. Specifically, the bursts of emotional messages occur, which obey the scaling laws and temporal correlations. Furthermore, a characteristic structure of the dialogs-based network was revealed as well as the dominance of the positive valence emotions. In order to investigate the mechanisms underlying the observed collective behaviors of users, in this work, we use agent-based modeling framework with the emotional agents (Schweitzer and Garcia 2010; Šuvakov et al. 2012b) and simulate the emotional influence between agents in the online social network.

Agent based approaches are gaining importance in quantitative study of emotions (Rodgers 2010; Kuppens et al. 2010). Our modeling framework has already proved its applicability for various online communications, e.g., emotional influence in chats (Garas et al. 2012), product reviews (Schweitzer and Garcia 2010; Garcia and Schweitzer 2011) and the dynamics on blogs (Mitrović and Tadić 2011; Mitrović and Tadić 2012b) and other online networked systems (Tadić and Šuvakov 2013; Tadić 2013). It is based on the concept of Brownian agents (Schweitzer 2003) which are described by two scalar variables, *valence* that describes the pleasure (attractiveness and averseness) associated with an emotion, and *arousal* that describes the activity level induced by the emotion. To quantify emotions by the valence and arousal components is motivated by Russel's model (Russell 1980; Coan and Allen 2007) from psychology.

The proposed model (Šuvakov et al. 2012b) is in various ways linked to the empirical observations in the MySpace network. Firstly, it takes some empirical findings as *input* for the computer simulations. In particular, the network of interactions, on which the model of agents is implemented, has been extracted from the empirical dataset of MySpace (Šuvakov et al. 2012a). To consider a realistic network structure is of a primary importance for the emotion dynamics, since it is known that hidden topology features, such as link correlations at next-neighborhood level, affect the spreading processes of information and other relaxation dynamics in complex networks (Roca et al. 2010; Tadić et al. 2007, 2005). Further important empirical input regards the interactivity pattern of the MySpace users. Such patterns appear across different communication media (Malmgren et al. 2009; Castellano et al. 2009; Crane et al. 2010; Mitrović and Tadić 2010, 2012a; Szell et al. 2010), representing a hallmark of a given communication system.

In specifying the model, we follow the agent-based framework of emotional influence outlined in Schweitzer and Garcia (2010), which was already applied to

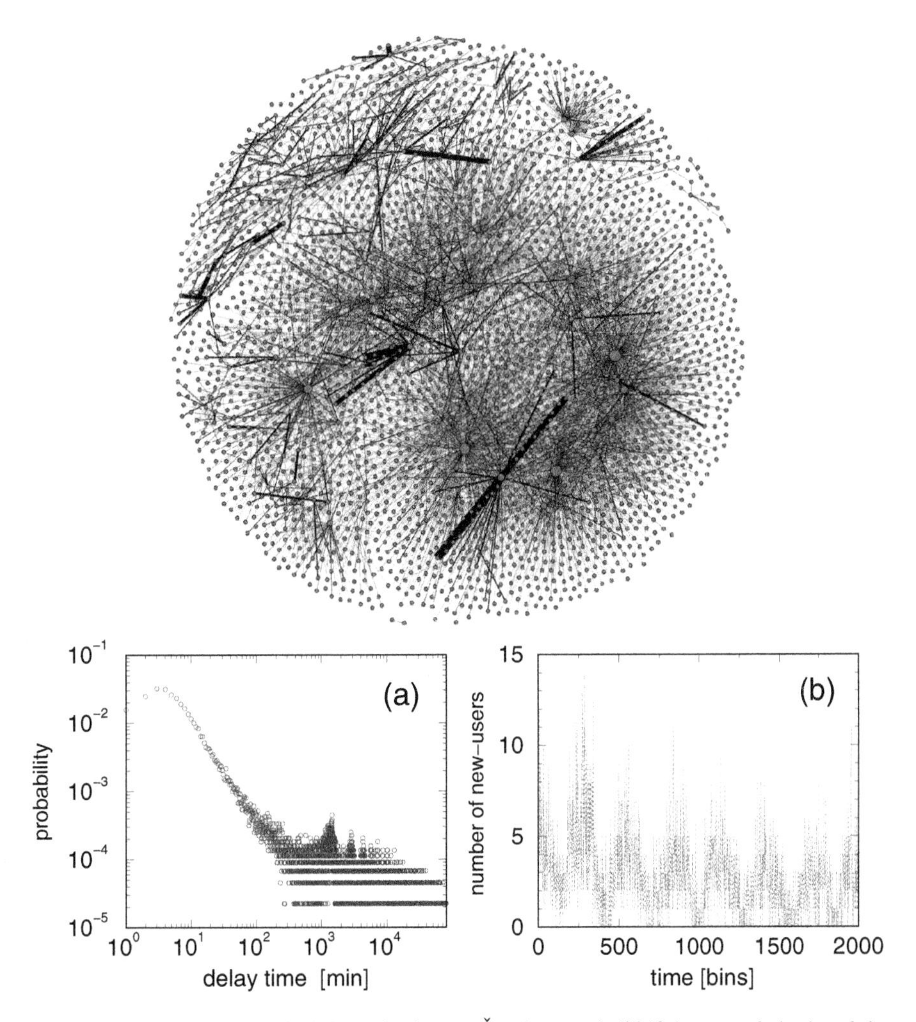

Fig. 11.1 Inputs from empirical data of reference Šuvakov et al. (2012a): network (*top*) and the histogram of raw data of delay times (**a**) and arrival of users time series (**b**)

product reviews (Garcia and Schweitzer 2011), chat rooms (Garas et al. 2012), and blog discussion (Mitrović and Tadić 2011). In this framework, the emotional state of an agent i is described by two variables, valence $v_i(t) \in [-1, 1]$, and arousal $a_i(t) \in [0, 1]$, both of which follow a stochastic dynamics. In contrast to the previous work (Garcia and Schweitzer 2011; Garas et al. 2012), where the stochasticity was modeled by simply adding a stochastic force, in the present model we assume that stochasticity may result from three sources, similarly to the model of agents on blogs (Mitrović and Tadić 2011). Specifically, these are:

- sampling from the empirical inter-activity time distribution $P(\Delta t)$, Fig. 11.1a;
- the empirical rate $p(t)$ at which new users enter the network, Fig. 11.1b;
- a spontaneous reset of both valence and arousal to a predefined value $(\tilde{v}, \tilde{a})$ with a probability p_0.

The mathematical complexity of the present model, with the *emotional agents on a fixed network*, lies in between two previously studied cases. On one side, we have the dynamics of blogs and chats with Bots, where the agent's actions cause the network on which they are situated to evolve (Mitrović and Tadić 2011; Mitrović and Tadić 2012b; Tadić and Šuvakov 2013); on the other end is the model of the product review communications, where the agents are not exposed to any geometrical constraints (Schweitzer and Garcia 2010). Numerical implementation of the present model and the preliminary results are available online (Šuvakov et al. 2012b).

11.2 Model of Emotional Agents in MySpace Social Network

11.2.1 Structure of the Model

To describe how an agent is affected by the emotional contents of perceived messages, three levels of aggregated influence are considered in our model:

- the aggregation of messages on the agent's wall, which is captured by a temporally varying arousal and valence fields $h_i^{a,v}(t)$;
- the aggregation of messages perceived on the friends' walls, resulting to additional arousal field $\overline{h}_i^a(t)$;
- a mean field $h_{\mathrm{mf}}(t)$, determined from all agents' walls, that captures a kind of "atmosphere" of the network.

In analogy to the empirical data, in the model each message carries a valence and arousal value, which give raise a valence and arousal component of the field, $h_i^v(t)$ and $h_i^a(t)$, respectively. Also in contrast to the previous modeling approaches in Schweitzer and Garcia (2010), Garcia and Schweitzer (2011), here, we assume that the agent's valence is affected solely by the valence field, while the agent's arousal and activity (e.g., in choosing conversation partners) are primarily affected by the arousal fields, but can also include the contribution of the valence fields, as it is explained below.

Using the discrete time dynamics, the emotional state of an agent i is described by the following maps:

$$v_i(t+1) = (1-\gamma_v)v_i(t) + \delta_{\theta_i,1}\mathscr{F}_v(t)\ , \tag{11.1}$$

$$a_i(t+1) = (1-\gamma_a)a_i(t) + \delta_{\theta_i,1}\mathscr{F}_a(t). \tag{11.2}$$

Here $\delta_{\theta_i,1}$ is the Kronecker delta which is one only if the agent is in an active state and zero otherwise. The nonlinear functions $\mathscr{F}$ capture how valence and arousal are affected by the information fields. A discussion supporting these forms of maps was provided in Schweitzer and Garcia (2010). Here, we use the following specific forms, which are suitable for the message exchange in the online social networks:

$$\mathscr{F}_v(t) = [(1-q)h_i^v(t) + qh_{mf}^v(t)] \times [c_1 + c_2(v_i(t) - v_i^3(t))][1 - |v_i(t)|] \, , \tag{11.3}$$

$$\mathscr{F}_a(t) = \{(1-q)[\epsilon h_i^a(t) + (1-\epsilon)\bar{h}_i^a(t)] + qh_{mf}^a(t)\} \times [1 - a_i(t)] \, . \tag{11.4}$$

Each of these functions consist of a term that depends on the information fields and a second term that depends on the arousal or the valence, respectively. For the latter, nonlinear assumptions are made in accordance with Schweitzer and Garcia (2010). The term $[1 - |x|]$ is added to confine both variables in the prescribed range of the phase space. The small parameter $q \in [0, 1]$ adjusts the fraction of the influence of the mean-field information, $h_{\text{mf}}(t)$, whereas the small parameter $\epsilon \in [0, 1]$ adjusts the influence of the emotional contents of the friends' walls, $\bar{h}_i^a(t)$, in relation to the own wall, $h_i^a(t)$ at time t. We assume that the arousal dynamics of an agent depends on the activity on the walls of its friends, that is captured by $\bar{h}_i^a$. However, we neglect such influence in the valence, assuming that the level of pleasure of an agent i should depend on the messages directed to i rather than on the messages that the agent i can see at the walls of its neighbors.

For our investigation in this work, we assume that the mean-field contribution can be neglected, i.e. we set $q = 0$. This corresponds to the actual situation in the online social networks like MySpace and Facebook. For comparison, in blogs dynamics (Mitrović and Tadić 2011; Mitrović and Tadić 2012b) some mean-field influence ($q = 0.4$) is necessary to match the empirical system; on the other hand, in product reviews (Garcia and Schweitzer 2011) or chat room conversation (Garas et al. 2012), the key role was attributed to the mean fields ($q = 1$). Moreover, in the case of social networks, *the main contribution of the information field comes from the agent's individual wall*, thus we set $\epsilon = 0.9$. We define the sequence of all messages from agent j to agent i as M_{ji}. The wall of an agent i contains the message sequences from all its friends. The influence of the aggregated messages on the agent's i emotion, however, decays in time with a rate γ^h. For the valence and arousal component of the influence field $h_i^v(t), h_i^a(t)$, we assume the following dynamics (here z stands either for valence v or arousal a)

$$h_i^z(t) = \frac{\sum_j \sum_{m \in M_{ji}} \theta(t, t_m) z_j(t_m) W_{ji} e^{-\gamma^h (t_{ji}^{lm} - t_m)}}{\sum_j \sum_{m \in M_{ji}} \theta(t, t_m) W_{ji} e^{-\gamma^h (t_{ji}^{lm} - t_m)}} e^{-\gamma^h (t - t_{ji}^{lm})} \tag{11.5}$$

where j runs over all neighbors of the node i in the network and m identifies each message from the streams of messages M_{ij} along the link from the neighbor j. For each messages m the creation time t_m is traced and the Heaviside step function $\theta[x]$ ensures that the influence of that message over time is correctly counted. The emotional content of the message is composed by the values of valence or arousal $z_j(t_m)$ of the neighboring agents j at time t_m. Its influence further depends on the weight W_{ji} of the directed link between the nodes, which is determined in the empirical network. The exponential terms indicate the aging messages and the decay of the entire influence filed with the rate γ^h, where t_{ji}^{lm} is the time of the last message arriving on the wall of agent i. The denominator of Eq. (11.5) plays the role of a normalization to keep the field values properly bounded.

In addition to the individual field component h_i, the influence of the friends' walls aggregated in $\bar{h}_i^a(t)$ is specified as follows:

$$\bar{h}_i^a(t) = \frac{\sum_j W_{ij} h_j^a(t)(1 + h_j^v(t)v_i(t))}{\sum_j W_{ij}(1 + h_j^v(t)v_i(t))} . \tag{11.6}$$

In contrast to the above defined fields, Eq. (11.5), the messages on the neighbor's wall can influence the agent's i arousal by two different mechanisms. Firstly, $\bar{h}_i^a$ is composed of the average over the weighted arousal fields on the friends' walls at time t. The weights W_{ij}, however, are modified by a term that takes into account the similarity between the valence $v_i(t)$ of the agent i and the valence fields of its friends walls h_j^v. The expression in the brackets in Eq. (11.6) indicates that the valence similarity enhances the importance of the friend's wall and vice versa, the valence dissimilarity reduces its contribution to the current arousal below the established width of the link W_{ji}. A psychological argument for this assumption is that users often search for information reinforcing their emotional state (Bradley 2009). Furthermore, there is a technical argument: to cope with information overload most social networking sites filter information such that content presented to the user is in line with its previous writing. We note that Eq. (11.6) captures the influence of every "friend-of-a-friend", who post messages on the walls of the friends of an agent. In other words, a next-nearest-neighbor influence occurs in the online social networks, that cannot be perceived in the same way in off-line social interactions. Our model captures this important feature of the online communication dynamics. When an agent is in the active state, $\theta_i = 1$, it writes a message with a probability ω_i that increases with its current arousal a_i. The proportionality, as mentioned before, depends on the global activity level, which is proxied by $p(t)$, and a strength parameter a_0, hence $\omega_i(t) = \delta_{\theta_i,1} a_0 p(t) a_i(t)$. The emotional content of the messages is given by the emotional state of the agent, $v_i(t)$ and $a_i(t)$, at the time of activity t. Finally, the recipient of the message is determined among the neighbor nodes. Instead of a uniform random choice, friend j of an agent will be chosen with a probability $s_j(t)$ that depends on (1) the aggregated information $h_{ji}(t)$ generated by j on the wall of i (whereas W_{ji} represents the strength of the social link between them), and (2) the importance of the wall of friend j to agent i. The rational behind this assumption is the following: when a user writes a message to another user, this can be a part of an ongoing conversation or an initiation of a new conversation. The former is reflected in the first term and the latter in the second term of the following equation:

$$s_j(t) = \mathscr{A}[\beta \frac{W_{ij} h_{ji}^a(t)}{\sum_k W_{ik}} + (1-\beta) \frac{W_{ij} h_j^a(t)(1 + h_j^v(t)v_i(t))}{\sum_k W_{ik}(1 + h_k^v(t)v_i(t))}] , \tag{11.7}$$

$\mathscr{A}$ is the inverse normalization constant and β is a parameter weighting between these two processes. For simplicity, we choose $\beta = \epsilon$, which weights the information in the neighbors' walls against the own wall. The aggregated information $h^a_{ji}(t)$ along the $j \rightarrow i$ link is assumed to depend only on the arousal component of the sender j in the following way:

$$h^a_{ji}(t) = \frac{\sum_{m \in M_{ji}} \theta(t, t_m) a_j(t_m) e^{-\gamma^h (t^{lm}_{ji} - t_m)}}{\sum_{m \in M_{ji}} \theta(t, t_m) e^{-\gamma^h (t^{lm}_{ji} - t_m)}} e^{-\gamma^h (t - t^{lm}_{ji})} . \quad (11.8)$$

Similar to Eq. (11.5), this aggregates a set M_{ji} of messages coming from the agent j on the wall of i.

11.2.2 Parameters of the Model and Input from the Empirical Data

As stated in Sect. 11.1, the network used for our simulations is taken from the empirical data of MySpace, which are collected and described in Šuvakov et al. (2012a). The network is a reduced version (termed Net3321) of the dialogs-based structure from 2-months dataset, which consists of $N = 3321$ nodes, connected by weighted directed links W_{ij} (Šuvakov et al. 2012a). The initial value of the agent's emotional state (i.e., when the agent appears for the first time in the process) are chosen uniformly at random from the intervals $a_i \in [0, 1]$, $v_i \in [-1, 1]$. Their activity pattern is drawn from the empirical inter-activity time distribution $P(\Delta t)$, shown in Fig. 11.1a. While we are sampling from a distribution observed at resolution of $t_{\text{res}} = 1$ min, the time scale t of our simulations is fixed by the driving signal $p(t)$ with $t_{bin} = 5$ min time bin, i.e., each time step in the simulations corresponds to one time bin of the real time. Hence, in case of sampling a value $\Delta t < 1$, the event happens at the current time step. Note that this includes five possible values of the delay time, to which different probability is assigned according to the high-resolution distribution in Fig. 11.1a. The global activity level on the social network is taken by the empirical signal $p(t)$. In Fig. 11.1b, the signal $p(t)$ that is used in the simulations is shown. Its resolution, here 5-min time bins, thus sets the time scale for all time series that are obtained from the simulated data in the following sections.

Apart from the empirical distribution function $P(\Delta t)$ and the time series $p(t)$ that we use as input for the simulations, the values for the control parameters of the model are specified in Table 11.1. We can distinguish between the parameters which control the emotional state of the agents ("internal", γ^a, γ^v, c_1, c_2), the parameters that affect communications between agents (γ^h, q, ϵ) and the parameters that control the activity of agents (p_0, a_0).

These parameters have a different role and, consequently, importance for the stochastic process. Specifically, the "internal" parameters shape the profile of

Table 11.1 Values of the control parameters and the input functions used in the simulations

Internal parameters	Decay rates	Influence	Driving	Global functions
$c_1 = 0.5$	$\gamma^a = \gamma^v = 0.05$	$q = 0$	$p_0 = 0.01$	$P(\Delta t)$
$c_2 = 0.5$	$\gamma^h = 0.01$	$\epsilon = 0.9$	$a_0 = 0.01$	$p(t)$

individual agents. In the absence of empirical data that yield the model of Eqs. (11.1) and (11.2), the parameters in Eqs. (11.3) and (11.4) are chosen such that the maps satisfy some formal conditions. In particular, $c_1 \neq 0$ and $c_2 > 0$ support additive field effects and the presence of fixed points in the relevant regions of the phase space, respectively. Similarly, the choice $\gamma_h < \gamma_a$ provides with a slower decay of the cumulative arousal of the field compared with the arousal of each individual message. The interaction parameter ϵ, the ratio between the weights of own wall and the friend's wall, is characteristic for the online social networks. Therefore, it should be nonzero. A plausible value, as shown in Table 11.1, is used for the simulations. In contrast, the mean-field fraction $q = 0$ is the most natural choice in the case of online social networks as MySpace since no network-wide knowledge is accessible to individual users. In principle, this parameter can be varied in the simulations, i.e. in analogy to online social systems on evolving networks (Tadić and Šuvakov 2013), where it can affect synchronization of the agents' activity. The parameter p_0 measures the probability at which the emotional state of an active agent is resetting by an external influence with the specified emotion components $(\tilde{v}, \tilde{a})$. In this respect, two different setups are discussed: (1) values $(\tilde{v}, \tilde{a})$ are chosen at random from a uniform distribution, and (2) $(\tilde{v}, \tilde{a})$ are fixed to one of the three sets representing a target emotion (a) "ashamed", (b) "enthusiastic", (c) "astonished". The first case corresponds to the null hypothesis of a completely unknown external influence that might come from any kind of influence to the agent's emotional state. The second case of fixed values of $(\tilde{v}, \tilde{a})$ will allow us to test the collective effects of large scale events, as for example, mass media, by assuming their influence in the emotional state of the users of MySpace.

11.3 Simulated Behavior of Agents and Comparison with the Empirical System

11.3.1 Agent's Trajectory in Phase Space

Before we discuss the aggregated output, we provide two examples of the dynamics of individual agents in terms of their emotional variables valence $v_i(t)$ and arousal $a_i(t)$. The simulation results which are shown in Fig. 11.2, demonstrate that states with high arousal can be built up due to the interaction of an agent with its neighborhood or, less often, by a single large input from the external environment. Repeated activities in short time intervals occur if an agent remains 'caught' by

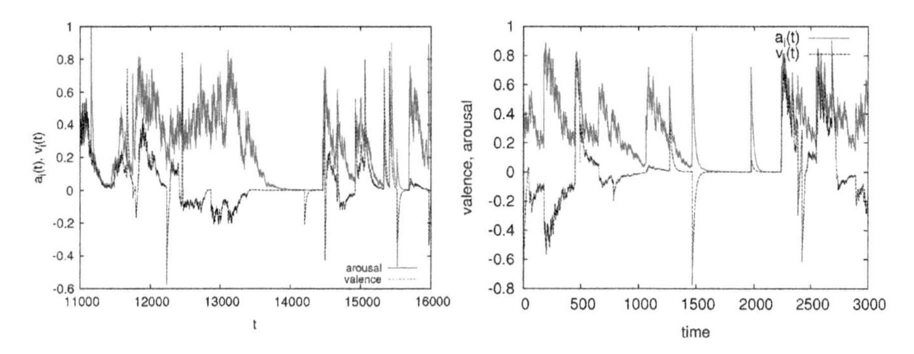

Fig. 11.2 Time series of the valence (*lower curve*) and the arousal (*upper curve*) of two active agents in a simulation of our MySpace model

an active neighborhood over long periods of time. In these time series, the peaks in both arousal and valence occur when the agent's emotional state was influenced by messages from its neighbors, or when its emotional components are reset by a large external input $(\tilde{v}, \tilde{a})$. In the absence of any action, however, both valence and arousal exponentially decayed towards zero. Further, one can notice that an agent's valence fluctuates being influenced either towards positive or towards negative values, without any specific preferences.

In the remaining parts of this section, we compare some simulated results with the corresponding results that we obtain from the empirical dataset in MySpace. As it was explained above, some empirical data, i.e., the driving signal $p(t)$ and the delay time distribution, $P(\Delta t)$, have been already used as the input parameters in the simulations. However, the activity patterns and emotional response of agents do not trivially follow from a driving signal $p(t)$ and a given network topology. Instead, only with the right assumptions about the agent's emotional interaction, we are able to reproduce the stylized facts as explained below. For the following simulation results, we sample the external influence $(\tilde{v}, \tilde{a})$ from a uniform distribution. The effects of a specified emotion input in the collective behavior of agents will be studied later. Here, we can compare the valence distribution of the messages generated by the emotional agents in the network with the distribution of valence of the emotional messages of users in the empirical data of (Šuvakov et al. 2012a). For this purpose, we first extract the corresponding emotional contents—valence and arousal—from the texts of messages in the empirical data.

11.3.2 Extraction of Emotional Content from Message Texts

For comparison of the simulated emotional behavior of agents with the empirical data in MySpace network, here we also analyze the emotional contents of the text messages in the empirical data of reference Šuvakov et al. (2012a). Specifically, we quantify emotions with respect to two dimensions, arousal and valence (Russell 1980). The latter indicates the pleasure associated with the emotion (positive, negative, neutral), the former the level activity that it induces.

In order to extract the emotional content from the messages, we use sentiment analysis by applying a standard procedure introduced in Dodds and Danforth (2010). It uses the ANEW dataset, a *lexicon of human ratings of valence and arousal* with about 1000 words (Bradley and Lang 1999), for which the emotional charge, or valence, v_i and the arousal a_i was determined. An algorithm then calculates the frequency f_i of such classified words in a given text message, to compute the valence and arousal of the text sequence as

$$v_{\text{text}} = \frac{\sum_{i=1}^{n} v_i f_i}{\sum_{i=1}^{n} f_i}\ ; \quad a_{\text{text}} = \frac{\sum_{i=1}^{n} a_i f_i}{\sum_{i=1}^{n} f_i}\ . \tag{11.9}$$

For the first time, here we apply this method to the `MySpace` dataset. Due to the limited size of the ANEW lexicon, the method should be preferably used on long texts because of statistical reasons. To overcome this limitation, we extracted the stem, or root form, of all words in the analyzed text. The stem contains most of the semantic information of a word (and thus its emotional content), and allows us to match similar words rather than exact matches, which eventually improves the statistics. To extract the stem, we used Porter's Stemming algorithm (van Rijsbergen et al. 1980), a technique that applies inverse generalized rules of linguistic deflection, mapping deflected words to the same stem. For example, the stem of the words "lovely" and "loving" by that method is "love", which matches the corresponding word in the ANEW lexicon. This way, the sentiment analysis covers a larger portion of the text and allows to calculate the emotional values for more than 60 % of the messages in the dataset.

The results of the analysis of emotional expression of the `MySpace` dataset are presented in Fig. 11.3. To be compatible with Russell's circumplex model (Russell 1980), we rescaled the output of Eq. (11.9) to the range $[-1, 1]$ and adopted the standard polar diagram for the quantitative representation of emotions by using the following transformation:

$$x' = x\sqrt{1 - \frac{y^2}{2}}\ ; \quad y' = y\sqrt{1 - \frac{x^2}{2}}\ . \tag{11.10}$$

Different points in this diagram are associated with different emotions. For comparison, the markers $1 - 19$ indicate examples of emotions which are known in psychology (Scherer 2005).

The distribution of valence values for the messages of our dataset is highly biased towards positive values, as the largest density in Fig. 11.3 is above 0. Figure 11.4 shows the distribution of valence for messages that contain at least one word from the ANEW dataset. We notice that the mode is at 0.5. On the other hand, we have shown (Garcia et al. 2012) that English written text is naturally biased towards positive emotions, with a mean $\mu_v = 0.31$ and a standard deviation $\sigma_v = 0.47$

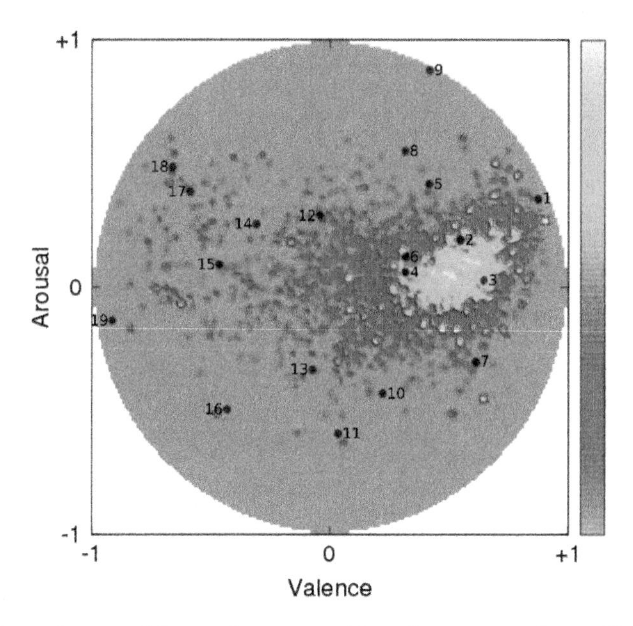

Fig. 11.3 Circumplex map of the emotions extracted from the MySpace dataset. The frequency is indicated by the *color* scale from 0 to 50, and the binning is done in a grid of 200 × 200 bins. *Markers* indicate different emotions: 1—"delighted", 2—"amused", 3—"interested", 4—"expectant", 5—"convinced", 6—"passionate", 7—"hopeful", 8—"feeling superior", 9—"astonished", 10—"longing", 11—"pensive", all on positive valence side, and 12—"impatient", 13—"worried", 14—"suspicious", 15—"distrustful", 16—"ashamed", 17—"frustrated", 18—"disgusted", 19—"miserable", on the negative valence side

for the valence. This needs to be taken into account to interpret Fig. 11.4, so we renormalise each valence value using $v' = (v - \mu_v)/\sigma_v/\sqrt{w}$, where v is the valence from Eq. (11.9) and w is the amount of ANEW words in the message. The renormalized valence distribution is shown in the inset of Fig. 11.4. We find that, despite this renormalization, there is still a large bias towards positive emotions in the messages of MySpace.

On the other hand, the distribution of the expressed arousal (vertical axis of Fig. 11.3) is quite concentrated around values close to 0, i.e. MySpace messages rarely contain words expressing strong arousal. This finding is in line with previous survey studies (Paltoglou et al. 2013; Thelwall et al. 2010) in which arousal from written texts showed a small variance.

11.3.3 Comparison of Simulated and Empirical Distributions

For the comparison of simulated agent's with the ones found in the empirical data, we consider the emotional valence of each agent in the moment of action. The

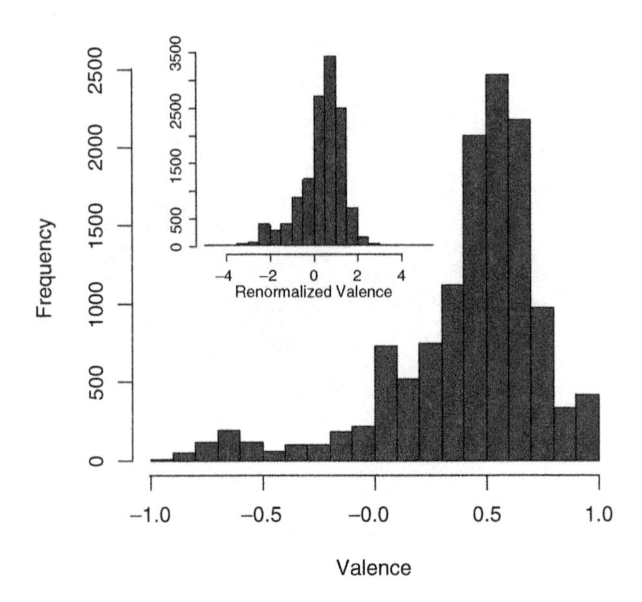

Fig. 11.4 Distribution of estimated valence of messages with at least one word from the ANEW dataset. *Inset*: distribution of the renormalized valence taking into account the amount of ANEW words in the message and the statistics of human expression on the Internet from (Garcia et al. 2012). Both histograms show a clear bias towards positive expression, even above the natural bias of human expression

histogram, averaged over all agents in the system, is shown in Fig. 11.5 together with the histogram of valences observed in the empirical data. The empirical distribution is equivalent to the one shown already in Fig. 11.4, except for the peak at $v = 0$ which contains all messages that did not contain any word from the ANEW dataset used for classification. We notice that both distributions have an obvious bias toward positive valence, which is stronger in the empirical data than in the simulations.

Here, rather than focusing on the quantitative comparison, we affirm that our model is in fact apt to generate this bias as the result of emotional interactions. Without any interaction, the agent's valence would relax towards zero because of the decay factor in Eqs. (11.1) and (11.2). However, the social interaction with other agents through the fields h on the network generates the positive bias in agreement with the empirical findings. This also supports previous work that argues about the social origin of the positive bias (Garcia et al. 2012; Rimé 2009). Further similarity between the simulated and the empirical system, in particular in the aggregated behavior of agents, is discussed Sect. 11.4.

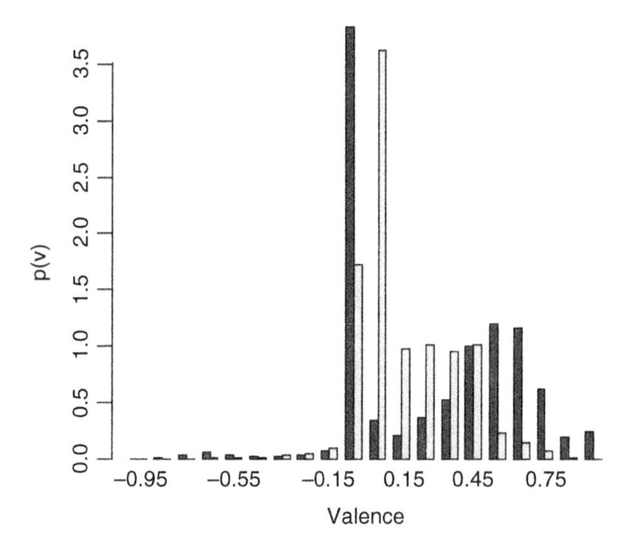

Fig. 11.5 Distribution of the valence of MySpace messages (*dark*), including the ones without any ANEW term in the bin at $v = 0$. In *grey*, the distribution of expressed valences in the simulation. Both distributions have a large bin of nonemotional expression close to zero, and the rest of the distribution shows a strong bias towards positive values

11.4 Simulated Network-Wide Emotional Activity of Agents

11.4.1 Temporal Correlations in Emotional Messages of Agents

As already explained in Sect. 11.2, in the simulations we use $P(\Delta t)$ and $p(t)$ as inputs in the computer code. Hence, other quantities, i.e., the time series of the number of all messages and the number of messages carrying positive/negative emotion, are reproduced in the simulations. Properties of these time series then can be compared with the respective time series of the empirical data studied in Šuvakov et al. (2012a). The simulated results are shown in Fig. 11.6 both in terms of the time series and the power spectrum $S(\nu) \sim \nu^{-\phi}$. While the driving signal $p(t)$ has the power-spectrum exponent $\phi_p = 0.67 \pm 0.11$, we obtain $\phi_{N_c} \approx 0.91 \pm 0.08$ for the driven signal, which exceeds the value $\phi_{N_c} \approx 0.65 \pm 0.12$ of the corresponding empirical data. This suggests that the emotional interactions among agents induce even stronger correlations. A similar observation holds for the range of the scaling region in the fluctuations, shown in Fig. 11.6c, although the Hurst exponent is the same as in the empirical data (see Šuvakov et al. 2012a for more details).

As discussed above, the simulated time series for $N_c(t)$ exhibit the long-term correlation, in analogy to the empirical system. However, the crossover between the uncorrelated and the correlated events occur at a larger time scale in the simulations, i.e., cascades on small time scales are not captured by the simulations. On the other

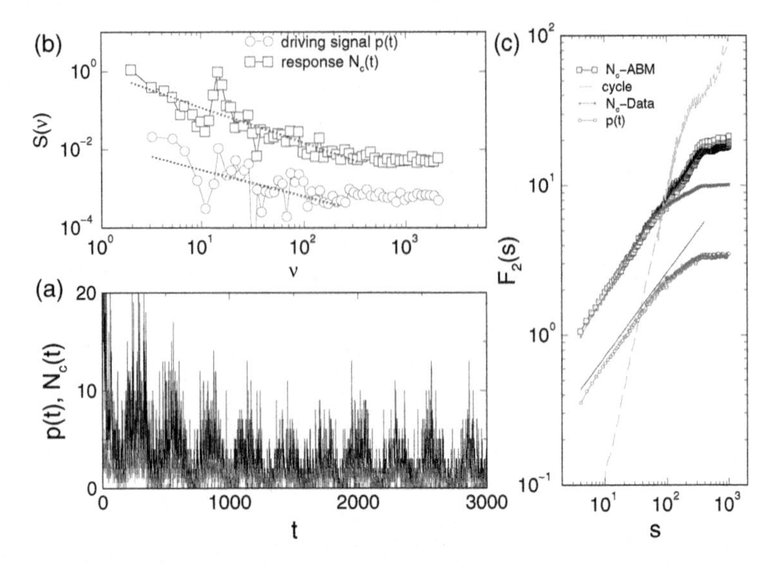

Fig. 11.6 Simulated time-series of the number of messages $N_c(t)$ in *black* and the driving signal $p(t)$ in *pink* of the emotional agents on the network `Net3321` (**a**), power spectrum (**b**) and scaled fluctuations of the simulated and empirical time series (**c**). The *straight lines* indicate the slopes of the correlated part of the power-spectrum in (**b**) and the scaling region in (**c**)

hand, the behavior on long time scales is well reproduced. Thus, our model is able to reproduce the emergence of long range correlations in the emotional expressions, which indicates the emergence of collective emotions.

Let us now discuss the importance of the spectral properties of the driving signal $p(t)$. Are the long-range correlation in the number of messages induced by the correlations in the driving signal? In order to test this, we performed simulations in which the driving signal is composed of white noise with a superimposed circadian component of the daily periodicity. The average value of the white noise is set to the mean of the empirical time series, $\langle p(t) \rangle$, but there are no long range correlations, in contrast to the original signal $p(t)$. Keeping all other parameters unchanged, we simulate the network response to this driving signal. Contrary to the results in Fig. 11.6, now $N_c(t)$ does not show any correlations, as displayed in Fig. 11.7, i.e., apart from the daily periodicity, the power spectrum retains the characteristics of a white noise. This suggests that the occurrence of circadian cycle of daily activity is not sufficient to create the long range correlations, which are characteristic for the real data. We comment on this in again in the conclusions.

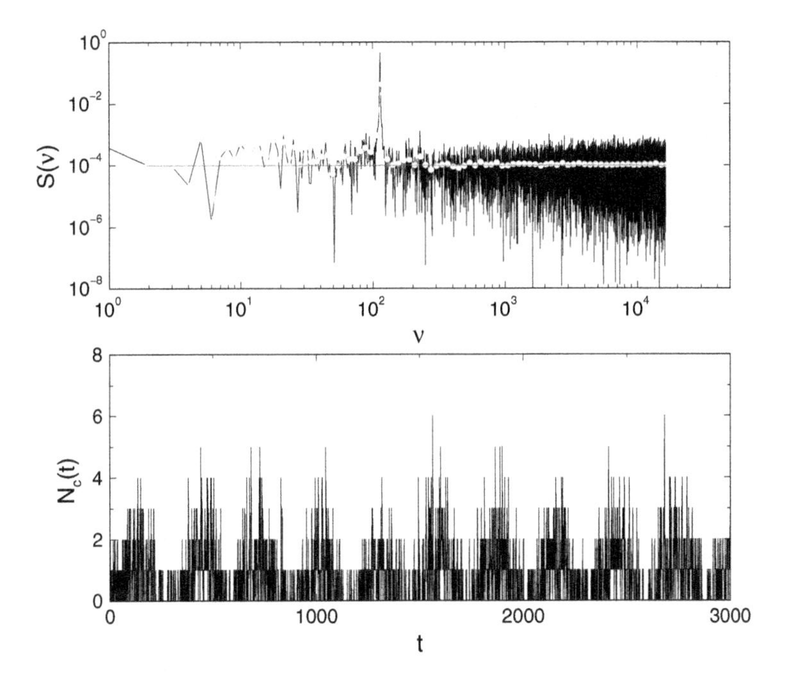

Fig. 11.7 Time series (*bottom*) and (*top*) power-spectrum of the number of messages $N_c(t)$ in a simulation as response to an artificial driving signal composed of white noise with super-imposed circadian cycles

11.4.2 Simulations of External Influence

To understand the influence of external events on the emergence of collective emotions, in our simulations we consider externally triggered resets of the emotional state of agents to a specified value for the valence and arousal, $(\tilde{v}, \tilde{a})$. As it was explained in Sect. 11.2.2, $(\tilde{v}, \tilde{a})$ can be either fixed or sampled from a uniform distribution. While the latter case is covered by the simulation results in Sect. 11.4.1, here we use fixed values for $(\tilde{v}, \tilde{a})$ which indicate three different emotional inputs, namely: (a) “astonished” (v=0.4, a=0.88), (b) “ashamed” (v= −0.44, a= −0.5), (c) “enthusiastic” (v=0.5, a=0.32). In the psychology literature (Scherer 2005), the emotions “astonished” and “ashamed” are believed to have different influence on the social interaction and emotional communication. In our quantitative model, these two emotional states are in the opposite parts of the circumplex map (see Fig. 11.3): “astonished” is a positive emotion with a high arousal value, while “ashamed” is a negative emotion with a low arousal value.

With the simulations, we examine how these fixed emotional inputs $(\tilde{v}, \tilde{a})$ influence the cascade of emotions on the social network. The simulated time series and power spectra and fluctuations of $N_c(t)$ are shown in Fig. 11.8, corresponding to “astonished” and “ashamed” emotional input. The power spectra of $N_c(t)$ in both

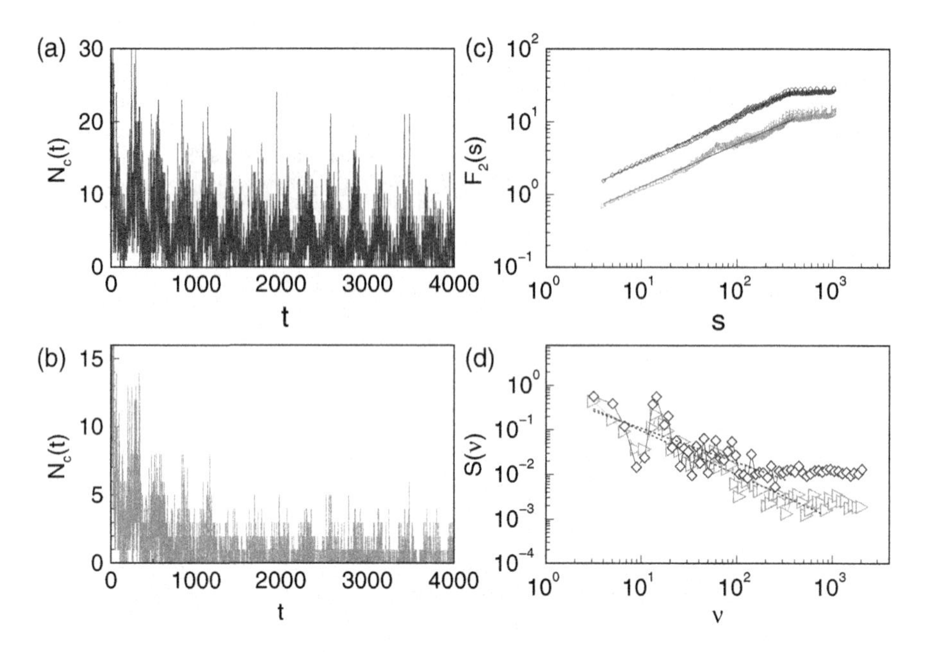

Fig. 11.8 Simulated time series $N_c(t)$ (**a**) and (**b**) in the case of fixed external influence $(\tilde{v}, \tilde{a})$ set to the point corresponding to "astonished" and "ashamed", respectively. In both cases the driving signal $p(t)$ was used. Fluctuations (**c**) and power spectrum (**d**) of these time series

cases show the same shape of $1/\nu^{\phi}$ with the exponents close to 0.76 ± 0.11 in the case "astonished", while 0.98 ± 0.05 approaching flicker noise, in the case "ashamed". These values are higher but not significantly different to the value found in the case of the uniformly sampled emotion components $(\tilde{v}, \tilde{a})$, shown in Fig. 11.6. But the correlations extend for a larger range in the case of "ashamed" (also noticeable in the lower value of $S(\nu)$ at high ν), even though it is a low-arousal negative emotion. On the other hand, the level of activity in the time series is higher in the case of the positive emotion "astonished", due to a higher arousal in these messages.

The collective emotional response of agents that is observed in the correlated time series, is built on the actions of individual emotional agents on the network. The evolution of the emotional state of an agent, like the examples shown in Fig. 11.2, can be seen as a trajectory in a circumplex (Ahn et al. 2010), similar to the circle map shown in Fig. 11.3. In this case, the circumplex is obtained by applying the transformation of Eq. (11.10) on the emotion variables arousal $a_i(t)$ and valence $v_i(t)$ of each agent's states over time. Hence, we can explore the collective effects of different types of external inputs by simply observing the emotional states which are "visited" by the agents on the circumplex.

In the color plots of Fig. 11.9 we show the histograms of different emotional states that were visited by the agents in our simulations. More precisely, we plot the

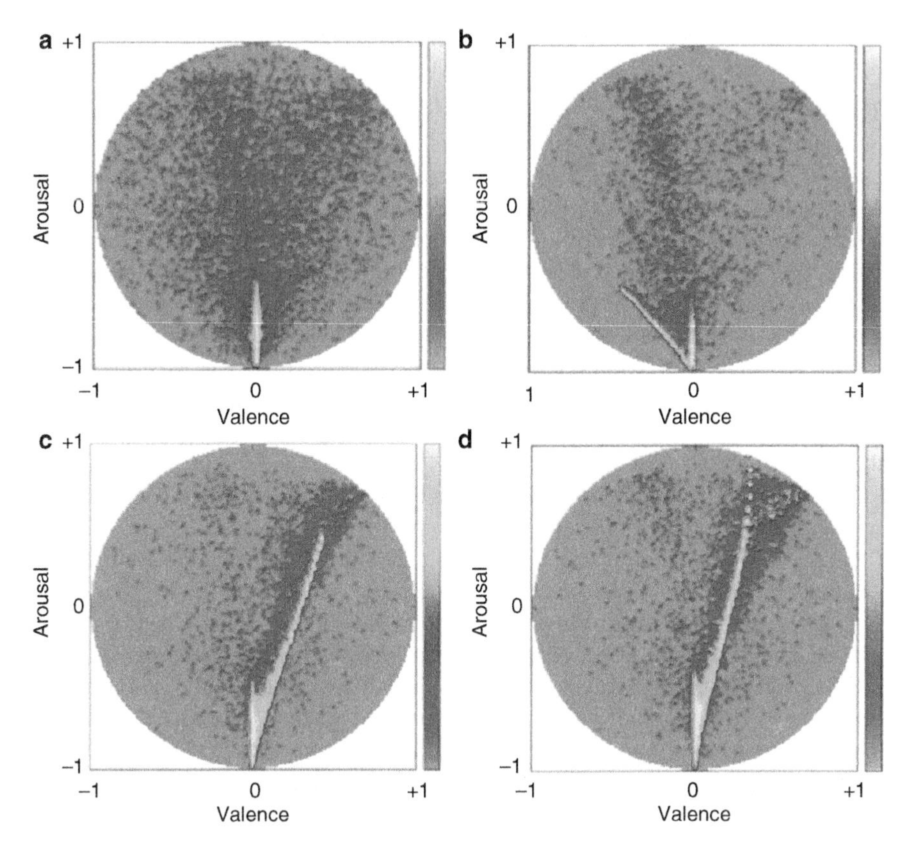

Fig. 11.9 Histograms of visited states in the circumplex for simulations with external influences sampled from a uniform distribution (**a**) and in fixed to "ashamed" (**b**), "enthusiastic" (**c**) and "astonished" (**d**). Arousal is rescaled to the interval $[-1, 1]$

emotion states of each agent i at every moment when the agent was active, i.e., $\theta_i = 1$. The level of activity among the agents varies a lot, depending on their location in the network and the events occurring in the agent's neighborhood. Consequently, an agent's contribution to these emotion histograms will vary according to their activity in a given simulation. The patterns shown in Fig. 11.9 represent situations with a random and the three specific external inputs: "ashamed", "astonished", and "enthusiastic". The histograms in the two upper panels of Fig. 11.9 show how high arousal states may arise starting from (a) uniformly distributed random input or (b) a low-arousal negative valence state like "ashamed". In the lower panels, the external influences are set to two different high arousal positive states: "astonished" and "enthusiastic". The influence of these states spreads through the networks by means of agent interactions through the field, giving raise to the asymmetrical V-shape pattern.

11.4.3 Propagation of Emotions in the Network

The agent-based model also allows us to understand how cascades of emotional influence propagate through the real social network of `MySpace`. Precisely, will positive emotions propagate along the same social links as negative ones? Obviously, agents may exchange positive or negative messages with different preferred neighbors. To what neighbor the message will be sent depends on the agent's past interaction along that link, the strength of the influence fields and the valence similarity with the wall of the recipient agent. The global problem of finding the pattern of most frequently used links on the entire network is adequately investigated by the *maximum-flow spanning tree* of that network. On these trees, each node is attached to the rest of the tree by its strongest link. Again, strength of a link between two agents is determined as the total amount of messages sent along that link during the simulation time.

Figure 11.10 shows these strongest links between the agents for two simulations with different external influence, i.e., with a positive ("enthusiastic") and a negative ("ashamed") input emotion. For comparison, the time series from the same simulation runs are shown in Fig. 11.8 and the patterns of visited areas in the phase space, in Fig. 11.9b, c. Obviously, the two flow patters, shown by the spanning trees for the entire network of $N = 3321$ nodes, differ considerably from each other.

Note that the spanning trees reflect the directedness of the links, i.e., an agent i may have its strongest link to j (in terms of messages exchanged), but not vice versa. The occurrence of strong hubs in the social network is common. The hubs appear in the spanning tree as the agents to whom many other agents have their strongest links.

Interestingly, in the case of the positive emotion with a high arousal, "enthusiastic", the large hubs occur along the central branch of the tree, and similarly, side branches contain smaller hubs of comparable size. On the other hand, in the case of negative emotions with a low arousal "ashamed", the tree splits in two major branches, and the hubs of different sizes appear along each branch. It should be stressed that these dynamical patterns, which emerge after long simulation time, are based on the emotional interactions between agents in the same network structure. Further simulations and a systematic analysis would be necessary to quantify differences between these patterns by suitable measures.

11.5 Conclusion

We have studied dynamics of the emotion-driven exchange of messages in online social networks with a particular emphasis on the emergence of collective behaviors of users. In the model, the users are represented by the agents, whose arousal and valence fluctuate in time, being influenced by internal and external inputs and "reactivity" of the network. In full analogy to the empirical data, a high-resolution

Fig. 11.10 Maximum-flow spanning trees for agent dynamics in our MySpace simulations with $(\tilde{v}, \tilde{a})$ fixed to the values corresponding to the state "enthusiastic" (*top*) and "ashamed" (*bottom*)

dynamics is maintained in the model. The rules of actions are motivated by realistic situations in the online social networks; the network structure and some of the parameters governing the dynamics are inferred from the dataset of MySpace dialogs.

Note that our model is directly applicable to other online social networks, which share the same interaction rules—message exchange through friend's walls—provided that the parameters are inferred from the related dataset of that network. Several other parameters, for instance decay time of emotion and network "reactivity", which can not be estimated from the available empirical data, have been kept within theoretically plausible limits. In addition, within the model we infer the *action-delay* and *circadian cycles* as generated by the real-world processes of MySpace users, which condition the pace of their actions and stepping into the world of the online social network, respectively. With this "native" set of control parameters, the simulation results enable us to derive several conclusions, in particular regarding the emotion spreading processes in MySpace, and point out the potential of the model for predicting user behaviors in hypothetical (experimental) situations.

Our main conclusions are summarized as follows:

- *Temporal correlations of user activity in* MySpace occur on long-time scale and are accompanied by high arousal and predominantly positive emotions.
- *Rhythms of users stepping from offline-to-online world* carry certain important features of the communication processes in the online social networks. Specifically, the temporal correlations in the online dynamics are built as a response to already correlated step-in processes. Otherwise, if not driven in a different way, the online social networks with their internal dynamics of the user-to-user contacts and restricted visibility of messages are not capable to generate correlations on a large temporal scale.
- *Patterns of emotion dynamics* in the online world of social networks are different for positive and for negative emotions. In the empirical data of MySpace the positive-valence emotions dominate. However, model simulations of spreading emotional states with different arousal–valence components and different social connotations, "enthusiastic" and "ashamed", for example, show different patterns in the phase space of the emotions involved as well as the social links used to spread the emotions on the network.
- *High-arousal states* in the dynamics are built on small noisy input for all initial emotion states in our simulations, which is reminiscent of "party"-like behavior of agents. According to our model this is a consequence of collective effects—repeated actions of an agent caught in the active network environment.

Quantitative comparison of the simulated data with the empirical data analysed in Šuvakov et al. (2012a) indicate similar results, for instance, the long-range correlations in the emotional time series in Fig. 11.6, and the range of the expressed valences, in Fig. 11.5. This suggests that the model of emotional agents can reproduce the stylized facts of the empirical data of MySpace dialogs, when the parameters are appropriately chosen. Moreover, within the model, genesis of the emergent behaviors—based on contributions of each user (agent) and its social connections—can be revealed. This makes the predictive value of the model. It is more subtle, however, to relate the predictions of the model which regard the individual agent's emotional state and its fluctuations with the "feelings change" observed in the psychology research of the online communications. In this respect, one can recognize that a characteristic area in the positive-valence high-arousal states recurrently being visited by the agents, may reflect the positive baselines of human valence and arousal found in Kuppens et al. (2010).

Moreover, the emergent asymmetrical V-shape patterns of Fig. 11.9 can be compared with the patterns of natural selective attention, that is often discovered in psychophysiological studies (Bradley and Lang 1999). For evolutionary reasons, humans have two modes of reaction to emotional content: appetitive and defensive motivation. Both tendencies can be seen in our simulations, opening the question of which one of them is predominant in the users' internal emotions. In this way, our agent based model provides testable hypotheses for further research, as the dynamics of the emotional reaction of users of online communities might depend on the emotional state of new members. An experimental setup, similar to the setup presented in Küster et al. (2011), can further provide a test whether the physiological reactions to new community members (arrivals) follow the patterns predicted by our model.

In conclusion, the presented agent-based simulations give a new insight into emotion dynamics in online social networks. In the model, the interaction rules, closely related to MySpace and other social network sites, take into account influence of the next-neighborhood on the agent's state—a salient feature of the online social networks, and the extended phase space where common emotions can be recognized. Hence, despite of its mathematical complexity, our model provides a "laboratory" for further experiments on the emotional agent's behavior (e.g., under different driving conditions, varied external inputs, and changed values of the parameters) and for a comparative analysis of online and offline social networks.

Acknowledgements The research leading to these results has received funding from the European Community's Seventh Framework Programme FP7-ICT-2008-3 under grant agreement n° 231323 and the project P-10044-3. B.T. is grateful for support from the national program P1-0044 of the Research Agency of the Republic of Slovenia and COST-TD1210 action. M.Š. also thanks the national research projects ON171037 and III41011 of the Republic of Serbia.

References

Ahn, J., Gobron, S., Silvestre, Q., Thalmann, D.: Asymmetrical facial expressions based on an advanced interpretation of two-dimensional Russell's emotional model. EPFL-CONF-164427. http://infoscience.epfl.ch/record/164427 (2010)

Amichai-Hamburger, Y.: Internet and personality. Comput. Hum. Behav. **18**(1), 1–10 (2002). doi:10.1016/S0747-5632(01)00034-6

Amichai-Hamburger, Y., Vinitzky, G.: Social network use and personality. Comput. Hum. Behav. **26**(6), 1289–1295 (2010). doi:10.1016/j.chb.2010.03.018

Bradley, M.M.: Natural selective attention: orienting and emotion. Psychophysiology **46**(1), 1–11 (2009). doi:10.1111/j.1469-8986.2008.00702.x

Bradley, M.M., Lang, P.J.: Affective norms for English words (ANEW): instruction manual and affective ratings. Technical Report C-1, University of Florida, Center for Research in Psychophysiology (1999)

Castellano, C., Fortunato, S., Loreto, V.: Statistical physics of social dynamics. Rev. Mod. Phys. **81**, 591–646 (2009). doi:10.1103/RevModPhys.81.591
Cheung, C.M.K., Chiu, P.Y., Lee, M.K.O.: Online social networks: why do students use facebook? Comput. Hum. Behav. **27**(4), 1337–1343 (2011). doi:10.1016/j.chb.2010.07.028
Chmiel, A., Sienkiewicz, J., Thelwall, M., Paltoglou, G., Buckley, K., Kappas, A., Hołyst, J.A.: Collective emotions online and their influence on community life. PLoS ONE **6**(7), e22207 (2011). doi:10.1371/journal.pone.0022207
Coan, J.A., Allen, J.J.B (eds.): The Handbook of Emotion Elicitation and Assessment. Series in Affective Science. Oxford University Press, Oxford (2007)
Crane, R., Schweitzer, F., Sornette, D.: New power law signature of media exposure in human response waiting time distributions. Phys. Rev. E **81**(5), 056101 (2010). doi:10.1103/PhysRevE.81.056101
DiMaggio, P., Hargittai, E., Neuman, W.R., Robinson, J.P.: Social implications of the internet. Annu. Rev. Sociol. **27**(1), 307–336 (2001). doi:10.1146/annurev.soc.27.1.307
Dodds, P., Danforth, C.: Measuring the happiness of large-scale written expression: songs, blogs, and presidents. J. Happiness Stud. **11**(4), 441–456 (2010). doi:10.1007/s10902-009-9150-9
Garas, A., Garcia, D., Skowron, M., Schweitzer, F.: Emotional persistence in online chatting communities. Sci. Rep. **2**, 402 (2012). doi:10.1038/srep00402
Garcia, D., Schweitzer, F.: Emotions in product reviews – empirics and models. In: Proceedings of 2011 IEEE International Conference on Privacy, Security, Risk, and Trust, and IEEE International Conference on Social Computing, PASSAT/SocialCom, pp. 483–488 (2011). doi:10.1109/PASSAT/SocialCom.2011.219)
Garcia, D., Garas, A., Schweitzer, F.: Positive words carry less information than negative words. EPJ Data Sci. **1**(1), 3 (2012). doi:0.1140/epjds3
Giles, M.: A world of connections - a special report on social networking. The Economist, p.16, January (2010)
Giles, J.: Social science lines up its biggest challenges. Nature **470**, 18–19 (2011). doi:0.1038/470018a
Gligorijević, V., Skowron, M., Tadić, B.: Structure and stability of online chat networks built on emotion-carrying links. Physica A **392**(3), 538–543 (2013). doi:10.1016/j.physa.2012.10.003
Guimerà, R., Danon, L., Díaz-Guilera, A., Giralt, F., Arenas, A.: Self-similar community structure in a network of human interactions. Phys. Rev. E **68**(6), 065103 (2003). doi:10.1103/PhysRevE.68.065103
Johnson, N.F., Xu, C., Zhao, Z., Ducheneaut, N., Yee, N., Tita, G., Hui, P.M.: Human group formation in online guilds and offline gangs driven by a common team dynamic. Phys. Rev. E **79**(6), 066117 (2009). doi:10.1103/PhysRevE.79.066117
Kleinberg, J.: The convergence of social and technological networks. Commun. ACM **51**(11), 66–72 (2008). doi:10.1145/1400214.1400232
Kuppens, P., Oravecz, Z., Tuerlincky, F.: Feelings change: accounting for individual differences in the temporal dynamics of affect. J. Pers. Soc. Psychol. **99**(6), 1042–60 (2010). doi:10.1037/a0020962
Küster, D., Tsankova, E., Theunis, M., Kappas, A.: Measuring cyberemotions: how do bodily responses relate to the digital world? In: 7th Conference of the Media Psychology Division of the Deutsche Gesellschaft für Psychologie, Bremen, August (2011)
Malmgren, R.D., Stouffer, C.B., Campanharo, A.S.L.O., Amaral, L.A.: On universality in human correspondence activity. Science **325**(5948), 1696–1700 (2009). doi:10.1126/science.1174562
Mitrović, M., Tadić, B.: Bloggers behavior and emergent communities in blog space. Eur. Phys. J. B **73**(2), 293–301 (2010). doi:10.1140/epjb/e2009-00431-9
Mitrović, M., Tadić, B.: Patterns of emotional blogging and emergence of communities: agent-based model on bipartite networks (2011). http://arxiv.org/abs/1110.5057
Mitrović, M., Tadić, B.: Emergence and structure of cybercommunities. In: Thai, M.M., Pardalos, P. (eds.) Handbook of Optimization in Complex Networks: Theory and Applications. Springer Optimization and its Applications, vol. 57, pp. 209–227. Springer, New York (2012a). doi:10.1007/978-1-4614-0754-6_8

Mitrović, M., Tadić, B.: Dynamics of bloggers' communities: bipartite networks from empirical data and agent-based modeling. Physica A **391**(21), 5264–5278 (2012b). doi:10.1016/j.physa.2012.06.004

Mitrović, M., Paltoglou, G., Tadić, B.: Networks and emotion-driven user communities at popular blogs. Eur. Phys. J. B **77**(4), 597–609 (2010). doi:10.1140/epjb/e2010-00279-x

Mitrović, M., Paltoglou, G., Tadić, B.: Quantitative analysis of bloggers' collective behavior powered by emotions. J. Stat. Mech. **2011**(2), P02005 (2011). doi:10.1088/1742-5468/2011/02/P02005

Paltoglou, G., Theunis, M., Kappas, A., Thelwall, M.: Predicting emotional responses to long informal text. IEEE. Trans. Affect. Comput. **4**(1), 107–115 (2013). doi:10.1109/T-AFFC.2012.26

Rimé, B.: Emotion elicits the social sharing of emotion: theory and empirical review. Emot. Rev. **1**(1), 60–85 (2009). doi:10.1177/1754073908097189

Roca, C.P., Lozano, S., Arenas, A., Sánchez, A.: Topological traps control flow on real networks: the case of coordination failures. PLoS ONE **5**(12), e15210 (2010). doi:10.1371/journal.pone.0015210

Rodgers, J.L.: The epistemology of mathematical and statistical modeling: a quiet methodological revolution. Am. Psychol. **65**(1), 1–12 (2010). doi:10.1037/a0018326

Russell, J.A.: A circumplex model of affect. J. Pers. Soc. Psychol. **39**(6), 1161–1178 (1980). doi:10.1037/h0077714

Ryan, T., Xsenos, S.: Who uses Facebook? An investigation into the relationship between the big five, shyness, narcissism, loneliness, and facebook usage. Comput. Hum. Behav. **27**(5), 1658–1664 (2011). doi:10.1016/j.chb.2011.02.004

Scherer, K.R.: What are emotions? And how can they be measured? Soc. Sci. Inf. **44**(4), 695–729 (2005). doi:10.1177/0539018405058216

Schweitzer, F.: Brownian Agents and Active Particles. Collective Dynamics in the Natural and Social Sciences, 1st edn. Springer Series in Synergetics. Springer, Berlin (2003). 10.1007/978-3-540-73845-9

Schweitzer, F., Garcia, D.: An agent-based model of collective emotions in online communities. Eur. Phys. J. B **77**(4), 533–545 (2010). doi:10.1140/epjb/e2010-00292-1

Šuvakov, M., Mitrović, M., Gligorijević, V., Tadić, B.: How the online social networks are used: dialogues-based structure of Myspace. J. R. Soc. Interface **10**(79), 20120819 (2012). doi:10.1098/rsif.2012.0819

Šuvakov, M., Garcia, D., Schweitzer, F., Tadić, B.: Agent-based simulations of emotion spreading in online social networks (2012). http://arxiv.org/abs/1205.6278

Szell, M., Thurner, S.: Measuring social dynamics in a massive multiplayer online game. Soc. Networks **32**(4), 313–329 (2010). doi:10.1016/j.socnet.2010.06.001

Szell, M., Lambiotte, R., Thurner, S.: Multirelational organization of large-scale social networks in an online world. Proc. Natl. Acad. Sci. USA **107**(31), 13636–13641 (2010). doi:10.1073/pnas.1004008107

Tadić, B.: Modeling behavior of Web users as agents with reason and sentiment. In: Kora, A.B. (ed.) Advances in Computational Modeling Research: Theory, Developments and Applications, pp. 177–186. Novapublishing, New York (2013)

Tadić, B., Šuvakov, M.: Can human-like bots control collective mood: agent-based simulations of online chats. J. Stat. Mech. Theory E **2013**(10), P10014 (2013). doi:10.1088/1742-5468/2013/10/P10014

Tadić, B., Malarz, K., Kułakowski, K.: Magnetization reversal in spin patterns with complex geometry. Phys. Rev. Lett. **94**, 137204 (2005). doi:10.1103/PhysRevLett.94.137204

Tadić, B., Rodgers, G.J., Thurner, S.: Transport on complex networks: flow, jamming and optimization. Int. J. Bifurcat. Chaos **17**(7), 2363–2385 (2007). doi:10.1142/S0218127407018452

Thelwall, M., Buckley, K., Paltoglou, G., Cai, D., Kappas, A.: Sentiment strength detection in short informal text. J. Am. Soc. Inf. Sci. Technol. **61**(12), 2544–2558 (2010). doi:10.1002/asi.21416

van Rijsbergen, C.J., Robertson, S.E., Porter, M.F.: New Models in Probabilistic Information Retrieval, Computer Laboratory. University of Cambridge, Cambridge (1980)

Part IV
Applications

Chapter 12
Does Sentiment Among Users in Online Social Networks Polarize or Balance Out? A Sociological Perspective Using Social Network Analysis

Matthias Trier and Robert Hillmann

12.1 Introduction

Users express and share sentiments electronically when they communicate within online social network applications. One way to analyze such interdependent data is focusing on the inter-user relationships by applying a *sociological perspective* based on *social network analysis*. Existing studies examined the existence or distribution of sentiments in online communication at a general level or in small observed groups. After a brief introduction into social network analysis, this chapter extends this research by studying ego-networks of focal actors (ego) and their immediate contacts in over 13,000 online social networks. Sentiment valence of all messages was determined with a trained and tested software algorithm. To explain sentiment-related patterns we draw from research on social influence and social attachment to develop theories of node polarization, balance effects and sentiment mirroring within communication dyads. Results from social network analysis support our theories and indicate that actors develop polarized sentiments towards individual peers but keep sentiment in balance on the ego-network level. Further, pairs of nodes tend to establish similar attitudes towards each other leading to stable and polarized positive or negative relationships. These results contribute to understanding the patterns of how sentiment propagates and resonates in large online groups.

M. Trier (✉)
Department of IT Management, Copenhagen Business School, Howitzvej 60, DK-2000 Frederiksberg, Denmark
e-mail: mt.itm@cbs.dk

R. Hillmann
Technische Universität Berlin, Einsteinufer 17, 10587 Berlin, Germany
e-mail: r.hillmann@tu-berlin.de

J.A. Hołyst (ed.), *Cyberemotions*, Understanding Complex Systems,
DOI 10.1007/978-3-319-43639-5_12

12.2 Sentiment in Online Communication

The fast adoption of social network applications by billions of users and the resulting digital data traces enable large-scale computer-supported analysis of human behavior in online environments. Based on this abundant data, researchers are now is a position to augment seminal research regarding the existence of sentiments or emotions in such communication channels. Rice and Love (1987) conducted an early study substantiating that computer-mediated communication does in fact allow for the exchange of emotions despite the inherent absence of non-verbal communication parts.

Subsequent research confirmed the existence of emotions and sentiments in various computer-mediated communication channels such as discussion boards, micro-blogging services and other social network applications (Belkin et al. 2006; Derks et al. 2007; Bollen et al. 2011b; Dodds and Danforth 2010; Thelwall and Wilkinson 2009). One typical tenet of subsequent research was that, compared to face-to-face setting, electronic interaction uses on few social cues and it hence takes longer as in the physical encounter to create similar interpersonal effects, such as trusted social relationships (Walther 1992).

We currently lack understanding of concrete interaction traces of micro-level interaction and the role of sentiment for triggering processes or even cascades of affective influence among users in online social networks. One way to address this gap is to adopt a social network perspective, modeling users as nodes and aggregate exchanged messages as links within complex social networks. Such a perspective enables explicating the micro-level processes of social influence that happen in actor communities and that might at least partially explain why actors (re-) act in a certain way.

In the context of this chapter, we are particularly concerned with affective influences brought about by the sentiment valence of messages to which an online actor is exposed. Such influences can bring about dissemination processes that could explain emerging effects like a main tonality of some online community or a separation of a group. This aspect of message exchange in communities can be studied by performing a classification of sentiments embedded in textual communication and by mapping their distribution onto network links. With this process we can trace the valence of unfolding relationships and actors in online social networks and the influence and exchange of sentiments. The focus of such an analysis lies on network nodes and their embeddedness within their ego-network.

An ego-network represents a subset of the complete network topology with a focal node in the center, the ego node, and all direct communication partners. The transmitted messages are analyzed regarding sentiment distribution at a local level of the relationship between communication pairs and at a global level within ego-networks.

We will now first review existing research on emotions in networks. We then introduce social network analysis, and in particular event-driven dynamic network analysis. Then we explore if there are important effects of node polarization, balance and sentiment mirroring within online communication dyads that might explain online sentiment distribution and diffusion. We can formulate the following guiding research question "What typical distributions of polarized and balanced relationships triggered by exchanged sentiments can be identified and what insights regarding human behavior in online environments can be derived?" Answers to this question can be used to gain insights about applicability of social science theories and support the understanding of communication mechanism in online environments.

12.3 Related Research to Explain Emotions in Networks

Despite the paucity of studies that focus on the impact and spread of emotions and sentiments in online networks, there has been research covering real life interactions. For example, Fowler and Christakis have identified happiness spreading effects among humans and the emergence of clusters of happy and unhappy people over time (Fowler and Christakis 2009).

Thelwall and Wilkinson analyzed the existence of sentiments and emotions in online social networking. Findings suggest that positive emotions are present in about two thirds of the analyzed comments. Negative emotion is much less present than positive emotion and is furthermore not associated with gender. It has been shown that Social Networking Sites are an emotion rich environment (Thelwall and Wilkinson 2009).

Based on the occurrence of sentiments in online networks, further research effort aimed at analyzing the consequences of affect in social networks. Bollen et al. (2011b) have addressed the research question dealing with collective emotional states and various socio-economic phenomena. Results imply that collective emotional states can be found in social networks and certain real world events have a measurable influence within social networks (Bollen et al. 2011a,b).

Research about the influence of emotions in online environments was done by Schweitzer and García (2010) who have developed an agent-based framework to model the emergence of collective emotions helping to understand and predict the emergence of collective emotions based on the interaction of agents with individual emotional states (see Chap. 10). Further research has shown the existence of emotion-driven user communities and collective emotional states and their influence on community life. Results corroborate the influence of sentiments on other nodes and the existence of emotional clusters with identical sentiment polarization (Chmiel et al. 2011; Mitrović et al. 2010) (see Chaps. 8 and 11).

12.4 Sociological Explanations of Interpersonal Relationships

For our analysis of the role of sentiment for affective social influence effects in online social networks, we can draw from a variety of academic contributions that theorized positive or negative network relationships among people (Belkin et al. 2006; Brzozowski et al. 2008; Huang 2009; Kunegis et al. 2009). Other research focused on the implications of affective polarized relationships on interpersonal relationships and emphasized the prevalence of positive or negative relationships in social networks (Labianca and Brass 2006). Such polarized relationships represent a certain actor's attitude towards a communication partner. Hence, the sentiments embedded within messages transmitted among users should exhibit a bias towards either positive or negative polarization. Based on this assumption and in addition to the one-sided view of actor's attitude, this study is focused on bi-directional network relationships by including reciprocal communication behavior and analyzing the distribution of reciprocal exchanged sentiments in communication dyads.

The interpersonal behavior theory deals with human behaviors that complement each other. The interpersonal circle by Kiesler outlines which behavioral patterns relate to each other. In the frame of his *Interpersonal Complementarity*, each action triggers a response and each response adapts to the previous action so that these will be repeated with high probability, e.g. friendliness invites friendliness (Kiesler 1983).

Emotional contagion describes the issue that human emotional states are influenced by other people through interaction and communication. People consciously and unconsciously adopt emotional states of their communication partners. Although emotional contagion theory is essentially based on the effect of interpretation of non-verbal parts of human communication which are suppressed in online communication, it has been shown that emotional contagion is also present in online environments (Hatfield et al. 1993; Belkin et al. 2006).

Miller et al. (2011) have analyzed sentiment dissemination through hyperlink networks and report that nodes are strongly influenced by their communication partners. In addition, they found that sentiment polarization is more likely to be present as the length of communication chains increases. Based on these findings regarding the influence of sentiments and the adoption of reciprocal attitudes, we can formulate:

H1	Communication partners mirror exchanged sentiments leading to reciprocal equally polarized relationships

Within the second part of the study, we further extend the analysis by looking at a node's complete set of communication partners. Our main emphasis lies on the relationship between the polarizations of individual communication links in respect to the expressed attitudes towards all communication partners especially on possible balance effects within a holistic analysis of the communication behavior.

To explain possible differences regarding the polarizations of affective ties, we can further apply attachment theory. This approach addresses the different levels of perceived importance of individual relationships between a focal actor and her contacts, e.g., the relationship between mother and their children (Cassidy and Shaver 2008). In our context, attachment theory suggests that polarization of relationships should be conceptualized as a property of the links or relationships rather than being a characteristic of the people themselves. This leads to the expectation that nodes within networks should, despite their polarized attitude towards specific communication partners, express a rather balanced pattern when looking at the whole set of all their links to contacts.

This notion of sentiment as a relational rather than a nodal attribute is in accordance with Heider's (1946) theory of social balance. It focuses on the distribution and arrangement of positive and negative relationships in social networks. Balance theory describes certain small structural patterns and common principles that lead to stable network structures on the basis of 'the friend of my friend is my friend' and 'the enemy of my friend is my enemy'. Heider's social balance theory describes the interplay of positive and negative relationships in network patterns among triplets of nodes and therefore supports expectations regarding possible balance effects within all communication links of a specific node.

Leskovec et al. (2010) have analyzed the distribution of positive and negative links within social networks. They found evidence that certain pattern configurations in accordance with social balance theory are present at a local network level and proved the applicability of Heider's theories in online environments. Based on attachment theory and social balance theory regarding local configurations of positive and negative ties, we expect a sentiment balancing effect when looking at ego's complete group of contacts. In other words, the polarized relationships of a focal actor are balancing out in ego-network (rather than being only negative or positive):

H2	Within the whole group of contacts, nodes tend to develop a balanced state regarding transmitted sentiments across all their communication links.

Social networks are known to belong to the class of scale-free networks showing scale free degree distributions which are dominated by few highly connected nodes and a majority of nodes with a small node degree (Barabási and Albert 2002; Barabási and Bonabeau 2003; Caldarelli 2007; Panzarasa et al. 2009). Hence, the communication activity of nodes within social network is unequally distributed—few nodes are responsible for the majority of network edges.

Results of our own exploratory study (manual reconstruction of ten online discussion threads) have shown, that in online discussions, topics are being discussed and people are sharing their opinions and express agreement or disagreement. Over time, implicit groups of people with same opinions are emerging, supporting each other and prolonging the discussion with people having different opinions. This mechanism was found early-on in different online discussion boards independent of the discussion lifetime and the resulting network size. Nodes with "relative" high communication activity are likely involved in different communication threads dealing with both agreement and disagreement and serving as some kind of moderator. Based on the presence of typical mixed interaction patterns indicating that nodes have the tendency to develop towards a balanced state within their personal communication network and including the communication mechanism described above, we expect a correlation of node degree and the expression of well-balanced states.

H3	Online social networks exhibit a correlation between node degree and the expression of a well-balanced state across all communication links independent of network size.

12.5 A Cross-Domain Study of 13,000 Online Social Networks

The data basis for the analysis covers online interaction data from various social network types including information from discussion forums, internet relay chats, micro-blogging services and newsgroups in the internet. The networks differ in terms of size, duration and communication intensity. Due to the abstract level of our propositions and the intention to derive insights about general communication mechanism, this data heterogeneity is appropriate.

The forum discussions were retrieved from the BBC website as well as the online portal `Digg.com`. The first dataset (DS I) covers all discussions from seven BBC message boards and covers a time span from 2005 till 2009. The Digg.com forum data (DS II) is a complete crawl of all story related discussions and covers the months February, March and April of the year 2009. The IRC dataset (DS III) includes chat interaction from the Ubuntu e-community and contains chat dialog recordings from 57 different communication channels and covers a period between summer 2005 and 2010. The micro-blogging dataset (DS IV) is based on Twitter posts from February 2010. The newsgroups dataset (DS V) includes a corpus of Usenet newsgroups with the complete posting history for several Austrian newsgroups from 1995 to summer 2010.

Each data source has a specific underlying data structure, determined by certain technical properties of the internet service in use. To overcome the disadvantages of heterogeneous source data structures, all data was transferred into a coherent event-based data model based on the approach of Trier (2008). The data model used for event-driven network analysis differs from other network data models and has a strong focus on network dynamics (Trier 2008). The data model consists of three main elements: networks, nodes and communication events, so called linkevents. The linkevents can be generally described as human interaction in social networks and are based on exchanged textual messages among users which are aggregated to network links among nodes (see Fig. 12.1).

Due to different quantities of the source data, the number of network entries per data set differs (see Table 12.1).

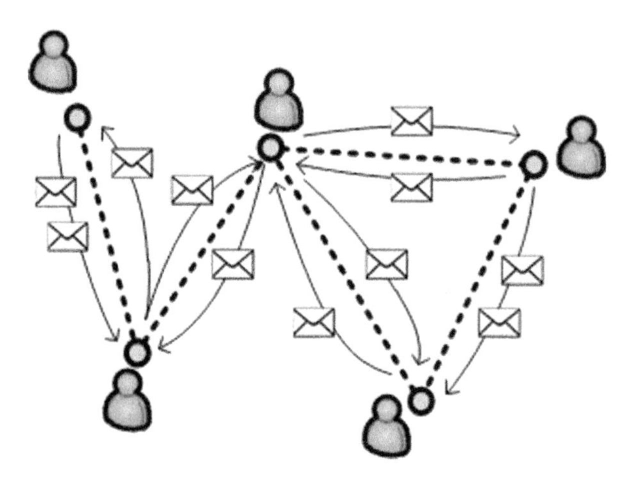

Fig. 12.1 Network links as aggregations of messages (as linkevents)

Table 12.1 Number of networks in database

Dataset	Number of networks
Forum dataset I	1657
Forum dataset II	8576
Chat dataset III	1925
Micro-blogging dataset IV	628
Newsgroup dataset V	1001

12.6 A Sociological Perspective on Social Network Analysis (SNA)

In the field of sociology, analytical methods of SNA can be traced by to the development of the sociogram by Moreno in the 1930s (Moreno 1934). SNA is applied for the quantified study of structural actor patterns. The method focuses on properties of network structure, which is generally defined as a set of actors connected via social relations (Wasserman and Faust 1994). Basic indicators for the description of networks, such as density, centrality, or diameter, are now firmly established in the network analyst toolset, which since the 1970s have been adopted in many other academic fields. Network metrics are further utilized in numerous studies that examine the relationships between network embeddedness of actors with external factors, e.g., the role of actor centrality for managerial performance in organizations (Brass 1984).

Conventional SNA analysis is based on data from a single cumulative collection of network data on a defined date (Wasserman and Faust 1994). In this case, studies hence evaluate the cumulative final structure displayed in static graphs. This structure-oriented approach is implicitly assuming that the investigated network is in equilibrium with continuous processes and thus the measured state is representative for the whole time period. These balance and homogeneity assumptions are not always correct and simplify the underlying unobserved network dynamics. This is particularly the case for online interaction which is subject to high volatility and fast dynamics (Trier 2008). The most central actors can quickly change and a summative figure is not likely to be representative.

Therefore, for our concern with online communication in the internet, analysis methods and theories that explicitly focus on the dynamics of networking processes and its stakeholders is a prerequisite for a more realistic and more detailed knowledge about social networks. In the description of methods for analysis of network dynamics, it is important to note, that the tools of the network analysts do not clearly distinguish between statics and dynamics. Rather, there is a transition, as many dynamic networking processes mirror the underlying static structures. In a developed friendship network, for example, there is an increasing occurrence of reciprocal relations in the network structure to be expected because, over time, the social processes are favored compared to alternative configurations. Often, different dynamic processes lead to the same structure (Trier 2008). In these cases, any cumulative snapshot of an evolving structure is not enabling clear conclusions on a possible dynamic process.

Approaches such as timeframe-based or event-driven approaches of the SNA capture small network periods over time and collect samples from the dynamic changes. These approaches are presented in detail below.

12.7 Dynamic Event-Driven Network Analysis

Relating to Doreian's and Stokman's (1997, p. 3) definition of a network process as a "series of events that create, sustain, and dissolve social structures", event-based network analysis disaggregates the network's relationships into smaller units, i.e. relational events. This approach has been designed for symmetric as well as asymmetric graphs derived from any kind of network data. The data model consists of a collection of actors (nodes), a collection of timed events, and a collection of timed relationships. This allows for more accurate dynamic graph visualization as well as measurement of network evolution in domains, where the overall relationship among actors is related to some underlying time-dependent unit, e.g. interaction frequency in online social networking (also cf. Trier 2008).

Examples for novel research applications based on event-based analysis include the visualization and analysis of group formation and stabilization over time, of actor's paths to central positions, or of process oriented activity patterns with a structural impact on the network. Explicit recognition of relational events is further able to capture the growth of relationships and the network's reaction to external events. Generally, the method provides multiple integrated levels of analysis by linking actor attributes (e.g., types), actors' activity patterns, to the resulting general network structure. By that, changes of the network can be tracked and traced back to individual sequences of networking events of participating actors. With this new method, networks become less a static phenomenon but can be perceived as a versatile structure in constant change and motion.

As individual events are the smallest unit of computation, event-based network analysis is utilized best with empirical network data that contains massive timed events. Such data can be retrieved from various interaction networks emerging from electronic communication, e.g., e-mail among employees. The comprehensive data model allows for capturing complete sequences of networking history over time. The intensity of the interaction and its relatedness to a defined context help to ensure that the network structure corresponds with real interaction and social networks.

Relational events can be different types of network activity including communication flow, information exchange, collaboration, document sharing, or status change. On a more abstract level, three types of events can be distinguished: timed interaction events, timed action events, and timed reaction events.

A timed interaction event $e_t(a_i, a_j)$ relates to an activity of two actors a_i and a_j that is either performed together, e.g. collaborating on a shared document, or performed by one actor with some effect on the other actor, e.g. writing an email. A timed action event $e_t(a_i)$ is an event performed by one actor without directly involving another actor. This relates to posting a message in a newsgroup with no reference to another posting, that is creating a new discussion thread.

A timed reaction event by actor a_i on actor a_j is then an activity related to some prior interaction event, denoted as $e_{t_2}(a_i, e_{t_1}(a_j, a_k))$, or to some prior action event, denoted as $e_{t_2}(a_i, e_{t_1}(a_j))$. An example for this is the citation of a co-authored paper or responding to some posting. The condition $t_1 < t_2$ demands that the referenced event has to be prior to the referencing event.

All events are either monadic (action events) or dyadic (interaction and reaction events). If it is necessary for the purpose of the analysis, several events can be aggregated to a group if they relate to the same context. For example, an email sent to more than one recipient will be designed as several dyadic events, $e_t^g(a_i, a_j), e_t^g(a_i, a_k), \ldots, e_t^g(a_j, a_n)$ that can be grouped. The group is then indicated by an index$g$. Interaction events with several initiators and reaction events referring to more than one action or interaction event and any combination of these examples can be designed as group events as well.

As timestamps can distinguish events by milliseconds it will rarely ever happen that two events involving the same actors occur exactly at the same time. However, if it is necessary one can number the events by an additional index.

Depending on the type of network the relationship between two actors will be established from either one type of events or a combination of them. Timed relationships include only events from a fixed period of time. They can be formalized as:

$$
\begin{aligned}
r_{t,t'}(a_i, a_j) = \{ & e_{t_1}, e_{t_2} \in E : e_{t_1}\left(a_i, a_j\right) \vee e_{t_2}\left(a_i, e_{t_1}\left(a_j\right)\right) \\
& \vee e_{r,t_2}\left(a_i, e_{t_2}\left(a_j, a_k\right)\right), t \leq t_1 < t_2 \leq t' \}
\end{aligned}
$$

Self-events with $i = j$ and therefore self-relationships are possible. With symmetric data, relationship $r_{t,t'}(a_i, a_j)$ and relationship $r_{t,t'}(a_j, a_i)$ with $i \neq j$ will not be distinguished in the graph and thus can be consolidated into one single relationship.

Each actor and each event can have properties, e.g. types of messages or ranks of actors. Relationships aggregate the properties of the events by some function, like the sum, the average, or the maximum. The properties of an actor can depend on the actor himself, e.g. gender, or name, or be related to the context of the network, e.g. position in the organizational hierarchy. Moreover, some properties can be derived from the position of the actor in the network, e.g. measures of centrality. This holds for event properties and relation properties as well.

If each property can be coded by a numerical value an actor a_i can then be described with a d-dimensional property vector $\mathbf{a}_i = (a_{i1}, a_{i2}, \ldots, a_{id})^T$. The same holds with events and relationships. The following Table 12.2 illustrates how the different event types can be applied.

In summary, the resulting data model of event-based dynamic network analysis consists of actors, actor properties, events, and event properties. For each event several properties are captured. For example, the time stamp of each message event is recorded as a message property. Hence, the sequence of messages and the change in relationship structure or strength is represented as a series of relational events in the data model.

Table 12.2 Examples of different types of events

Network	Actor	Interaction event	Action event	Reaction event
Email exchange	Any person identified by email address	Email to one or more recipients	–	–
Newsgroup	Any person identified by user account	–	Posting new message (unrelated)	Posting new message (related to a prior posting)
Citation	Any person identified by full name	Paper with several authors (co-authorship)	Paper with one author	Citation of any paper

12.8 Commetrix as a Software-Support for Event-Driven Dynamic Network Analysis

The above method of modeling and measuring dynamic network properties has been implemented using the dynamic network analysis software framework Commetrix.[1] It interprets the above data models and manages the massive set of computations for a large number of time-windows, nodes, and shortest paths. It allows for computing time window measures and additionally provides very sophisticated functionality for animating the community evolvement as an evolving graph to visually inspect the actors' activities. This helps to actually represent and visually trace change in a network and adds additional insight to the quantitative results. Complex options for time, actor, and relationship filtering help to select relevant actors, relationships or time periods. For any time period and for any selection, typical social network measures can be computed, analyzed and exported. With this process, we derived weekly measurements of the evolving network and of the environments (i.e., ego-networks) of identified actors with high centrality and high impact on the final network structure.

As in the sociogram (i.e. the social network graph), the software visualizes actors with nodes and edges represent the relationships as flexible aggregations of message events. The sociogram has been extended with additional means for information visualization and with the capability to adapt to longitudinal network change. These additions yield a dynamic graph termed 'communigraph'. Utilizing Bertin's (1983) concept of visual variables to encode information, properties can be visualized by label, node size, node color (brightness, transparency), or a number of rings around the node. Relationship properties are graphically represented using colors, thickness, length, and labels.

[1] cf. http://www.commetrix.net.

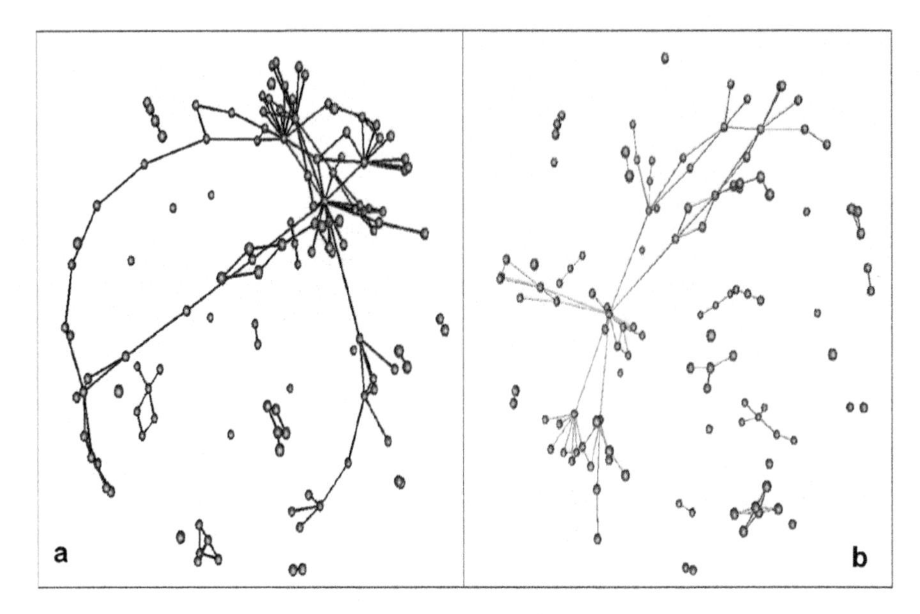

Fig. 12.2 Visualizations of the negative (**a**) versus the positive (**b**) subnetworks of an online discussion in `Digg.com`

For the analysis of sentiment dissemination in online interaction networks, a special viewer has been built, based on the Commetrix framework (available from the authors of this chapter). Figure 12.2 shows a screenshot of the network visualization provided by the Commetrix framework.

12.9 Analysis with Event-Driven Network Analysis

The event-based data model allows for the detailed reconstruction of online social networks. The methodology described below is tailored to this data model utilizing the aggregation of single messages to network links to enable a fine grained analysis of the relationship of sentiments embedded in textual messages and communication links.

As a first step, the exchanged linkevents in all networks have been analyzed regarding their embedded sentiments. That was done with the specialized sentiment detection software *SentiStrength* (introduced in a separate chapter in this book). The used classifiers represent the state-of-the-art in sentiment classification and were developed as part of the European FP7 project "Collective Emotions in Cyberspace" (Paltoglou et al. 2010; Thelwall et al. 2010, 2012). The classification consists of two phases. In the training phase, the classifiers are using a set of human annotated documents from which the algorithms learn the characteristics of the sentiments. In the second phase, the software classified the content of each linkevent regarding the

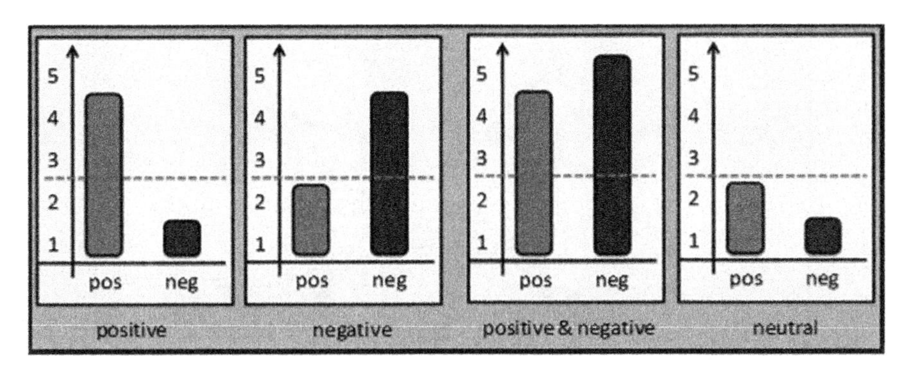

Fig. 12.3 Transformation of sentiment data into four disjoint categories. *Horizontal dashed line* marks the threshold level

sentiments of the text. Each analyzed message is rated regarding included positive and negative sentiments with a scale ranging from 1 to 5 for positive and (−1)–(−5) for negative sentiments. To overcome some limitations regarding the accuracy of the classification results (Thelwall et al. 2012), the sentiment spectrum of the classifier was converted into four disjoint categories according to a specified scheme (see Fig. 12.3).

The basic schema differentiates four disjoint categories of sentiment classification. Every textual message can either be of positive, negative, both positive and negative or neutral nature. Due to the computational reason that each message contains at least a sentiment value of one, a threshold level is defined. A classification result with a positive result above the threshold of three and a negative result of less than three is categorized as positive and vice versa. If both the positive and the negative results are above the threshold level, the message is categorized as both positive and negative. If both results are below the threshold level, the linkevent is classified as neutral expressing the fact, that the text contains neither positive nor negative sentiments at a meaningful level.

This analysis is focused on influence and dissemination of sentiments and resulting polarized links within ego-networks. An ego-network represents a subset of the network topology and is defined as a node in the centre, the ego node, and all of its communication partners while the rest of the network is suppressed (see Fig. 12.4).

Each of the aforementioned hypotheses H1–H3 is addressed with a specific analysis part. To derive meaningful statements about link and ego-network polarization, certain minimum requirements for links among nodes and ego-networks are defined which determine the interpretability of the data. Ego-nodes have to transmit a minimum of four messages towards a specific node to create an adequate link due to the four different sentiment types available. Nodes must have more than three different communication partners to represent a meaningful ego-network. Both requirements are necessary to exclude trivial cases of polarization effects.

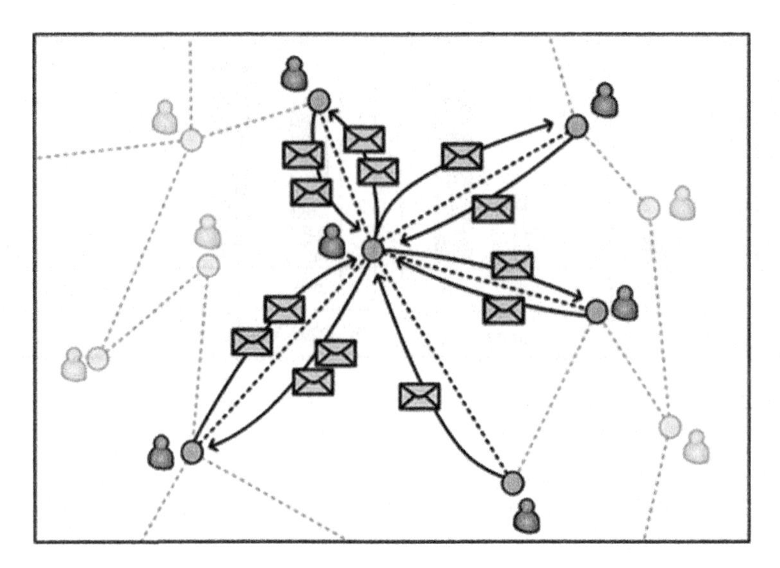

Fig. 12.4 Ego-network extraction

As a preliminary step, each of the outgoing network links within ego-networks is analysed regarding a possible link polarization. The following formula (12.1) is used to determine the polarization of nodes towards communication partners on link i. Hence, only messages transmitted from the ego-node across link i are taken into account.

$$pol_i = \frac{\#messages_{positive} - \#messages_{negative}}{\#messages_{total}} \tag{12.1}$$

The number of positive and negative messages is subtracted and the result is divided by the total number of messages. If the annotation algorithm found both positive and negative sentiment in messages (see Fig. 12.5 for the percentage share in the different datasets) then these were excluded, as they do not contribute to our analysis. The distribution of the results (values between -1 and $+1$) are sorted into 0.1 intervals.

The analysis regarding H1 is focussed on polarization effects of reciprocal connections. The polarization of the ego-node n towards the communication partner m is determined together with the opposite attitude of node m towards node n with formula (12.1). The absolute difference (distance) of both polarization values is computed and the distribution is determined for intervals between exact match (0) and maximal opposite polarization (2).

Regarding H2 and H3, analyzing balance effects within ego-networks and correlation between node degree and ego-network polarization, the same formula (12.1) is used. However, all messages transmitted from the ego-node to all communication partners are simultaneously incorporated in the analysis.

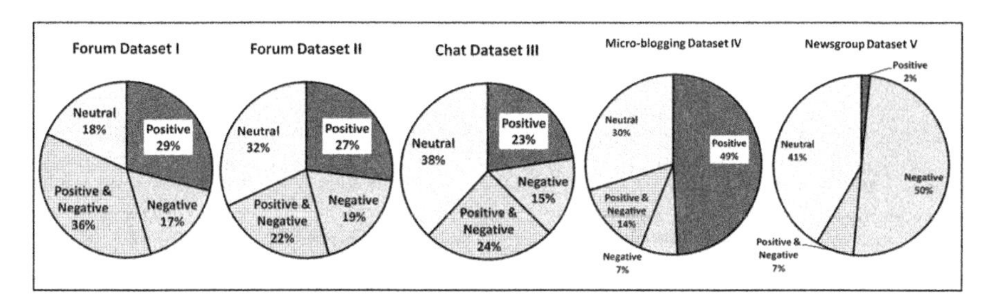

Fig. 12.5 Sentiment distribution

For the analysis of H3, the out-degree of all nodes is computed using an adjacency matrix. Due to the heterogeneous quantitative characteristics of social networks, the degree threshold is calculated for each network separately. The 50 % best connected nodes within networks are interpreted as having a "high" node degree; all other nodes are interpreted as having a "low" node degree. For each node, the two features node degree and ego-network polarization are determined and stored in a 2-dim matrix to perform a chi-square test.

The aforementioned calculations are repeated for the five datasets including all networks and nodes together with their ego-networks under consideration of the quantitative requirements mentioned above.

12.10 Polarizing or Balancing Out?

As a preliminary step we have analyzed the descriptive sentiment distribution within the datasets (see Fig. 12.5). The datasets differ in terms of their sentiment distribution. The positive fraction ranges from 2 to 49 %. The highest share is within the micro-blogging dataset. The negative sentiment linkevents have a share from 7 to 50 % with a peak at the newsgroup dataset. The fraction of both positive and negative linkevents ranges from 7 % in the newsgroup dataset to 36 % in the forum dataset 1. The percentage of neutral messages has the lowest value in forum dataset 1 and the highest proportion in the newsgroup dataset.

Figure 12.6 includes the results of the analysis regarding the share of polarized links within the data. The distributions differ among the five datasets. The highest share can be found within the micro-blogging dataset, the lowest part is present in the chat dataset. The distribution of polarized links is separated according to 0.1 range intervals from strong negative polarization (-1) to strong positive polarization ($+1$) as a result of formula (12.1). Except the chat dataset, a tendency towards polarization having peaks at the positive and negative extremes can be seen. The link-based polarizations are not normal-distributed exhibiting a U or W shape.

Figure 12.7 includes an overview about the results that focus on sentiment mirroring effects. As already mentioned above, this part is focused on reciprocal

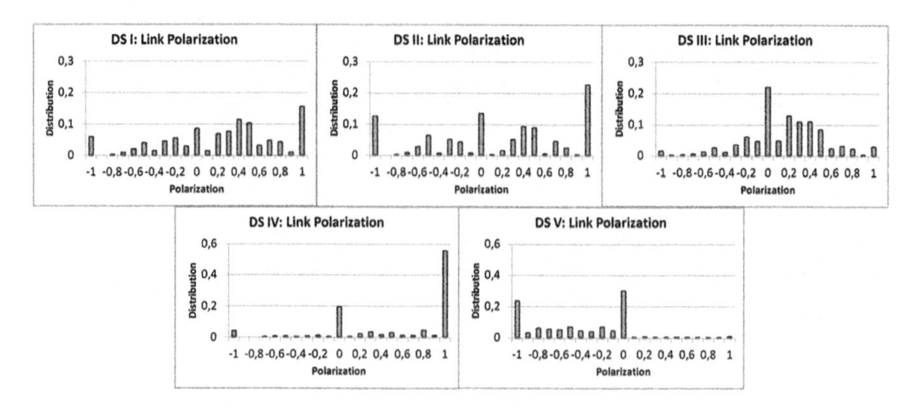

Fig. 12.6 Link-polarization distribution

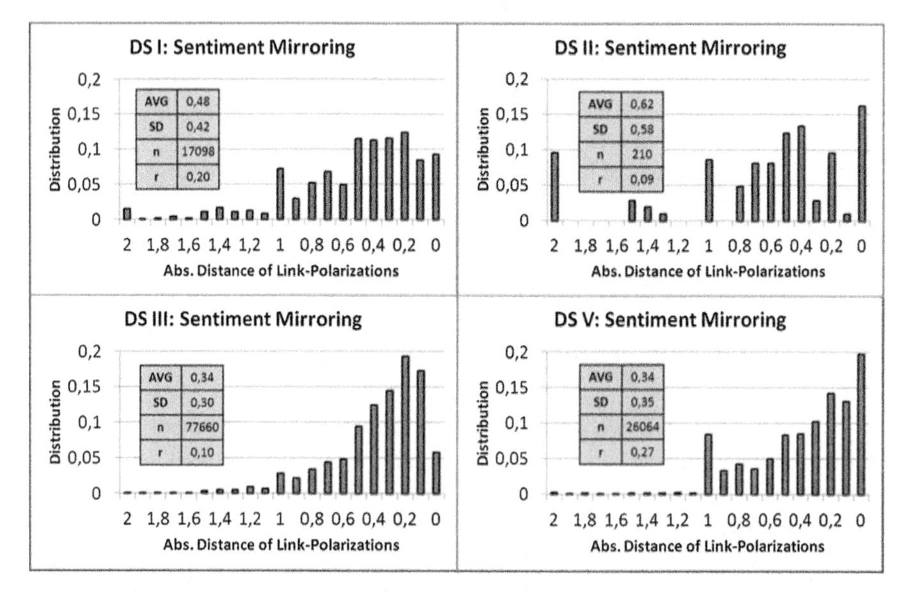

Fig. 12.7 Sentiment mirroring effects

links. The polarization value is calculated for both directions separately. The distance between both values is calculated and presented together with the number of samples, the average distance, standard derivation and the correlation coefficient (DS IV could not be used due to insufficient number of valid results).

Results exhibit correlations regarding the exchanged sentiments on reciprocal network links. The average distance ranges from 0.34 to 0.62. The majority of network relationships only exhibit a small distance between both polarization values. Between 57 and 78 % of all analyzed links express reciprocal similar sentiments with results = 0.6. H3 predicting equally polarized network links can therefore be supported.

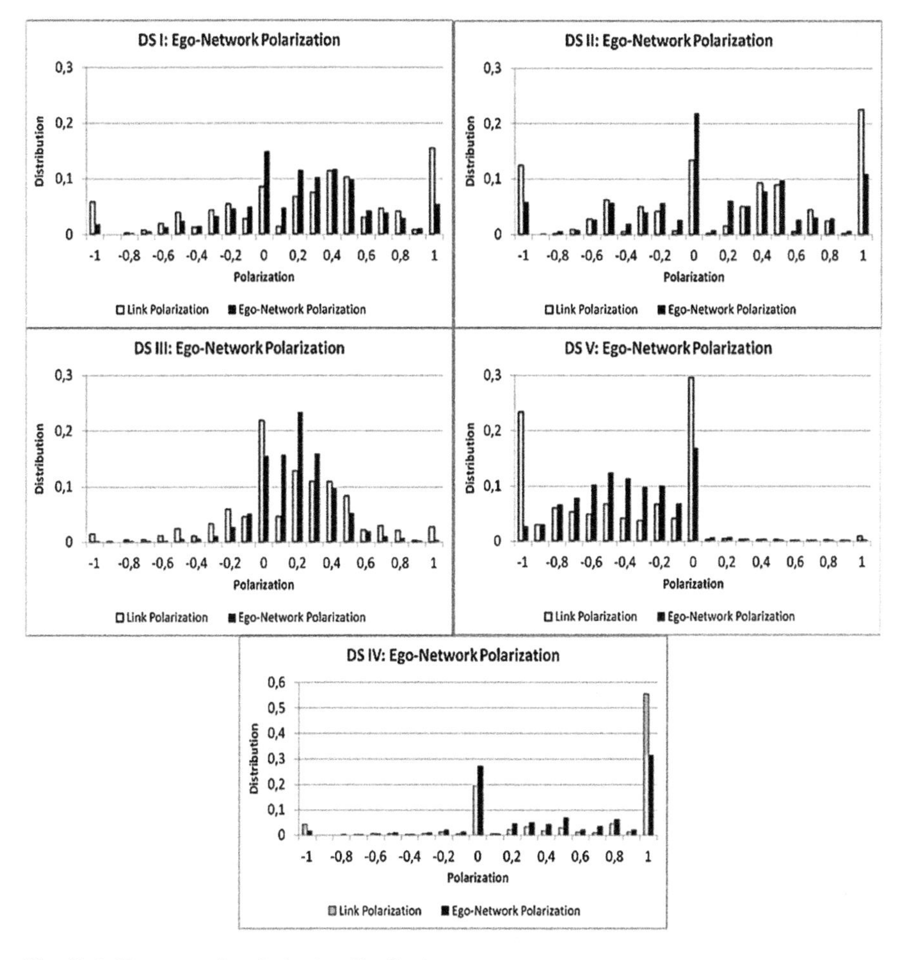

Fig. 12.8 Ego-network polarization distribution

Figure 12.8 shows the polarization results for both ego-networks (black bars) and individual links (grey bars). The bars represent the percentage share of polarizations according to formula (12.1) separated by 0.1 intervals from (-1) to $(+1)$.

Compared with the link polarization (grey), a tendency towards balanced ego-node states can be seen. Except the dataset IV, the graphs do not exhibit a U or W shape but instead having a peak in the middle expressing a slightly positive tendency. In case of the newsgroup dataset, the peak is slightly negative between $I = [-0.6, -0.4]$ which is in accordance with the sentiment distribution of the population with a strong share of negative sentiment. In comparison with the link polarization results, ego-networks exhibit a tendency towards a well-balanced state regarding expressed sentiments supporting H2.

Table 12.3 Correlation between node degree and polarization: datasets I–V

Data	Degree	Polarization			
		Low[a]	Medium[b]	High[c]	Σ
Dataset I*	Low	1910	1515	1053	4478
	High	4446	2343	620	7409
	Σ	6356	3858	1673	11,887
Dataset II*	Low	2075	1856	2071	6002
	High	7319	5121	2932	15,372
	Σ	9394	6977	5003	21,374
Dataset III*	Low	6249	2338	400	8987
	High	9907	986	47	10,940
	Σ	16,156	3324	447	19,927
Dataset IV**	Low	45	20	60	125
	High	227	68	223	518
	Σ	272	88	283	643
Dataset V*	Low	1901	1550	1382	4833
	High	2036	2622	1262	5920
	Σ	3937	4172	2644	10,753

*Statistical significance $p < 0.01$, **Statistical significance $p = 0.27$
[a] $(p < 0.33)$
[b] $(0.33 \leq p < 0.66)$
[c] $(0.66 \leq p)$

Table 12.3 conveys the results of the analysis regarding H3. The analyzed ego-networks have been categorized according to their outgoing node degree and their ego-network polarization. The absolute polarization values are sorted according to three intervals from low to high.

Results exhibit a significant correlation between node degree and ego-network polarization supporting H3. Except dataset IV, the independence of both features needs to be rejected. The distributions exhibit a correlation showing low polarization in combination with high-node degree. Due to a large scattering of network sizes used for this analysis, the effect is independent of specific network size or node degree levels.

12.11 It Is Not the Actor—It Is the Relation

Our data suggests the existence of links with polarized (positive or negative) sentiments. However, a large part of network links express a balanced state, either with the exchange of an equal amount of positive and negative or neutral messages. One can conclude that people participating in online networks either do take a certain positive or negative attitude towards people or maintain a rather balanced position towards their communication partners. Our theories predicting polarized relationships among users of social interaction can be supported by analyzing sentiments embedded within exchanged messages. However, subject to the specific context, neutral or balanced relationships are also present in social networks.

Our research extended the one-sided view of nodes in ego-networks and their outgoing communication by focusing on bi-directional relationships. Results show that communication partners are likely to respond in a similar way expressing the same attitude towards the ego-node. Social networks exhibit a certain effect to establish links with a corresponding view or attitude between both users leading to reciprocal identical polarized relationships.

In contrast to the presence of polarized network links, sentiments in ego-networks developed towards a balanced distribution. A polarized attitude of users towards other nodes can therefore be seen as a property of the relationships themselves rather than as a node attribute. The existence of polarized users in general cannot be supported. Instead, the users express a balanced state at a global level together with specific positive and negative relationships at the local level within dyads. Within the third part of the analysis, a significant correlation between node degree and the expression of a well-balanced state can be found indicating a certain moderating role of central actors independent of network size or communication activity.

12.12 Conclusions

This empirical study focused on identifying and explaining patterns of polarization effects in online social networks. The comprehensive data basis allows for the discovery of general regularities independent of specific communication types or application contexts. Our results support the role of previous social science theories in online network environments and serve as a basis for future research.

Our data supported the existence of polarized network links, balance tendencies within ego-networks and sentiment mirroring effects in bi-directional communication. In accordance with a variety of social science studies, relationships with an expression of positive or negative attitudes can be found by applying a sentiment analysis of message exchange. This polarization can be seen as an attribute of the network links rather than being an actor property due to balanced sentiment distributions within ego-networks. A significant correlation between node degree and the expression of a well-balanced state is found emphasizing a moderating role of central actors.

Despite this local polarization compared to global balance and in accordance with Kiesler's *Interpersonal Complementarity Theory* and *Emotional Contagion* theory, users are likely to mirror exchanged sentiments in reciprocal message exchange leading to stable and polarized relationships.

This study analyzed sentiments in messages without information about the actual content of the communication. Further research could extend this scope by integrating a content perspective and combining sentiment with content or topic dissemination. The analysis scope could be further extended from links to network motifs and the influence of social network effects such as reciprocity or transitivity in social networks.

Acknowledgements This work was supported by EU FP7; Theme 3: Science of complex systems for socially intelligent ICT: Project Collective Emotions in Cyberspace—CYBEREMOTIONS.

References

Barabási, A.-L., Albert, R.: Statistical mechanics of complex networks. Rev. Mod. Phys. **74**(1), 47 (2002). doi:10.1103/RevModPhys.74.47

Barabási, A.-L., Bonabeau, E.: Scale-free networks. Sci. Am. **288**, 50–59 (2003). doi:10.1038/scientificamerican0503-60

Belkin, L.Y., Kurtzberg, T.R., Naquin, C.E.: Emotional contagion in the online environment: investigating the dynamics and implications of emotional encounters in mixed-motive situations in the electronic context. Presentation at 19th International Association of Conflict Management Annual Meeting, Montreal, Canada, June (2006). doi:10.2139/ssrn.913774. Available from http://ssrn.com/abstract=913774

Bertin, J.: Semiology of Graphics. The University of Wisconsin Press, Madison (1983)

Bollen, J., Gonçalves, B., Ruan, G., Mao, H.: Happiness is assortative in online social networks. Artif. Life **17**(3), 237–251 (2011a). doi:10.1162/artl_a _00034

Bollen, J., Mao, H., Pepe, A.: Modeling public mood and emotion: twitter sentiment and socioeconomic phenomena. In: Adamic, L.A., Baeza-Yates, R.A., Counts, S. (eds.) Proceedings of the Fifth International AAAI Conference on Weblogs and Social Media. The AAAI Press, Menlo Park, CA (2011b). Also arXiv:0911.1583

Brass, D.J.: Being in the right place: a structural analysis of individual influence in an organization. Adm. Sci. Q. **29**(4), 518–539 (1984). doi:10.2307/2392937

Brzozowski, M.J., Hogg, T., Szabo, G.: Friends and foes: ideological social networking. In: CHI '08 Proceedings of the SIGCHI Conference on Human Factors in Computing Systems, pp. 817–820. ACM, New York (2008). doi:10.1145/1357054.1357183

Caldarelli, G.: Scale-Free Networks: Complex Webs in Nature and Technology. Oxford University Press, Oxford (2007)

Cassidy, J., Shaver, P.R. (eds.): Handbook of Attachment - Theory, Research and Clinical Applications. The Guilford Press, New York (2008)

Chmiel, A., Sienkiewicz, J., Thelwall, M., Paltoglou, G., Buckley, K., Kappas, A., Hołyst, J.A.: Collective emotions online and their influence on community life. PLoS ONE **6**(7), e22207 (2011). doi:10.1371/journal.pone.0022207

Derks, D., Fischer, A.H., Bos, A.E.R.: The role of emotion in computer-mediated communication: a review. Comput. Hum. Behav. **24**(3), 766–785 (2007). doi:10.1016/j.chb.2007.04.004

Dodds, P., Danforth, C.: Measuring the happiness of large-scale written expression: songs, blogs, and presidents. J. Happiness Stud. **11**(4), 441–456 (2010). doi:10.1007/s10902-009-9150-9

Doreian, P., Stokman, F.N.: The dynamics and evolution of social networks. In: Doreian, P., Stokman, F.N. (eds.) Evolution of Social Networks, pp. 1–17. Gordon & Breachp, New York (1997)

Fowler, J.H., Christakis, N.A.: Dynamic spread of happiness in a large social network: longitudinal analysis of the Framingham Heart Study social network. Br. Med. J. **338** (2009). doi:10.1136/bmj.a2338

Hatfield, E., Cacioppo, J.T., Rapson, R.L.: Emotional contagion. Curr. Dir. Psychol. Sci. **2**(3), 96–99 (1993). doi:10.1111/1467-8721.ep10770953

Heider, F.: Attitudes and cognitive organization. J. Psychol. **21**, 107–112 (1946). doi:10.1080/00223980.1946.9917275

Huang, M.: A conceptual framework of the effects of positive affect and affective relationships on group knowledge networks. Small Group Res. **40**(3), 323–346 (2009). doi:10.1177/1046496409332441

Kiesler, D.J.: The 1982 interpersonal circle: a taxonomy for complementarity in human transactions. Psychol. Rev. **90**(3), 185–214 (1983). doi:10.1037/0033-295X.90.3.185

Kunegis, J., Lommatzsch, A., Bauckhage, C.: The slashdot zoo: mining a social network with negative edges. In: Proceedings of the 18th International Conference on World Wide Web, pp. 741–750. ACM, New York (2009). doi:10.1145/1526709.1526809

Labianca, G., Brass, D.J.: Exploring the social ledger: negative relationships and negative asymmetry in social networks in organizations. Acad. Manag. Rev. **31**(3), 596–614 (2006). doi:10.5465/AMR.2006.21318920

Leskovec. J., Huttenlocher, D., Kleinberg, J.: Predicting positive and negative links in online social networks. In: Proceedings of the 19th International Conference on World Wide Web, pp. 641–650. ACM, New York (2010). doi:10.1145/1772690.1772756

Miller, M., Sathi, C., Wiesenthal, D., Leskovec, J., Potts, C.: Sentiment flow through hyperlink networks. In: Adamic, L.A, Baeza-Yates, R.A., Counts, S. (eds.) Proceedings of the Fifth International AAAI Conference on Weblogs and Social Media. The AAAI Press, Menlo Park, CA (2011)

Mitrović, M., Paltoglou, G., Tadić, B.: Networks and emotion-driven user communities at popular blogs. Eur. Phys. J. B **77**(4), 597–609 (2010). doi:10.1140/epjb/e2010-00279-x

Moreno, J.L.: Who Shall Survive? Nervous and Mental Disease Publishing Company, Washington, DC (1934)

Paltoglou, G., Gobron, S., Skowron, M., Thelwall, M., Thalmann, D.: Sentiment analysis of informal textual communication in cyberspace. LNCS State-of-the-Art Survey, pp. 13–25 (2010)

Panzarasa, P., Opsahl, T., Carley, K.M.: Patterns and dynamics of users' behavior and interaction: network analysis of an online community. J. Am. Soc. Inf. Sci. Technol. **60**(5), 911–932 (2009). doi:10.1002/asi.21015

Rice. R.E., Love, G.: Electronic emotion: socioemotional content in a computer-mediated communication network. Commun. Res. **14**(1), 85–108 (1987). doi:10.1177/009365087014001005

Schweitzer, F., García, D.: An agent-based model of collective emotions in online communities. Eur. Phys. J. B **77**(4), 533–545 (2010). doi:10.1140/epjb/e2010-00292-1

Thelwall, M., Wilkinson, D.: Data mining emotion in social network communication: gender differences in MySpace. J. Am. Soc. Inf. Sci. Technol. **61**(1), 190–199 (2009). 10.1002/asi.21180

Thelwall, M., Buckley, K., Paltoglou, G., Cai, D., Kappas, A.: Sentiment in short strength detection informal text. J. Am. Soc. Inf. Sci. Technol. **61**(12), 2544–2558 (2010). doi: 10.1002/asi.21416

Thelwall, M., Buckley, K., Paltoglou, G.: Sentiment strength detection for the social Web. J. Am. Soc. Inf. Sci. Technol. **63**(1), 163–173 (2012). doi:10.1002/asi.21662

Trier, M.: Towards dynamic visualization for understanding evolution of digital communication networks. Inf. Syst. Res. **19**(3), 335–350 (2008)

Walther, J.B.: Interpersonal effects in computer-mediated interaction: a relational perspective. Commun. Res. **19**(1), 52–90 (1992). doi:10.1177/009365092019001003

Wasserman, S., Faust, K.: Social Network Analysis: Methods and Applications. Cambridge University Press, Cambridge (1994)

Chapter 13
Towards the Instantaneous Expression of Emotions with Avatars

Ronan Boulic, Junghyun Ahn, Stéphane Gobron, Nan Wang, Quentin Silvestre, and Daniel Thalmann

13.1 Introduction

In the present chapter we describe how to provide an instantaneous visual feedback that faithfully conveys the emotions detected in a conversational text exchange as it happens over time, e.g. through a chat system. Indeed, many on-line communities still rely on text messages as their privileged medium of communication and we believe it will remain so even if it can be complemented by images and video. This is mainly due to the effectiveness of the written form of language for conveying meaning with a small number of signs hence making it the best choice for establishing a fluid dialog among remotely connected users. However, despite being such dominant means of communication, text exchanges filter out the large range of non-verbal communication channels used by human beings (Morris 1978). As a consequence it can introduce ambiguity in case the text content lacks redundancy

R. Boulic (✉) • N. Wang
Immersive Interaction Group, Ecole Polytechnique Fédérale de Lausanne, Switzerland
e-mail: Ronan.Boulic@epfl.ch; Nan.Wang@epfl.ch

J. Ahn
Technicolor, Rennes, Cesson-Sévigné, France
e-mail: junghyun.ahn@technicolor.com

S. Gobron
Haute Ecole ARC/HES-SO, Neuchatel, Switzerland
e-mail: stephane.gobron@gmail.com; stephane.gobron@he-arc.ch

Q. Silvestre
Genview, Lausanne, Switzerland
e-mail: quentin.silvestre@gmail.com

D. Thalmann
Institute for Media Innovation, Nanyang Technological University, Singapore
e-mail: danielthalmann@ntu.edu.sg; daniel.thalmann@epfl.ch

J.A. Hołyst (ed.), *Cyberemotions*, Understanding Complex Systems,
DOI 10.1007/978-3-319-43639-5_13

about the expressed meaning. Therefore we advocate the use of a complementary visual communication channel in the form of an animated 3D avatar to convey the potential emotions carried by the text messages. This choice is motivated by the observation from Darwin himself (Darwin 1872) that human beings and animals (unconsciously) communicate their current state of mind through facial expressions and body gestures. Since that early contribution, it has been shown that facial expression plays an important role in the process of empathy (Preston and de Waal 2002), and emotional contagion (Hatfield et al. 1993), i.e. unconscious sharing of the emotions of conversation members. Nevertheless, the exact relationship between non-verbal communication and emotions is still difficult to define even by specialists (Ekman 2004; Kappas 2010). In our 3D-chatting context Non-Verbal Communication (NVC) encompasses the speechless process of communication that consists of the following animation components: gaze, facial expressions, head and body orientation, gestures and idle movements (Gobron et al. 2012). It is critical that this visual feedback update is automated for two reasons. First, as already noted by earlier authors, NVC is mostly an unconscious process; it would be a heavy cognitive burden for users to explicitly define the current facial and body expression again and again over time. Second, such an additional task would request some time and introduce an intrinsic delay in the interaction, hence disturbing the communication dynamics (Gobron et al. 2011). For these reasons the main purpose of our contributions is to provide an automated mapping of an emotion model to real-time facial expressions and body movements (Fig. 13.1).

The next section recalls the key references in the synthesis of NVC for conversational agents and avatars. It is followed by a presentation of our first approach relying on a 2D Valence-Arousal emotion model and its evaluation through a

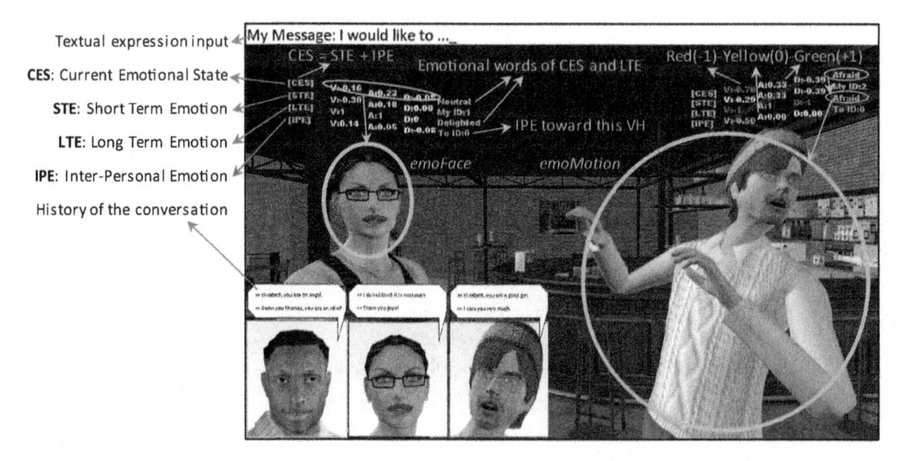

Fig. 13.1 Displaying non-verbal facial expressions and body movement to convey emotions detected from the text of the messages with the SentiStrength tool (Thelwall et al. 2010) [Chap. 7]. The short and long term temporal dimension of the emotion display is managed through the emotion dynamics model presented in Ahn et al. (2012a)

user test. The following section describes the model extension to integrate a 3D Valence-Arousal-Dominance (VAD) representation of emotions for producing facial expressions. We then justify the introduction of some degree of asymmetry in the facial expression to produce a wider range of emotions, especially ambivalent ones; this section includes some preliminary evaluation results. Facial expression is complemented with the description of a set of body movements in the next section. The following two sections provide an overview of the software architecture and the network management when exploiting the UNITY3D framework. The result section illustrates the performances of our approach while the conclusion elaborates on future research directions.

13.2 Non-verbal Communication of Emotions

Non-verbal communication has been deeply studied in psychology (Wiener et al. 1972). Transposing these prior findings to produce plausible NVC for virtual agents is still an active research topic (Krämer et al. 2007). Pelachaud and Poggi provide a rich overview of a large set of expressive means to convey an emotional state (Pelachaud and Poggi 2002) including head orientation. A number of researches about 3D chatting or agent conversational system have been presented so far. For example, a behavior expression animation toolkit entitled "BEAT" that allows animators to input typed text to be spoken by an animated human figure was described in Cassell et al. (2001). A few years later, a model of emotional dynamics for conversational agent has been presented in Becker and Wachsmuth (2004). Similarly to our approach, their architecture of an agent called "Max" used an advanced representation of emotions. Instead of a restricted set of predefined emotions, a dimensional representation was exploited with the three dimensions Valence-Arousal-Dominance (VAD). Later, an improved version of agent "Max" has been also presented as a museum guide (Kopp et al. 2005) and as a gaming opponent (Becker et al. 2005). A model of behavior expressivity using a set of six parameters that act as modulation of behavioral animation has been developed in Pelachaud (2009) as well. Niewiadomski et al. (2011) then proposed to introduce a constraint-based approach for the generation of multimodal emotional displays. Those works offer a good insight about how to derive sequential behavior including emotions. Very recently Lee et al. (2015) presented encouraging results regarding the acceptance of an emotionally augmented storytelling agent that was integrating nonverbal behaviors based on text analysis.

For what concerns the cooperation between agents and humans, it was found in Melo et al. (2009) that users appreciate to cooperate with a machine when the agent expresses gratitude by means of facial expressions. For this reason, adding emotional NVC to virtual environments would not only enhance user experience, but also foster collaboration and participation in online communities. From an application point of view the specificity of our approach is to give more weight to the NVC emotional dynamics among multiple chatting users supported by a real-time model of emotion expression within a robust network management architecture.

13.3 Valence and Arousal Model

In this section we describe a first emotional model only based on two emotional axes, i.e. valence and arousal $\{v, a\}$. To understand the need and structure of this solution the following Sect. 13.3.1 describes the architecture enabling chatting in 3D virtual environment between avatars and also autonomous agent. Then, to validate this model, Sect. 13.3.2 presents a user-test study (Gobron et al. 2013) demonstrating how the 3D graphical facial expression enhances the perception of the chatting—see Fig. 13.2.

To describe the different types of conversation, we use a specific vocabulary in this section. Users chat with a "virtual" bartender also called "affect" bartender. When the machine is actually a computer, therefore an agent, users chat with a "chatting system" also called "conversational system".

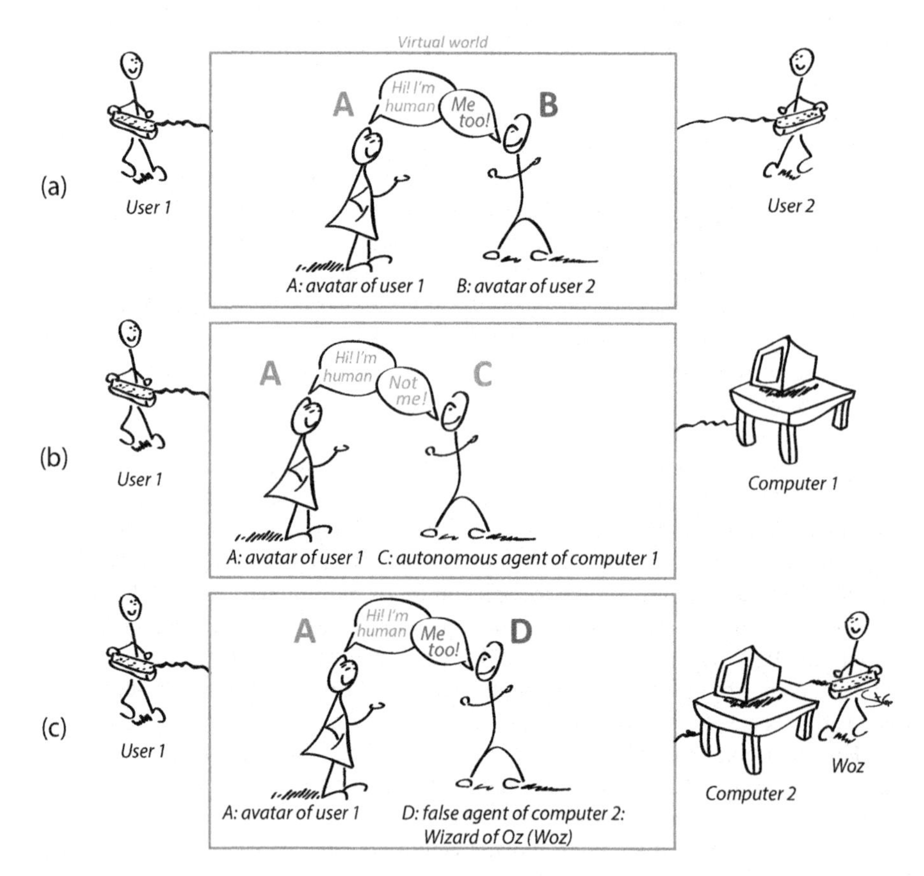

Fig. 13.2 Sketch of semantic and nonverbal communication in virtual environment: (**a**) VH (avatar) "A" chatting with VH (avatar) "B"; (**b**) "A" chatting with an agent "C"—i.e. a VH controlled by a computer, whether autonomous or task oriented; (**c**) "A" believing to chat with an agent but in reality chatting with another VH (avatar) "D", i.e. a Wizard of Oz with the acronym (*Woz*)

13.3.1 Architecture

We focus here on the components allowing virtual humans (VH) to produce conversations with consistent facial expression. For that purpose three types of conversations between two users have been identified and tested: (a) through their avatars, (b) between an avatar and an agent (i.e. a virtual human fully controlled by a computer), and (c) between an avatar and a Wizard of Oz (Woz). To this end, the following key features are required to design and implement this system entitled 3D-emoChatting:

- a global architecture that combines components from several research fields;
- a real-time analysis and management of emotions that allows interactive dialogs with non-verbal communication;
- a model of a virtual emotional mind called emoMind that allows to simulate individual emotional characteristics.

This VR architecture—illustrated in Fig. 13.3—is enabling chatting dialogs with semantic (i.e. text utterances) including emotional communication based on valence and arousal emotional dimensions. The main steps of this model are as follows:

- An avatar or an agent can start a conversation, and every text utterance is a new event that enters the event engine and is stored in a queue;
- The sentence is analyzed by classifiers to extract potential emotional parameters, which are then refined to produce a multi-dimensional *probabilistic emotional histogram* (PEH) (Gobron et al. 2010), that define the heart of the emotion model for each VH. In our approach, we concentrate on "wider displays of emotion" driven by a Valence-Arousal $\{v,a\}$ plane, which brings various facial expressions from any given VA coordinate;
- This generic PEH is then *personalized* depending on the character (e.g. for an optimist this would introduce a shift towards higher valence);
- Then a new current $\{v,a\}$ (emotional coordinate axes valence and arousal) state is selected among the VH mind state;
- The resulting emotional $\{v,a\}$ values are transmitted to the interlocutor;
- If this is a conversational system (such as the *affect bartender* used in Sect. 13.3.2), then it produces a text response potentially influenced by the emotion;
- In parallel, and depending on the events, different animated postures are selected (e.g. idle, thinking, speaking motions);
- This process continues until the end of the dialog.

As shown in the Fig. 13.3, all interdisciplinary aspects of verbal and non-verbal communications are included (i.e. data mining, VR, CG, distant protocols, artificial intelligence, and psychology), this model to can potentially be of a real practical use for entertainment applications involving virtual societies. Details relative to this architecture model are presented by Gobron et al. (2011).

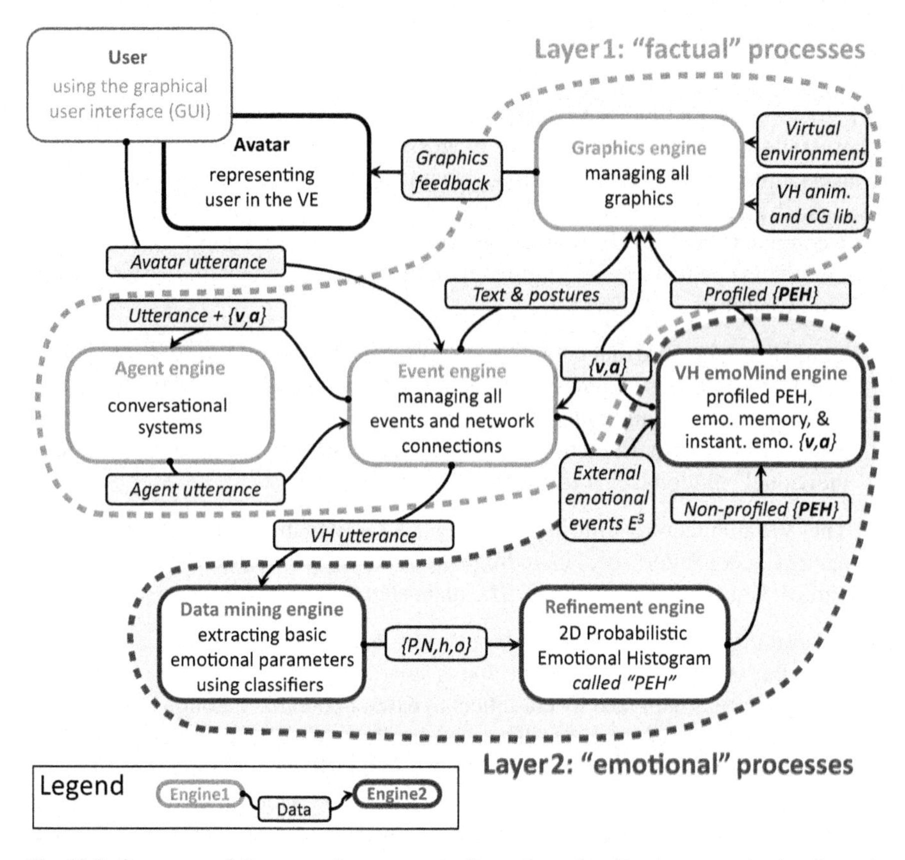

Fig. 13.3 Summary of the general process pipeline where the direct communication layer is represented by the three *green* engines and the non-verbal engines in the shaded area (*red*). *Arrows* represent the data transfer between different engines. Details of each engine can be found in Gobron et al. (2011)

The following section describes a user-test study on the impact of facial expressions automatically induced from text dialogs.

13.3.2 User Test

Two research issues motivated the use of a user-test tool: first, evaluating the impact of no-emotional and emotion facial expression simulation on the chatting experience and second, understanding whether people like chatting within a 3D graphical environment. For that purpose, we identified the following questions relative to two

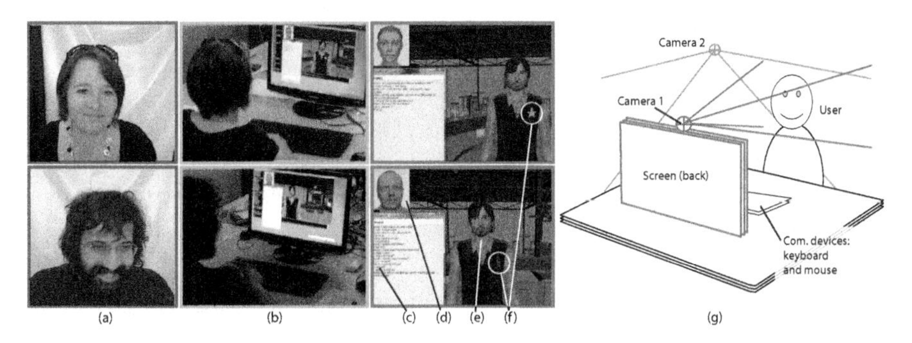

Fig. 13.4 Settings of the user-test. (**a**) View of camera 1, (**b**) view of camera 2, (**c**)–(**f**) screen display

fundamental aspects of chatting through VHs:

- What is the influence of facial expressions that are induced from text dialog?
- What is the impact of nonverbal cues in conversation with a chatbot or with a human operator with a *wizard of oz* approach?

Examples of users performing the test with different bartender conditions are proposed in Fig. 13.4. (a) Users enjoying our chatting system (user face camera 1); (b) Second view at the same moment showing the user interface (global view camera 2); (c) to (f) Close up of the user interface: (c) Message window, where user type input and receive an answer from an agent or a *Woz*; (d) Facial expression of user's avatar generated by our graphics engine; (e) Bartender who could be an agent or a *Woz*; (f) A tag showing different condition of the bartender; (g) Setup during user-test where two cameras were recording the chatting.

The study involved 40 participants. We proposed only subtle facial expressions and head movements induced from spontaneous chatting between VHs and the participants. All of them performed four experiments: with or without facial emotion and with a conversational system or a Wizard of Oz—see Fig. 13.5. Notice that the color code used in this figure is also used in the charts in Fig. 13.6.

In the two top rows of Fig. 13.6, only three of the six graphs provide statistically significant differences to conclude: when chatting without facial expressions, chatting is more enjoyable and emotionally connected with a machine compared to a human; and if facial expressions are rendered, users have a stronger emotional connection when chatting with a human. In the two bottom rows of Fig. 13.6, only graphs representing conversations with the human Wizard of Oz provide statistically significant differences to allow a conclusion: when chatting with facial expressions, the chatting enjoyment, the emotional connection, and even the text consistency are perceived by users to be improved.

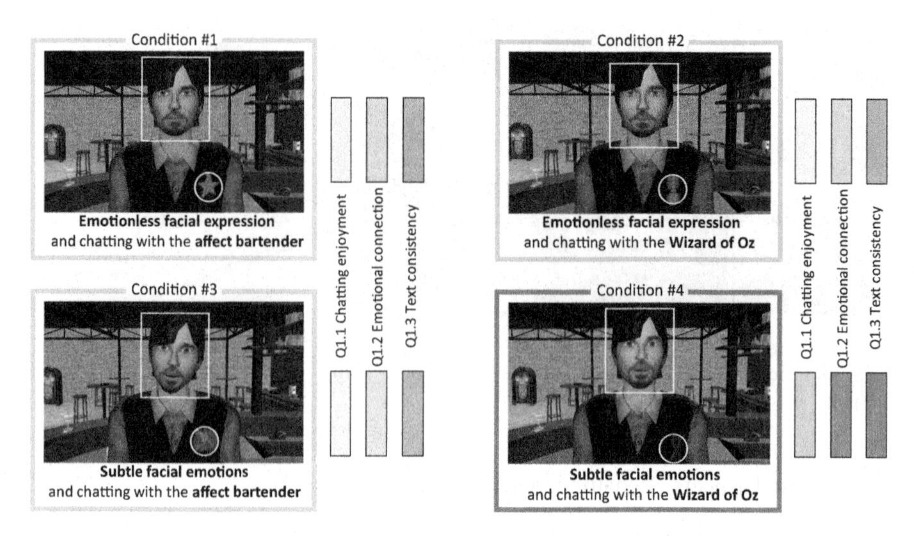

Fig. 13.5 The four conditions of the user-test obtained by combining the facial expression factor (without on *top row*, with on *bottom row*) and the dialog partner factor (chatbot agent in *left column*, human Woz in *right column*). Each condition is identified by specific patch textures on the bartender: a *yellow star* (*top left*), a *blue medal* (*top right*), a *green tag* (*bottom left*), and a *red ribbon* (*bottom right*). As conditions where randomly ordered, users were told to pay attention to these textures and eventually to take notes to help them answering the questionnaire

In summary, this test demonstrated that the influence of 3D graphical facial expression enhances the perception of the chatting enjoyment, the emotional connection, and even the dialog quality when chatting with another human. This implies that nonverbal communication simulation is promising for commercial applications such as advertisement and entertainment.

Furthermore, the test setup that was created provides a solid foundation: first, for further experiments and statistical evaluations at individual level (i.e. one to one VH communication); second, for extension to crowd interaction, a kind of "virtual society" level, e.g. MMO applications including realistic emotional reaction of user's avatars and AI system's agents.

We believe such a study (revealing clues on how people react to virtual communication) can be of an interest before designing applications involving VHs interaction in virtual worlds. Details on this user test can be found in Gobron et al. (2013).

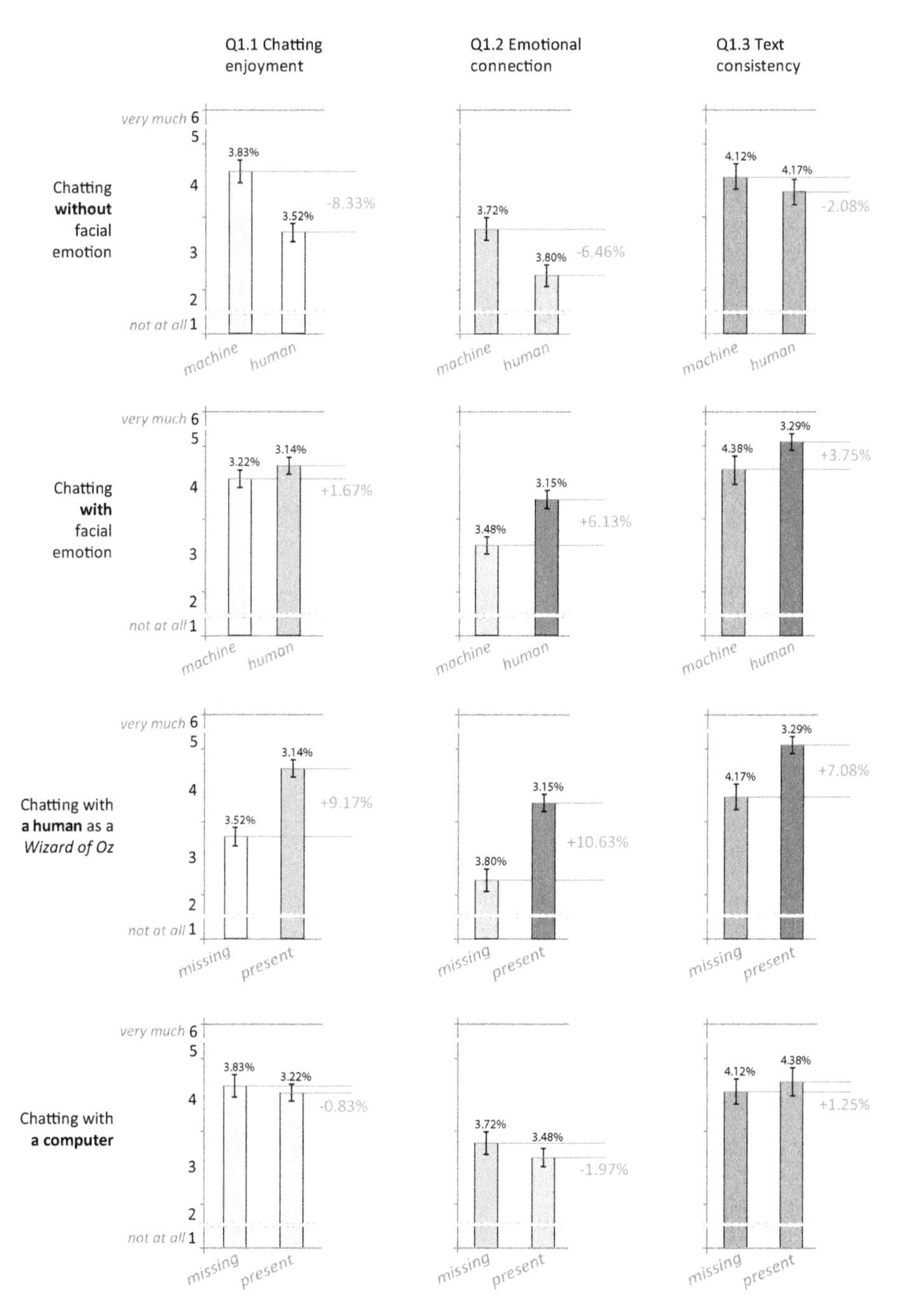

Fig. 13.6 Questionnaire analysis: chatting enjoyment (*left column*), emotional connection (*middle column*), text consistency (*right column*). First (*two top rows*) comparison of user appreciation and influence of the subtle facial expressions when *missing* or *present* (the chatbot agent is labelled with 'machine', whereas the Wizard of Oz is labelled with 'human'). Second (*two bottom rows*) comparison of user appreciation: when chatting with a computer (*machine*) or a human (*Woz*). The *vertical bars* indicate the facial expression factor (missing/present)

13.4 Integrating the Valence-Arousal-Dominance Model of Emotions

The motivation for adopting an emotion space spanned by three independent dimensions is best illustrated in Fig. 13.7. One can evaluate how the third dimension, indicating the feeling of power of the emotion (Dominance), significantly influences the facial expression for a sampling of nine emotions in the Valence-Arousal space. This approach has been proposed in the mid 1970s with slightly different terms depending on the authors (Bush 1973; Russell 1980; Scherer 2005). We stick to the Valence-Arousal-Dominance (VAD) terminology to be consistent with and take advantage of the ANEW set that assesses the affective meaning of 1034 English words in the VAD space (Bradley and Lang 1999).

Based on the set of Action Units (AU) identified in Ekman and Friesen (1978) we have chosen to produce facial expressions through the action of 12 Facial Part Action (FPA) corresponding to antagonist muscle groups (Fig. 13.8). We modelled

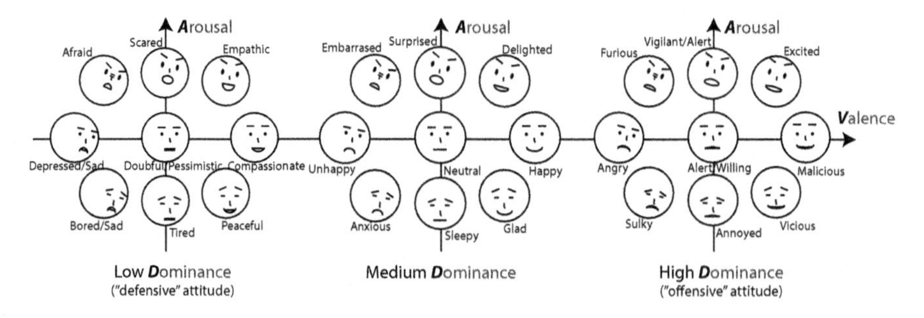

Fig. 13.7 Keywords and iconic representation for a sampling of the Valence-Arousal-Dominance (VAD) emotion space. The first axis (Valence) represents the positivity or negativity of an emotion. The second one (Arousal) describes the degree of energy of the emotion. Finally, the third axis (Dominance) indicates the feeling of power of the emotion (Ahn et al. 2012a)

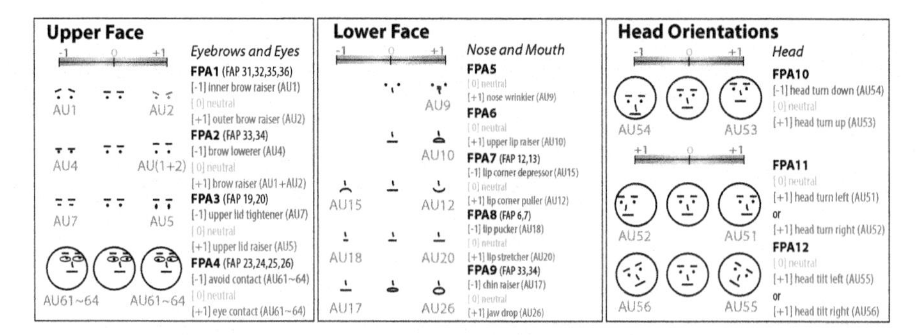

Fig. 13.8 The 12 Facial Part Actions (FPA) retained for producing facial expressions. Intensity is normalized mostly within [−1,1] except for FPA5 and FPA6 [0,1]. The related Action Units (AU) from Ekman and Friesen (1978) appear under the iconic representation of the muscle group action

the normalized activity of each FPA as a linear function of the three VAD emotion parameters to minimize the computing cost at run-time. Briefly stated, each FPA activity being a linear function of Valence, Arousal and Dominance, we only have to identify one coefficient for each dimension. We consider that the origin of the VAD space is the neutral expression for which all FPA activities are null, hence all the linear functions pass through the origin.

The linear mapping has been calibrated on a sample of 18 emotion keywords listed in Table 13.1. These 18 emotions include the 6 basic emotions (fear, anger, sadness, surprise, joy, disgust) and 12 additional emotion key-words from the ANEW set (Bradley and Lang 1999) to ensure a homogeneous coverage of the VAD space. Table 13.1 displays the VAD coordinates of each keyword and the corresponding FPA normalized intensities (Fig. 13.9) to produce the associated facial expression. The optimized linear coefficients appear in the last three lines of the table. For example, the normalized intensity y of FPA7, responsible of the smiling muscles (Fig. 13.8), is expressed as $y = 0.55v - 0.08a + 0.37d$, where v, a and d represent respectively the normalized amplitude of the valence, arousal and dominance (Fig. 13.7).

13.5 Revisiting the Facial Expression of Emotion Through Asymmetry

In the field of real-time applications, economic constraints of character design cost, run-time computing cost and memory footprint have often resulted into the production of symmetric faces. Likewise the VAD emotion model is associated by default to the synthesis of symmetric facial expressions. However, if one takes a look at the emotional expressions from pictures, movies or the real world, there is ample evidence that neither a face, nor its expression are always perfectly symmetric (Schatz 2006). This observation is supported by numerous studies in psychology such as those reviewed by Borod et al. (1997), and by recent findings in psychophysiology stating the possible simultaneous activation of emotions with positive and negative valence (Norman et al. 2011).

Our major motivation for considering asymmetry in the facial expression of emotion is its ability to convey a wider spectrum of complex emotions, in particular those simultaneously displaying conflicting emotions, also known as *ambivalent* emotions (Ahn et al. 2012b). It is our conviction that integrating the asymmetry dimension into the real-time facial expression model increases the believability of avatars and virtual human agents.

By definition a symmetric facial expression displays the same 12 FPA intensities on both sides of the face. In order to produce an asymmetric expression we allow five FPA to have a different intensity on the left and right sides of the face. These asymmetric FPAs are related to corrugator, orbicularis, and zygomatic muscles, which are mostly observed for left/right asymmetric facial expression from EMG

Table 13.1 Set of 18 calibrating emotions and optimized coefficients for the 12 FPA linear mapping

Emotions	**V**	**A**	**D**	FPA1	FPA2	FPA3	FPA4	FPA5	FPA6	FPA7	FPA8	FPA9	FPA10	FPA11	FPA12
Afraid	−0.75	0.42	−0.26	−0.5	1.0	1.0	−0.5	0.5	0.0	−0.5	0.0	1.0	1.0	1.0	0.0
Angry	−0.54	0.54	0.14	1.0	−1.0	0.5	1.0	1.0	1.0	−1.0	0.0	−0.5	1.0	0.0	0.0
Disgusted	−0.64	0.11	−0.17	1.0	−1.0	−1.0	−0.5	1.0	1.0	−1.0	0.5	0.5	1.0	1.0	0.0
Happy	0.80	0.37	0.41	−0.5	0.0	0.0	0.5	0.0	0.5	1.0	1.0	0.0	−0.5	0.0	1.0
Sad	−0.85	−0.22	−0.39	−0.5	0.0	−0.5	−0.5	0.0	0.0	−0.5	−0.5	−0.5	−1.0	0.0	0.0
Surprised	0.62	0.62	0.28	−0.5	1.0	1.0	0.5	0.5	0.5	0.0	0.0	0.5	1.0	0.0	0.0
Anxious	−0.05	0.48	0.08	0.5	−0.5	0.0	−0.5	0.0	0.5	−0.5	0.0	0.5	0.0	0.5	0.0
Bored	−0.51	−0.54	−0.22	−0.5	0.0	−0.5	−0.5	0.0	0.0	−0.5	0.0	0.5	−0.5	0.5	0.5
Consoled	0.20	−0.12	−0.14	−0.5	0.5	0.0	0.0	0.0	0.0	0.0	0.5	0.5	−0.5	0.0	0.5
Gloomy	−0.78	−0.29	−0.36	−0.5	0.0	−0.5	−0.5	0.0	0.0	−0.5	0.0	0.5	−1.0	0.5	0.0
Hopeful	0.53	0.20	0.10	−0.5	0.5	0.5	0.5	0.0	0.0	0.5	0.5	0.0	0.5	0.0	0.5
Indifferent	−0.10	−0.46	−0.04	−0.5	0.0	−0.5	−1.0	0.0	0.0	0.0	0.5	0.0	0.0	0.5	0.5
Inspired	0.54	0.26	0.42	0.0	0.5	0.5	1.0	0.5	1.0	0.0	0.5	0.5	0.5	0.0	0.0
Overwhelmed	−0.20	0.50	−0.28	0.5	−0.5	0.0	−0.5	0.5	0.5	0.0	−0.5	0.5	−0.5	0.5	0.5
Peaceful	0.68	−0.51	0.11	−0.5	0.0	−0.5	0.0	0.0	0.0	0.5	0.5	0.0	−0.5	0.0	1.0
Pleasant	0.82	0.19	0.29	−1.0	1.0	0.5	0.5	0.0	1.0	1.0	1.0	0.5	0.5	0.0	0.5
Relaxed	0.50	−0.65	0.14	−1.0	0.5	−0.5	0.0	0.0	0.0	0.5	0.5	0.0	−0.5	0.0	0.5
Shy	−0.09	−0.31	−0.39	0.0	0.0	−0.5	−1.0	0.5	0.0	0.0	−0.5	1.0	−0.5	0.5	0.0
β_{vi}				−0.42	0.63	0.19	0.29	−0.32	0.13	0.55	0.48	−0.07	0.07	−0.46	0.52
β_{ai}				0.46	−0.02	0.49	0.25	0.59	0.51	−0.08	−0.00	0.35	0.56	0.06	−0.22
β_{di}				−0.12	0.35	0.32	0.46	−0.09	0.36	0.37	0.52	−0.58	0.37	−0.48	0.26

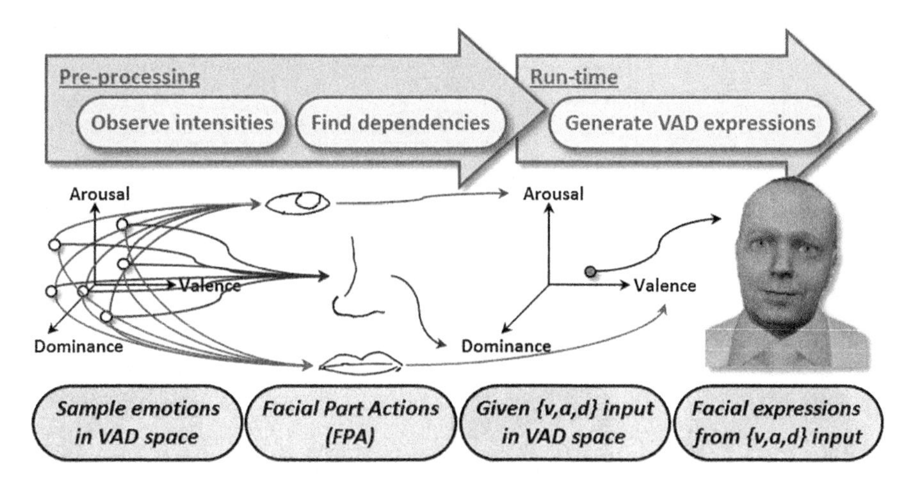

Fig. 13.9 (*left*) Preprocess to identify the VAD to FPA activity mapping; (*right*) exploiting the mapping to generate facial expression at run-time for a given emotions in the VAD space (Ahn et al. 2012b)

experiments (Schwartz et al. 1979; Dimberg and Petterson 2000; Wyler et al. 1987):

- Corrugator supercilii: facial muscle placed on the inner side of the eyebrow (FPA1 and FPA2)
- Orbicularis oculi: facial muscle that closes the eyelids (FPA3)
- Zygomatic major: facial muscle which draws the angle of the mouth upward and laterally (FPA7 and FPA8).

The proposed editing framework is described in Fig. 13.10. It offers two different types of run-time input edited at pre-processing stage: (1) a pair of VAD values adjusted with the face-edit parameters described in Fig. 13.10 left; or (2) a pair of VAD flows manually designed by an animator. For an efficient face edit process, the proposed interface provides a list of 117 emotional keywords, each linked to a VAD value. These keywords were chosen among the 1035 words provided in the ANEW Table (Bradley and Lang 1999). The face edit operators are necessary to adjust the two component emotions as their default VAD coordinates rarely match to produce a believable asymmetric expression. The principle of these operators is to allow a local displacement in the VAD space along the following independent directions: Valence, Arousal and Dominance axis, and/or applying a 3D scaling factor W and/or introducing *a* bias *b* controlling the inter-face weight I between the two component emotions (Fig. 13.10 left). This process is illustrated on Fig. 13.11 with two examples of complex emotions.

We have conducted an experimental study with 58 subjects to evaluate the proposed VAD framework and the impact of asymmetric facial expressions (Ahn et al. 2013). Subject were shown 64 pairs of static facial expressions, one symmetric and one asymmetric, illustrating eight emotions (three basic and five complex ones)

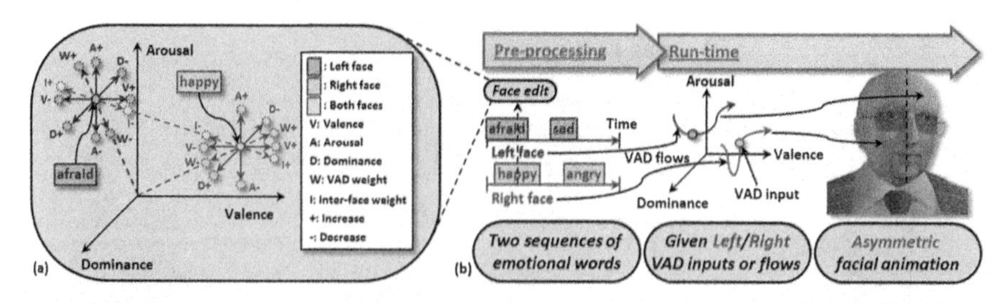

Fig. 13.10 (**a**) Asymmetric face edit parameters; (**b**) asymmetric facial expression pipeline

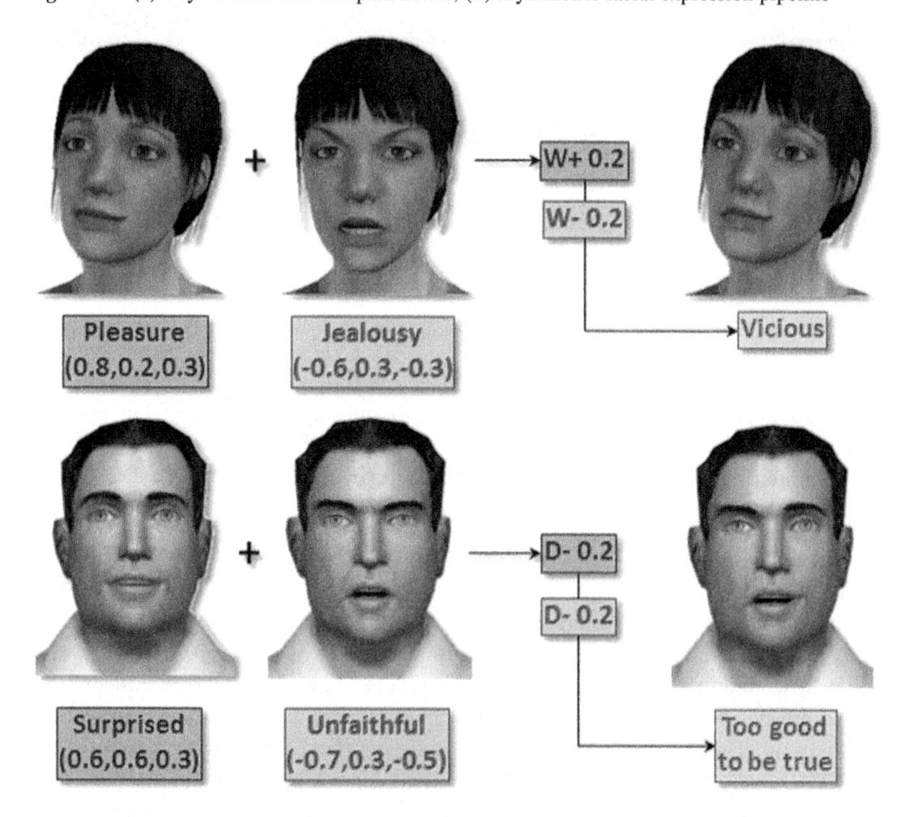

Fig. 13.11 The emotion of the *left column* (resp. *middle column*) appears on the *left* (resp. *right*) *side* of the composite facial expression after the action of the face edit operators W and D

on two virtual characters, one male and one female. They were asked to grade each facial expression on a continuous scale about its ability to correctly represent the associated emotion. The analysis of the results is still going on but we can give a first overview in Fig. 13.12. Although not always the case, results tend to confirm the preference for symmetric face for expressing basic emotions (sadness, peaceful, fear) whereas asymmetric faces are clearly preferred for a subset of the studied

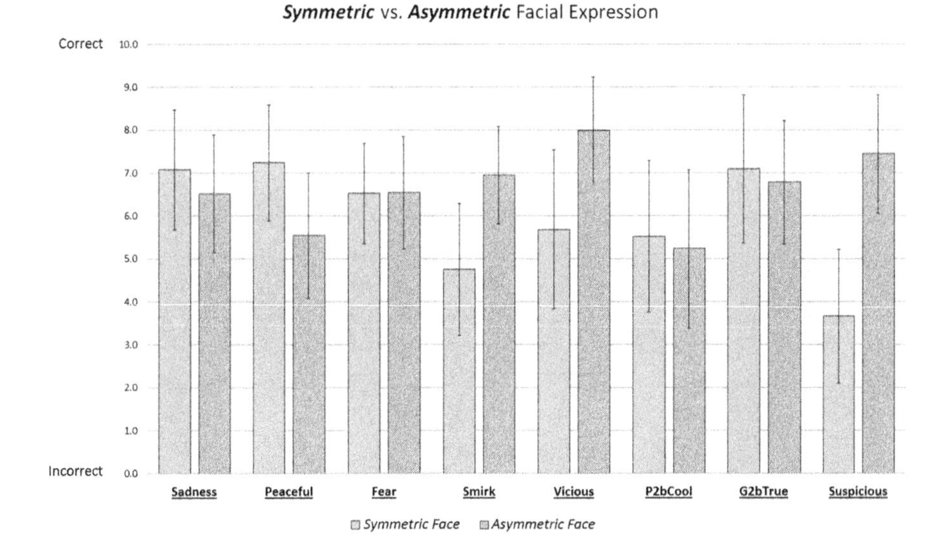

Fig. 13.12 Preliminary results about the comparative study of symmetric and asymmetric facial expressions for expressing basic (sadness, peaceful, fear) and complex emotions (smirk, vicious, pretend to be cool, too good to be true, suspicious). More details can be found in Ahn et al. (2013)

complex emotions (smirk, vicious, suspicious). These results tend to confirm the subject sensitivity to facial expression displaying the co-existence of conflicting emotions. However the presented asymmetric expressions did not always convince subjects, such as for "Too good to be true" also visible in Fig. 13.11 and "pretend to be cool". It might come partly from the static nature of the image medium or from the chosen combination of conflicting emotions. More experimental studies are necessary that integrate the temporal dimension to better take advantage of the richer expressive space made possible through asymmetric expressions.

13.6 Combining Facial Expressions and Body Movements

When associating full body movements to the facial expression of emotions it is important to model first the influence of the emotions on a background movement which is essential to ensure the believability of the 3D avatar: breathing. The choice of the VAD emotion space simplifies the modeling problem as, by construction, the Arousal dimension encapsulates the energy of the emotion which is much more correlated to movement than Valence or Dominance (Fig. 13.7). For this reason we expressed the breathing amplitude and frequency as a function of the sole Arousal value of the current emotion (Ahn et al. 2012a).

In addition to the breathing movement pattern, we captured 27 emotional body movements to fit all the areas of the VAD model illustrated in Fig. 13.7. The

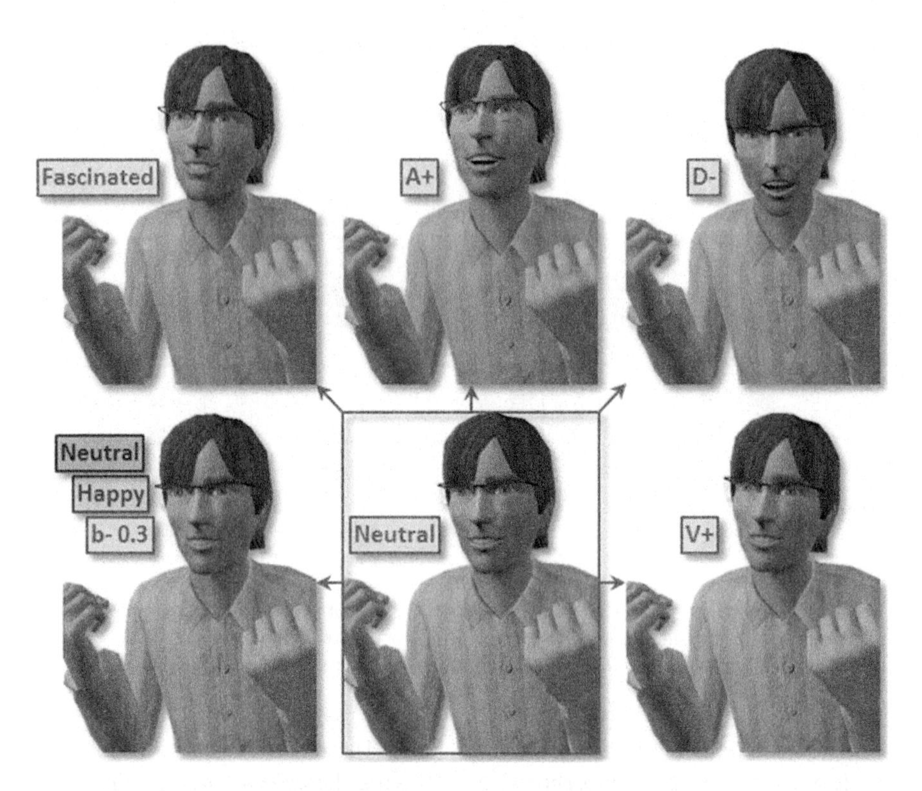

Fig. 13.13 The captured emotional body movement (here the excited movement) can be associated with a large variety of facial expressions owing to the face edit parameters

parameters presented in Fig. 13.10 left (V, A, D, W, b and I) and a selection of predefined VAD keywords allow to produce a wide range of facial modulations for the same emotional body animation as can be seen on Fig. 13.13. However unlike the dynamic model handling the facial expressions, a full-body animation is only triggered from the analysis of emotions over a longer temporal window (e.g. 10 s) (Ahn et al. 2012a).

13.7 3D Chatting System Overview

As discussed in the introduction section, users of text-based chatting systems may not anticipate potential ambiguities or misinterpretations of their messages due to the very limited redundancy of such type of communication. Our goal is to enrich and disambiguate this type of communication with a visual channel displaying an instantaneous expression of emotions with avatars. Such a goal introduces various technical challenges considering network protocols, operating system, and various technical issues (Gobron et al. 2012).

Our latest 3D-chatting visualization system uses the Unity3D platform as a develop environment. The advanced features (especially network architecture and pipeline) provided by Unity3D bring the possibility of having a reliable affective and emotional communication system. Our chatting system allows users to (1) select their virtual human avatar, (2) move their avatar locally in the 3D environment or (3) change the 3D scene during the conversation.

To enhance the user experience during the conversation, we enable real-time facial and full body animation. When a user types a sentence, not only is the text-based message sent to the receiver(s), but also the related emotional parameters, such as Valence, Arousal and Dominance. The emotional state of a user's text messages are detected based on the SentiStrength tool (Chap. 7). Then according to the VAD values, the receiver client performs facial and body animation based on facial and full body joints control algorithm. There are three ways to trigger facial and full body animations:

- User's own messages
- Messages from other users
- User explicit emotion activation with the interface

Unlike other online chat systems, our emotional chat system supports a wide range of platforms and environment configurations.

Cross Platform Users using different platforms can seamlessly chat with each other effectively, such as using iOS (iPhone iPad), Android, Linux, Mac, and Windows. We also provide an online web-based emotional chat application that supports chatting with other possible platforms (Fig. 13.14).

Cross Environment Our chat system not only supports classical PC-to-PC chatting, but also allows multi-environment chatting, such as a user A using a PC

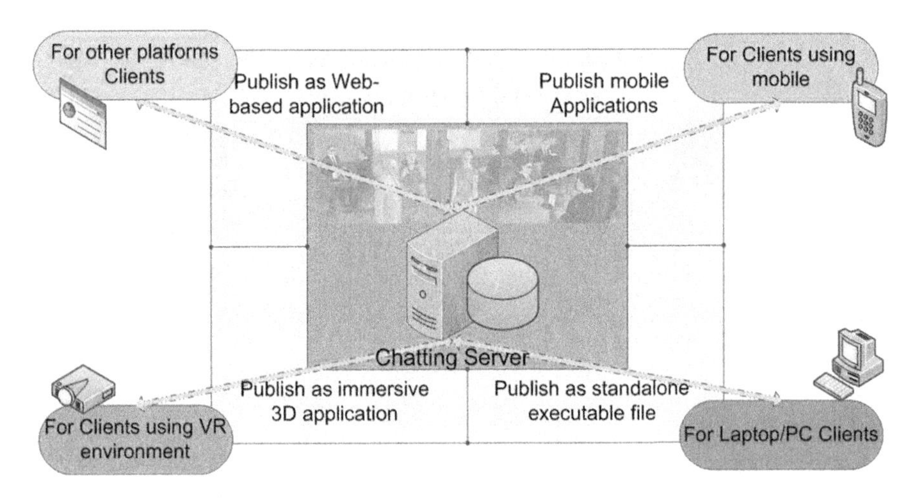

Fig. 13.14 Cross-platform and cross-environment system approach with UNITY3D

can communicate with a user B who is standing in a Virtual Reality interactive environment wearing a stereo glasses (e.g. CAVE in Fig. 13.18 left). During the chat, both users can move within the 3D scene by using their dedicated interface.

13.8 Network Architecture

Real-time networking is a complex issue because of the numerous types of network protocols, variation of network connections, network programming, data transmission format, etc. Our system provides two different types of network connections based on different network environments and research requirements: Direct Connection (DC) and Internet Connection (IC).

The Direct Connection can be applied to a simple network environment, where two potential clients know each other's local IP address. The following steps should be applied before starting the communication (Fig. 13.15):

- One of the clients (Client *A*) has to create a chat room. This will open a network port on Client *A*'s computer; then wait for the connection from another client.
- In order to join the chat room which is created by Client *A*, Client *B* has to specify the local IP address which is used by Client *A* and the network port which is opened for this connection.

The Internet Connection allows a widespread connection through Internet with multi-connection. The following essential steps have to be performed in order to setup a connection with remote users (Fig. 13.16):

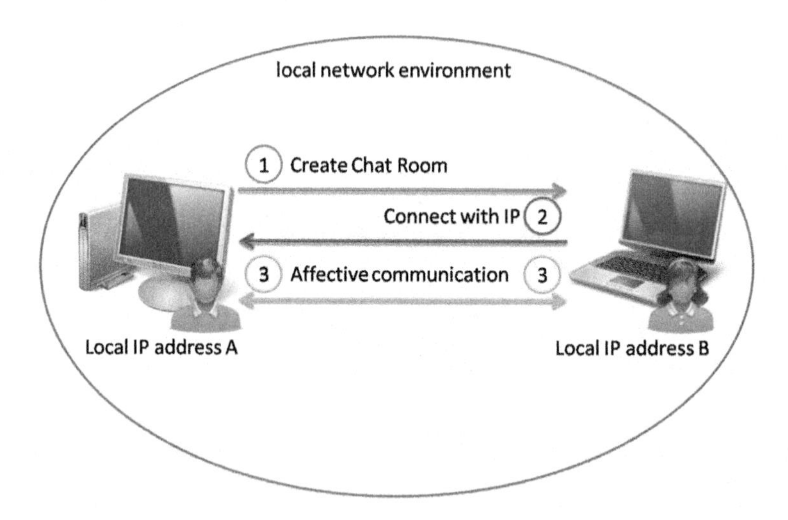

Fig. 13.15 Direct connection suited for a simple network environment

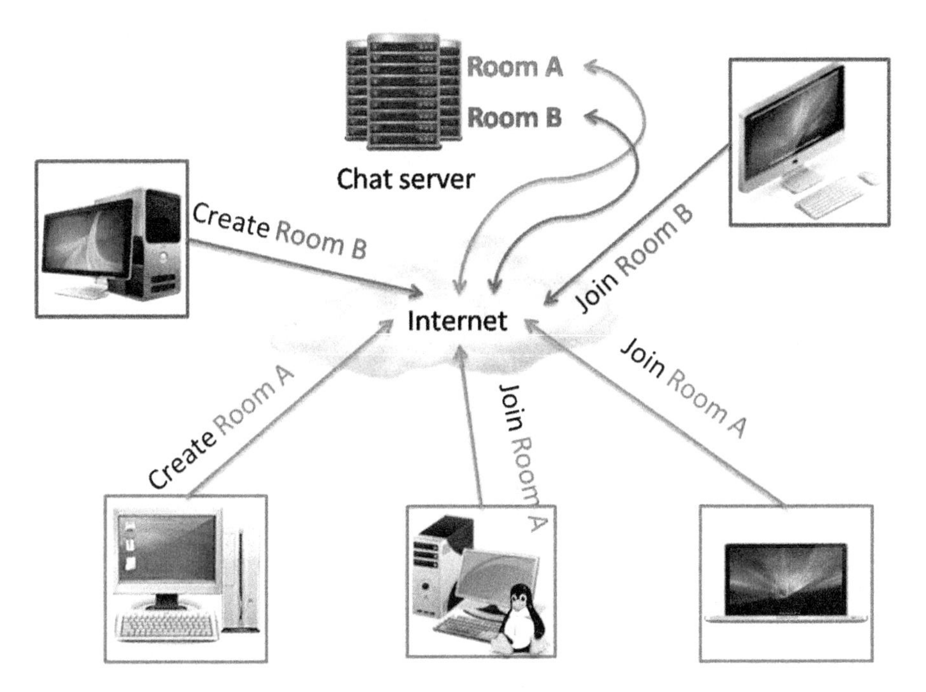

Fig. 13.16 The indirect connection through chat rooms dynamically created on a server

- In order to connect multi-client which are geographically separated anywhere on the Internet, a chat server must be set up first.
- One of the potential users has to connect to a chat server and create a chat room on the server (e.g. Room A).
- Other remote users have to join into the chat room (e.g. Room A) in order to communicate with each others.

13.9 Results

The emotion dynamic model embedded into the chatting system version corresponding to Fig. 13.1 is demonstrated in the video http://youtu.be/UGbW8nDNO24 associated to Ahn et al. (2012a). It illustrates how emotions evolve over time according to the emotional content detected with SentiStrength when three users exchange messages. The example deliberately triggers various emotional movements when interacting with the user represented by the Dorothy avatar.

The FPA interface is demonstrated in http://youtu.be/ycUoZ4jAU_E (Ahn et al. 2012b); it is used to produce instantaneous facial expressions, including asymmetric ones. Performance-wise this second video also demonstrates the display rate of a

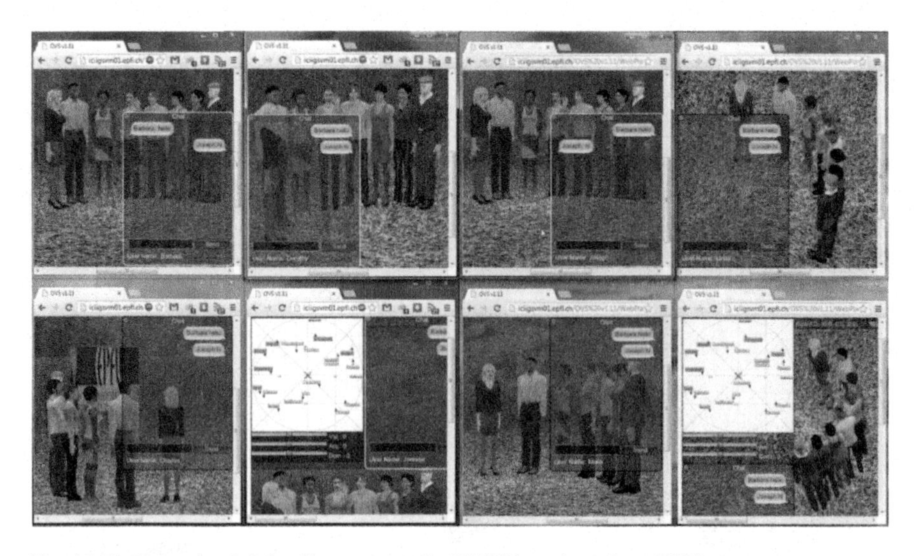

Fig. 13.17 Example of eight client users to the UNITY version of our NVC system

scene with 200 characters with continuously changing emotions. With an NVidia GTX 460 1 GB hardware, we were able to simulate and edit facial expressions at >60 fps. This confirms the suitability of our approach to handle a potentially large number of users simultaneously connected to a single chat room.

The network engine we used in our UNITY-based configuration is RakNet. RakNet is a cross platform open source networking engine for game programmers. A lower level framework powers higher-level features such as game object replication, voice communication and patching. Using new features of RakNet, a set of issues have been solved and improved, such as: secure affective communication connections, robust communication and message ordering on multiple channels and peer to peer application if necessary. With a basic free version Raknet, connection with 1000 clients is possible. As we did not have enough computer resource and clients, at the stage of system evaluation we only conducted a group affective communication with eight clients (Fig. 13.17).

Figure 13.18 (left) illustrates the use of a CAVE as one environment supported by our NVC chat system with UNITY3D. Finally an independent study of crowd evaluation within the CAVE (Ahn et al. 2012c) has highlighted the positive increase, although not significant, of the believability of the crowd when crowd agents were displaying random emotions instead of a neutral face (Fig. 13.18 right).

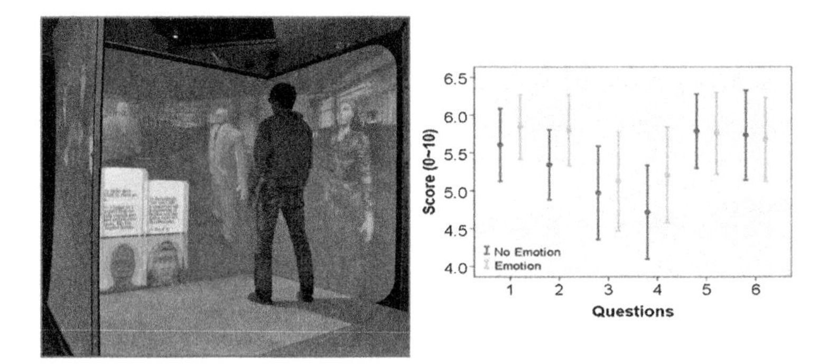

Fig. 13.18 (*left*) The CAVE can be used as a chatting environment, (*right*) positive influence of emotions on believability questions while evaluating real-time crowds in a CAVE (Ahn et al. 2012c)

13.10 Conclusion

Through this chapter we have demonstrated the effectiveness of an architecture allowing to support instantaneous Non-Verbal Communication as an additional visual feedback for multiple simultaneous chatting users. Results from a first user study with a Valence-Arousal emotion model have shown that the influence of 3D graphical facial expression enhances the chatting enjoyment, the emotional connection, and even the dialog quality when chatting with another human. Based on these results, we have extended the system to integrate the 3D Valence-Arousal-Dominance emotion model. At its core lays an animation engine that relies on a fast linear mapping of emotions on facial expressions and body movements. We have also recalled how asymmetric facial expressions can help to enlarge the spectrum of displayed emotions, especially ambivalent ones. A first evaluation study has helped to outline when this new class of facial expressions should be used and when not. We have found that users are very sensitive to ambivalent emotions and that only an asymmetric facial expression is able to convey it successfully. On the contrary, the successful understanding of basic emotions such as fear, sadness and joy is more successfully conveyed with symmetric facial expressions on virtual characters. More studies should be performed to be able to fully take advantage of this enlarged range of emotions in chatting systems.

In summary, we proposed a multi-platform system with the UNITY3D environment to make possible the creation of a 3D virtual society interconnecting a potentially large number of simultaneous users. The visualization software, which simulates this society, allows connecting people from different types of hardware platforms such as desktop PC, portable high-end laptop, and even from an immersive CAVE system. This environment allows to sustain the performance needed for a large number of simultaneous avatars and agents. It includes the SentiStrength tool and the emotional dynamics from Ahn et al. (2012a).

Acknowledgements This work was supported by a European Union grant by the 7th Framework Programme, Theme 3: Science of complex systems for socially intelligent ICT. It is part of the CyberEmotions project (contract 231323).

References

Ahn, J., Gobron, S., Garcia, D., Silvestre, Q., Thalmann, D., Boulic, R.: An NVC emotional model for conversational virtual humans in a 3d chatting environment. In: Perales Lopez, F.J., Fisher, R.B., Moeslund, T.B. (eds.) Articulated Motion and Deformable Objects: 7th International Conference, AMDO 2012, Port d'Andratx, Mallorca, July 11–13, 2012, Proceedings. Lecture Notes in Computer Science (Image Processing, Computer Vision, Pattern Recognition, and Graphics), vol. 7378, pp. 47–57. Springer, Berlin (2012a). doi:10.1007/978-3-642-31567-1_5

Ahn, J., Gobron, S., Thalmann, D., Boulic, R.: Conveying real-time ambivalent feelings through asymmetric facial expressions. In: Kallmann, M., Bekris, K. (eds.) Motion in Games: 5th International Conference, MIG 2012, Rennes, November 15–17, 2012 Proceedings. Lecture Notes in Computer Science, vol. 7660, pp. 122–133. Springer, Berlin (2012b). doi:10.1007/978-3-642-34710-8_12

Ahn, J., Wang, N., Thalmann, D., Boulic, R.: Within-crowd immersive evaluation of collision avoidance behaviors. In: Thalmann, D., Wu, E., Pan, Z., El Rhalibi, A., Magnenat-Thalmann, N., Adcock, M. (eds.) Proceedings of the 11th ACM SIGGRAPH International Conference on Virtual-Reality Continuum and Its Applications in Industry, pp 231–238. ACM, New York (2012c). doi:10.1145/2407516.2407573

Ahn, J., Gobron, S., Thalmann, D., Boulic, R.: Asymmetric facial expressions: revealing richer emotions for embodied conversational agents. Comput. Anim. Virtual Worlds **24**(6), 539–551 (2013). doi:10.1002/cav.1539

Becker, C., Kopp, S., Wachsmuth, I.: Simulating the emotion dynamics of a multimodal conversational agent. In: André, E., Dybkjær, L., Minker, W., Heisterkamp, P. (eds.) Affective Dialogue Systems: Tutorial and Research Workshop, ADS 2004, Kloster Irsee, June 2004. Lecture Notes in Computer Science (Lecture Notes in Artificial Intelligence), vol. 3068, pp. 154–165. Springer, Berlin (2004). doi:10.1007/978-3-540-24842-2_15

Becker, C., Nakasone, A., Prendinger, H., Ishizuka, M., Wachsmuth, I.: Physiologically interactive gaming with the 3D agent Max. Presented at the JSAI 2005 Workshop on Conversational Informatics in Conjunction with the 19th Annual Conference of the Japan Society for Artificial Intelligence, June 2005 (2005)

Borod, J.C., Haywood, C.S., Koff, E.: Neuropsychological aspects of facial asymmetry during emotional expression: a review of the normal adult literature. Neuropsychol. Rev. **7**(1), 41–60 (1997). doi:10.1007/BF02876972

Bradley, M.M., Lang, P.J.: Affective norms for English words (ANEW): instruction manual and affective ratings. Technical Report C-1, University of Florida: Center for Research in Psychophysiology (1999)

Bush, L.E.: Individual differences multidimensional scaling of adjectives denoting feelings. J. Pers. Soc. Psychol. **25**(1), 50–57 (1973). doi:10.1037/h0034274

Cassell, J., Vilhjálmsson, H.H., Bickmore, T.: BEAT: the behavior expression animation toolkit. In: Pocock L (ed.) The 28th International Conference on Computer Graphics and Interactive Techniques SIGGRAPH '01, pp. 477–486. ACM, New York (2001). doi:10.1145/383259.383315

Darwin, C.: The Expression of the Emotions in Man and Animals. Murray, London (1872)

Dimberg, U., Petterson, M.: Facial reactions to happy and angry facial expressions: evidence for right hemisphere dominance. Psychophysiology **37**(5), 693–696 (2000). doi:10.1111/1469-8986.3750693

Ekman, P.: Emotions Revealed. Henry Holt and Company, LLC, New York (2004)

Ekman, P., Friesen, W.: Facial Action Coding System: A Technique for the Measurement of Facial Movement. Consulting Psychologists Press, Palo Alto (1978)

Gobron, S., Ahn, J., Paltoglou, G., Thelwall, M., Thalmann, D.: From sentence to emotion: a real-time three-dimensional graphics metaphor of emotions extracted from text. Vis. Comput. **26**(6–8), 505–519 (2010). doi:10.1007/s00371-010-0446-x

Gobron, S., Ahn, J., Silvestre, Q., Thalmann, D., Rank, S., Skowron, M., Paltoglou, G., Thelwall, M.: An interdisciplinary VR-architecture for 3D chatting with non-verbal communication. In: Coquillart, S., Steed, A., Welch, G. (eds.) JVRC11: Joint Virtual Reality Conference of EGVE - EuroVR, pp 87–94. Eurographics Association (2011). doi:10.2312/EGVE/JVRC11/087-094

Gobron, S., Ahn, J., García, D., Silvestre, Q., Thalmann, D., Boulic, R.: An event-based architecture to manage virtual human non-verbal communication in 3D chatting environment. In: Perales Lopez, F.J., Fisher, R.B., Moeslund, T.B. (eds.) Articulated Motion and Deformable Objects: 7th International Conference, AMDO 2012, Port d'Andratx, Mallorca, July 11–13, 2012, Proceedings. Lecture Notes in Computer Science (Image Processing, Computer Vision, Pattern Recognition, and Graphics), vol. 7378, pp. 58–68. Springer, Berlin (2012). doi:10.1007/978-3-642-31567-1_6

Gobron, S., Ahn, J., Thalmann, D., Skowron, M., Kappas, A.: Impact study of nonverbal facial cues on spontaneous chatting with virtual humans. J. Virtual Reality Broadcast. **19(2013)**(6) (2013). doi:10.20385/1860-2037/10.2013.6. https://www.jvrb.org/past-issues/10.2013/3823/?searchterm=gobron

Hatfield, E., Cacioppo, J.T., Rapson, R.L.: Emotional contagion. Curr. Dir. Psychol. Sci. **2**(3), 96–99 (1993). doi:10.1111/1467-8721.ep10770953

Kappas, A.: Smile when you read this, whether you like it or not: conceptual challenges to affect detection. IEEE Trans. Affect. Comput. **1**(1), 38–41 (2010). doi:10.1109/T-AFFC.2010.6

Kopp, S., Gesellensetter, L., Krämer, N.C., Wachsmuth, I.: A conversational agent as museum guide — design and evaluation of a real-world application. In: Panayiotopoulos, T., Gratch, J., Aylett, R., Ballin, D., Olivier, P., Rist, T. (eds.) Intelligent Virtual Agents: 5th International Working Conference, IVA 2005, Kos, September 2005. Lecture Notes in Computer Science, vol. 3661, pp. 329–343. Springer, Berlin (2005). doi:10.1007/11550617_28

Krämer, N., Simons, N., Kopp, S.: The effects of an embodied conversational agent's nonverbal behavior on user's evaluation and behavioral mimicry. In: Pelachaud, C., Martin, J.-C., André, E., Chollet, G., Karpouzis, K., Pelé, D. (eds.) Intelligent Virtual Agents: 7th International Conference, IVA 2007 Paris, September 2007 Proceedings. Lecture Notes in Computer Science, vol. 4722, pp. 238–251. Springer, Berlin (2007). doi:10.1007/978-3-540-74997-4_22

Lee, S., Johnson, A.E., Leigh, J., Renambot, L., Jones, S., Di Eugenio, B.: Emotionally augmented storytelling agent. In: Brinkman, W.-P., Broekens, J., Heylen, D. (eds.) Intelligent Virtual Agents: 15th International Conference, IVA 2015, Delft, August 2015, Proceedings. Lecture Notes in Computer Science, vol. 9238, pp. 483–487. Springer, Berlin (2015). doi:10.1007/978-3-319-21996-7_53

Melo, C.M., Zheng, L., Gratch, J.: Expression of moral emotions in cooperating agents. In: Ruttkay, Z., Kipp, M., Nijholt, A., Vilhjálmsson, H.H. (eds.) Intelligent Virtual Agents: 9th International Conference, IVA 2009 Amsterdam, September 2009 Proceedings. Lecture Notes in Computer Science, vol. 5773, pp 301–307. Springer, Berlin (2009). doi:10.1007/978-3-642-04380-2_32

Morris, D.: Manwatching: A Field Guide To Human Behaviour. Triad Panther, London (1978)

Niewiadomski, R., Hyniewska, S., Pelachaud, C.: Constraint-based model for synthesis of multimodal sequential expressions of emotions. IEEE Trans. Affect. Comput. **2**(3), 134–146 (2011) doi:10.1109/T-AFFC.2011.5

Norman, G.J., Norris, C.J., Gollan, J., Ito, T.A., Hawkley, L.C., Larsen, J.T., Cacioppo, J.T., Berntson, G.G.: The neurobiology of evaluative bivalence. Emot. Rev. **3**(3), 349–359 (2011). doi:10.1177/1754073911402403

Pelachaud, C.: Studies on gesture expressivity for a virtual agent. Speech Commun. **51**(7), 630–639 (2009). doi:10.1016/j.specom.2008.04.009

Pelachaud, C., Poggi, I.: Subtleties of facial expressions in embodied agents. J. Vis. Comput. Animat. **13**(5), 301–312 (2002). doi:10.1002/vis.299

Preston, S.D., de Waal, F.B.M.: Empathy: its ultimate and proximate bases. Behav. Brain Sci. **25**(01), 1–20 (2002). doi:10.1017/S0140525X02000018

Russell, J.A.: A circumplex model of affect. J. Pers. Soc. Psychol. **39**(6), 1161–1178 (1980). doi:10.1037/h0077714

Schatz, H.: In Character: Actors Acting. Bulfinch, New York (2006)

Scherer, K.R.: What are emotions? And how can they be measured? Soc. Sci. Inf. **44**(4), 695–729 (2005). doi:10.1177/0539018405058216

Schwartz, G.E., Ahern, G.L., Brown, S.L.: Lateralized facial muscle response to positive and negative emotional stimuli. Psychophysiology **16**(6), 561–571 (1979). doi:10.1111/j.1469-8986.1979.tb01521.x

Thelwall, M., Buckley, K., Paltoglou, G., Cai, D., Kappas, A.: Sentiment strength detection in short informal text. J. Am. Soc. Inf. Sci. Technol. **61**(12), 2544–2558 (2010). doi:10.1002/asi.21416

Wiener, M., Devoe, S., Rubinow, S., Geller, J.: Nonverbal behavior and nonverbal communication. Psychol. Rev. **79**(3), 185–214 (1972). doi:10.1037/h0032710

Wyler, F., Graves, R., Landis, T.: Cognitive task influence on relative hemispheric motor control: mouth asymmetry and lateral eye movements. J. Clin. Exp. Neuropsychol. **9**(2), 105–116 (1987) doi:10.1080/01688638708405351

Chapter 14
Zooming in: Studying Collective Emotions with Interactive Affective Systems

Marcin Skowron, Stefan Rank, David Garcia, and Janusz A. Hołyst

14.1 Introduction

Computer-mediated communication between humans is at the center of the formation of collective emotions on the Internet. In parallel, the influence of artificial systems and services on the pace of information propagation and the selection of topics is growing and this needs to be accounted for in a holistic analysis of the emergence of collective emotions on networks. Interactive artificial systems can play an important role in studying the influences on the formation of collective emotions in cyberspace.

Numerous online services rely on computer algorithms for selecting and prioritizing content displayed to users. This enables websites that collect millions of units of information, constantly updated and served to millions of users in a timely and often personalized fashion. The scope of potential applications of such systems and their effects on personal and collective states can easily be underestimated by the average Internet user. Artificial systems can also actively search and establish contact with users and provide content. This ranges from the

M. Skowron (✉)
Austrian Research Institute for Artificial Intelligence, Vienna, Austria
e-mail: marcin.skowron@ofai.at

S. Rank
Department of Digital Media, Drexel University, Philadelphia, PA, USA
e-mail: stefan.rank@drexel.edu

D. Garcia
ETH Zurich, Zurich, Switzerland
e-mail: dgarcia@ethz.ch

J.A. Hołyst
Faculty of Physics, Warsaw University of Technology, Koszykowa 75, 00-662 Warsaw, Poland
e-mail: jholyst@if.pw.edu.pl

J.A. Hołyst (ed.), *Cyberemotions*, Understanding Complex Systems,
DOI 10.1007/978-3-319-43639-5_14

occasional dispatch of email or twitter messages to synchronous natural-language-based chats or task-specific dialogues in a written or oral form. The artificial systems have different functions and identities: from operators and supportive bots on internet communication channels, personal assistants on mobile devices, over 'non-playable characters' (NPCs) in online games and 'chat bots' applied as means of entertainment, to a range of service and information providers in commercial, health-care and educational contexts. The multi-layer impact of systems with different characteristics of their interactive behavior on the spread of information and emotions, on the formation of social bonds between users as well as on decisions concerning personal or group choices clearly reach beyond the online world.

In this chapter we focus on a subset of artificial systems which can be characterized by their strongly interactive nature: specifically involving direct, natural language based communication with users, paired with the ability to perceive, model, and elicit emotions. We elaborate here that such interactive affective systems (1) play an important role in the scientific exploration of the factors which influence the formation of individual and collective emotions in online setups, (2) show great potential for practical applications and (3) necessitate a discussion on ethical aspects related to their deployment in real world e-communities.

The rest of this chapter is structured as follows: First, we present the motivation for interactive studies of collective emotions and introduce practical examples of application scenarios for artificial interactive systems endowed with an awareness of affective phenomena. We also describe how such systems can be applied for interactive studies of processes related to the formation of collective emotions and provide an overview of the main components related to the computational awareness of emotions in artificial systems. The following section presents the experimental results obtained from a series of experiments with a specific realization of interactive affective systems. Next we enumerate challenges related to application of IASs in multiple users environment and present an overview of a theoretical framework and specific agent-based modelling approach which support the systems' decision making and influence their interactive and affective characteristics in such environments. Finally, we discuss ethical implications of the application of interactive affective systems for supporting online e-communities.

14.2 An Interactive Approach to Collective Emotions

Recent studies of collective emotions in online communication are focused on post-hoc analysis or modelling of valence exchanged in textual messages and the related interaction patterns in e-communities (Mitrović and Tadić 2011; Chmiel et al. 2011a; Thelwall et al. 2011, 2013; Hillmann and Trier 2012; Garas et al. 2012), as well as physiological responses related to the perception and generation of emotionally charged online content (Kappas et al. 2010). The interactive study of hboxe-communities complements these lines of work by employing affective interactive systems to provide experimental setups in order to (a) support studies on how interactions with artificial and human agents influence us in online virtual settings (Skowron 2010); (b) gather supplementary data on specified topics from selected groups, enabling the acquisition of relevant background information that is not available to post-hoc analysis (Skowron et al. 2011c); (c) evaluate theoretical models based on quantitative analyses of large data-sets by experimentally reproducing posited effects online (Rank et al. 2013). The approach also enables direct querying of users to discover a range of variables that are likely to influence affective reception, thus extending the scope of analysis to e.g. motivations and stances of particular users.

14.2.1 Examples of Application Scenarios for Interactive Affective Systems

Interactive affective systems (IAS) aim to model the affective dimensions of multiparty interactions, to simulate potential future changes in group dynamics, and, in part based on that information, to suitably respond to utterances both on the content and the affect level (Skowron and Rank 2014). In the following, we outline two general examples of realised application scenarios of IAS to illustrate the range of foreseen activities for such systems. The *Affective Interaction Analyser (AIA)* focuses on the analysis of interaction patterns, especially in multi-user environments. It is only concerned with the affective content of the exchanged textual messages and the tracking of group-level characteristics and is inactive in terms of interactions with casual users, except for infrequent messages provided to selected users such as e.g. Internet Relay Chat (IRC) channel operators. The *Affective Supporter and Content Contributor (ASCC)* participates to a moderate extent in ongoing discussions by providing new content related to the discussed topic or the results of affective group dynamic analysis and real-time simulations, communicating with the whole group.

14.2.2 Role of the Method

The role of the proposed method, i.e., interactive studies of e-communities, rests on the interplay between quantitative and qualitative research. This mutually beneficial relation is similar to practice in psychology or sociology: questionnaires and field studies complement statistical data and enable to (a) zoom in to particular groups of subjects, aspects of interactions, or topics of interest; and (b) investigate various hypotheses in direct interactions with users. Current post-hoc methods are limited to the acquisition of large data-sets of online communication paired with the application of algorithms for detecting sentiment expression. The non-interactive approach to studying collective emotions, by its very nature, can only acquire a limited amount of information, bounded by the state of the art in sentiment classification, information extraction, natural language processing and data mining. Additionally, information posted on the Internet often presents only a distilled summary of users' opinions on a given topic of interest, frequently missing any explicit statement of affect due to social convention or shared background knowledge, specific to a given community.

The validity of applying artificial IAS in experimental online setups in order to overcome those limitations is grounded in the influence that emotions exert on communication processes even in virtual setups. It has been argued that people treat interactions with certain media (including virtual agents) in the same way as interactions with real humans (Reeves and Nass 1996). Further, for example Forgas (2011) demonstrated that mood influences the disclosure of personal information both in real and virtual setups. Such disclosure is an essential part of human relationship formation (Altman and Taylor 1973). These studies provide the foundation for transferring experimental results obtained in human-computer interaction setups to human-human interaction. Other aspects relevant to the practical implication of experimental results, presented in this chapter and privy to this particular method, relate to the experimental evidence on people's increased willingness to cooperate with agents that are more human-like [mechanisms similar to kin selection (Hamilton 1964)], and are more engaged in interactions with agents that express emotions (Bates 1994; Lester et al. 1997).

A prerequisite for IAS usable for this method and that can further potentially affect collective states are mechanisms to assess the impact of collective emotions on online and offline processes, described in the following section.

14.3 Computational Awareness of Collective Emotions

Several aspects of the role of emotions and collective emotions in offline communities have been identified, such as: an increased potential for responding to opportunities or threats (Kemper 1991; Preston and Waal 2002); the benefit of emotional similarity for perceiving others' intentions and motivations (Hatfield et al. 1994; Levenson and Ruef 1992); or validating one's feelings and appraisals (Locke and Horowitz 1990; Rosenblatt and Greenberg 1991). The implementation of an awareness of the affective dimension of online interactions has significant benefits for the development of artificial systems. The ability to recognise these complex dimensions of human behaviour and related interaction patterns allows for an increased adaptability to social, real-world interactive settings. An understanding of the fundamental aspects of human emotions, caring about specific states of the world, and the subjective assessment of the relevance of any changes in that regard (Ellsworth and Scherer 2003; Petta 2003; Frijda 2007), is a step towards achieving beneficial human qualities such as cooperation, empathy, fairness, or reciprocity that rely on the concern of the well-being of others in artificial systems (Marsella et al. 2010).

In principle, interactive affective systems can communicate with users directly, provide new content to a group of users, and provide reports on group activities identifying interaction patterns. While the interaction is, in this sense, unrestricted, the chosen domain for an IAS is constrained in so far as the systems are concerned mainly with the emotional states of individual users as well as the dynamics of group and collective emotions and employ suitable strategies to keep the conversation going. This restriction is based on the strong influence of emotion in group dynamics, as mentioned above, as well as by consideration of practical feasibility of system complexity. The system, thus, analyses affective aspects of group interactions, utilising information about the network structure, state of collective emotions perceived in an e-community, and the contributions of individuals, and provides reports on the simulation outcomes for the group. In the following, we outline the components that are necessary to achieve this computational awareness of emotion. More detailed description of the components for computational awareness of emotion in IAS and specific methods and tools used for their detection, representation and modelling is presented in Skowron and Rank (2014).

14.3.1 Components for Computational Awareness of Emotion in Interactive Systems

A prerequisite for IAS that intentionally study and affect collective states are mechanisms to assess the impact of collective emotions on online and offline processes. There are methods for relating textual expressions to a range of affective dimensions but, additionally, the modelling of large e-communities at different levels that account for the role of emotions has been undertaken. The proposed realisation of computational awareness of collective emotions for IAS is based on these abilities to automatically detect and categorize emotional expressions, to model the dynamics of emotion exchanges in e-communities, and to interact with users directly including affect generation. We distinguish two (overlapping) layers of competencies for IAS: (a) Direct interactions, i.e. dialog analysis in 1-on-1 and multi-user settings, where analysis is focused on the perception of affective dimensions of the dialog and the identification of related aspects such as timing, style, novelty and coherence of contributions; and (b) Social/network interactions, i.e. perceiving, modelling and simulating interactive and affective dynamics in online groups, which includes accounting for the network structure and the roles of individual nodes as well as, for the ASCC scenario, the ability to select peers for interaction. The following list provides an overview of system components relevant for the scenarios outlined above organised along these two layers of competencies.

- Direct Interactions
  - Modelling of conversational partners
  - Modelling of self in 1-on-1 interactions

  - Affect detection
  - Affect generation and affective dialog management
  - Management of long-term interactions
- Social and Network Aspects
  - Modelling groups of individuals
  - Modelling of self in online social networks
  - Anticipation of effects of interactions in multi-user environments
  - Modelling and analysis of collective emotions

The development of systems capable to model and potentially influence emotional dynamics in groups based on such an computational awareness of emotion presents new opportunities to support e-communities. Making information on affective dynamics explicit and presenting this information or directly interacting with the group, e.g., by posting relevant novel content, has the potential to increase cooperation in a group or to counteract negative tendencies. In the following, we present the interaction scenarios and the experimental results which were primarily focused on the evaluation of such systems and studies of their impacts on users.

14.4 Interaction Scenarios for Experiments with an Affective Dialog System

The commonality between the experimental setups presented here is that the variants of the Affective Dialog System communicate with users in a predominantly textual modality, rely on integrated affective components for computational awareness of emotion, and use the acquired information to aid generation and selection of responses. The system variants interact with users via a range of communication channels and interfaces that share common characteristics of online chatting. In study 1 this consisted of a three-dimensional virtual reality setting, in studies 2 and 3 a web-chat like interface was used. Detailed description of the experimental systems is presented in the following publications: (Skowron et al. 2011b, 2013, 2014) Here we provide an overview of the experimental setups and main findings from the conducted tests.

14.4.1 Study 1: Virtual Reality WOZ Setting

The focus of the first study was the evaluation of the system in comparison with a Wizard-of-Oz (WOZ) setting as a prerequisite for the following studies. Further, we investigated the practical use of affective cues detected in user input for the generation of affective responses. As setting for the experiment, a Virtual Reality (VR) bar was created: a furnished virtual bar room and a virtual bartender with facial expressions and a chat interface, see Gobron et al. (2011) regarding the technical aspects of the VR setting.

The experimental setting consisted of the user, represented by an avatar (male or female according to the user's gender), interacting with a virtual human (male bartender). Each participant interacted four times, 5 min each, randomized order, in 2×2 conditions: The conversational partner was either a variant of the system called Affect Bartender (AB) or a Wizard of Oz (WOZ),[1] and the generation of emotional facial expressions was either active or not. In the AB condition, a simulation of thinking and typing speed was introduced to prevent an influence of differences in the response delivery time between the system and the human operator.

14.4.2 Study 2: Distinct Affective Profiles

In this study, an *artificial affective profile* was defined as a coarse-grained simulation of affective characteristics of an individual, corresponding to dominant, observable affective traits, that can be consistently demonstrated by a system during the course of its interactions with users (Skowron et al. 2011c). In this round of experiments, three distinct affective profiles were implemented for the dialog system—labeled as positive, negative and neutral—limiting variations to baseline levels of positive and negative affectivity in personality (Watson and Tellegen 1985). Each affective profile aims at a consistent demonstration of character traits of the system that are described as, respectively:

- polite, cooperative, empathic, supporting, focusing on similarities with a user;
- conflicting, confronting, focusing on differences with a user;
- professional, focused on the job, not responding to expressions of affect.

The study directly addressed research questions about the artificial system's ability to consistently simulate affective profiles.

14.4.3 Study 3: Key Social Processes

Affective profiles simulate human dispositional variations in baseline valence, sometimes referred to as "mood" (Watson and Tellegen 1985) or "affective home

[1]Participants were led to believe that they communicate with a dialog system, while responses are actually provided by a human operator. In the presented experiments, the operator adhered to general guidelines stating the objectives that needed to be achieved during the interactions. These included: providing realistic and coherent responses to the users' utterances and avoiding utterances that demonstrate an unusual sense of humor or eloquence. Before initiating the experiments, the operator conducted several rounds of pre-test interactions which helped to test and assure a consistency of his communication patterns, in line with the general instructions.

base" (Kuppens et al. 2010) that impact behavior patterns in different communication scenarios and are an important part of implementing different personalities. For this study, we chose two of those patterns in social human-human communication as reference points: getting acquainted and sharing of emotions (Skowron et al. 2013). When two previously unacquainted people meet for the first time, whether online or face-to-face, they try to reduce uncertainty by "getting to know" one another. They exchange personal information, making themselves known to one another, a process called *acquaintance* (Altman and Taylor 1973). Another important process at play when two people meet and converse is called *social sharing of emotion* (Rimé 2009). It consists in exchanging information about what happened to one another, and more specifically about the events which elicited emotional responses: e.g., anger over a recent public transport strike or happiness about one's latest personal success. Both processes—acquaintance and social sharing of emotion—are core components of everyday conversations, fostering relationship development and maintenance. They have been studied for several decades and can be elicited experimentally.

14.5 Experimental Results

In the following, we give a short overview of the main experimental results for each of the three studies introduced above.

14.5.1 Study 1

Interactions in all four experimental settings ($2 \times$ WOZ, $2 \times$ System) were completed by 35 participants (13 female, 22 male), aged between 20 and 50, resulting in 140 interaction logs. After each of the experimental interactions, lasting 5 min, participants were asked the following questions for assessing the conversational system (VH = Virtual Human, the label used for the graphical bartender avatar):

1. Did you find the dialog with the VH to be realistic?
2. How did you enjoy chatting with the VH?
3. Did you find a kind of emotional connection between you and the VH?

The participants provided their ratings on a six-point scale, i.e., from 1 = *not at all* to 6 = *very much*. Figure 14.1 presents the aggregated results obtained for the experimental settings with the Affect Bartender and for those with a Wizard-of-Oz.[2] In all three tasks, the results achieved by the conversational system match those

[2] In all figures, data are normalized by the number of utterances emitted by a user in a given interaction. Asterisks indicate significant differences at $p < 0.05$. Error bars indicate ± 1 standard error.

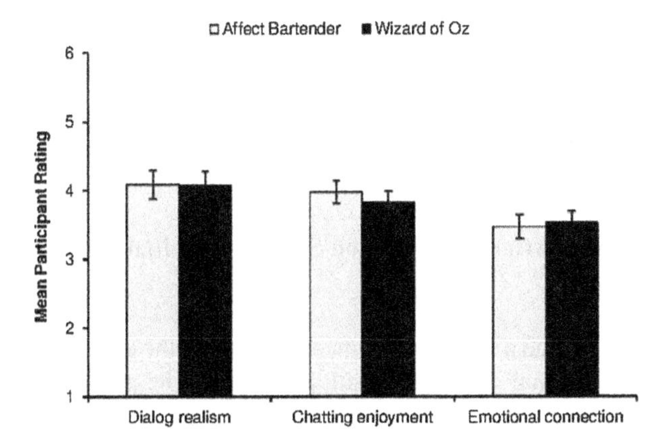

Fig. 14.1 Study 1: System vs. WOZ—evaluation results. Figure adapted from Skowron et al. (2011b)

obtained for the WOZ. In particular, the correlation coefficients for the aggregated AB and WOZ ratings varied between 0.95 (chatting enjoyment), 0.96 (emotional connection) and 0.97 (dialog realism). All these correlations differ from 0 at a significance level of 0.001. A repeated measures analysis of variance showed no main effect of the setting (AB vs. WOZ) on the three dependent measures (all Fs $(1, 34) < 0.50$, ps > 0.49). Pairwise comparisons with Bonferroni correction (Holm 1979) confirm the absence of significant differences between the two settings on the perception of dialog realism, chatting enjoyment, and subjective feeling of emotional connection with the system.

14.5.1.1 Comparisons Between System and WOZ Data

Comparisons between system and WOZ utterances demonstrate success in conveying interest in feelings and concerns of users, potentially *connecting* with the user. Specifically, examined with a lexicon-based sentiment classifier (Paltoglou et al. 2010), system utterances were both more positive, and more negative compared to WOZ utterances (repeated measures analysis of variance, Fs$(1, 34) > 36.61$, ps < 0.001). Simply put, system utterances were more emotionally loaded than WOZ utterances, both in the positive and in the negative orientation. Additionally, when examining utterances with LIWC personal concerns categories, we find that the system generated significantly more words related to work, home, money, religion, and death, compared to the WOZ (Fs$(1, 34) > 11.59$, ps < 0.01). In short, the system was able to talk more about emotions as well as potential user concerns, in an attempt to relate to users' feelings and interests.

14.5.2 *Study 2*

Interactions were completed by 91 participants (33 female, 58 male), aged between 18 and 52, in all three experimental settings resulting in 273 interaction logs.

14.5.2.1 Effects of Affective Profile on System's Evaluation and Emotional Changes

The affective profile had a series of significant effects on the evaluation of the system and on users' emotional changes. Detailed results of the analyses performed are presented in Skowron et al. (2011c,a). Concerning evaluation, the affective profile had significant effects on all dependent measures (see Fig. 14.2). The largest effect sizes were found on statements 5 and 6 (positive and negative emotional change, respectively). As expected, affective profiles successfully induced corresponding emotional changes in users, affecting perception of dialog realism and coherence only to a smaller extent.

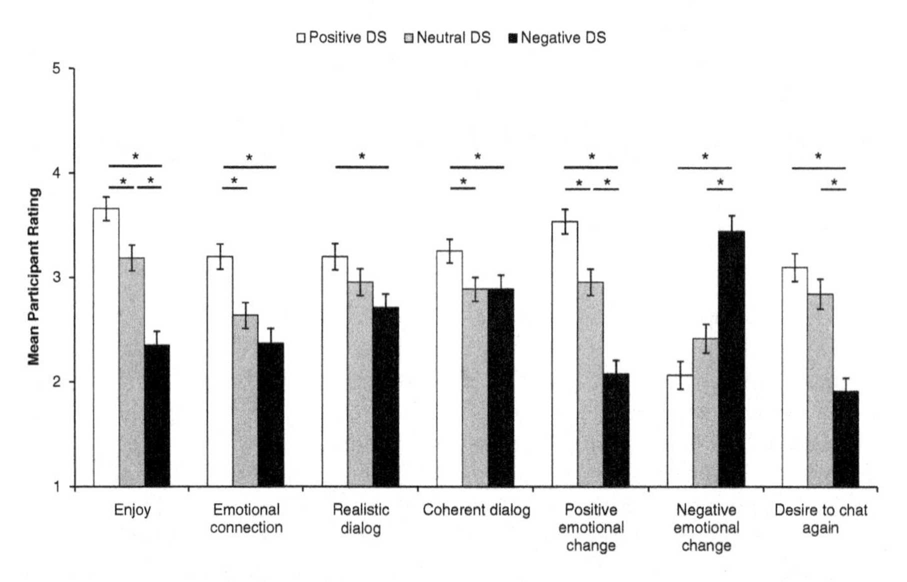

Fig. 14.2 Participant's mean ratings on all dependent variables, of their interactions with the dialog system (DS) with three different affective profiles (positive, neutral, negative) (Skowron et al. 2011c)

14.5.2.2 Effects of Affective Profile on Users' Interaction Style

Users were equally fast in replying to different affective profiles. Specifically, when analyzing the whole interactions, there were no differences in the participants' average response time to a number of letters generated. They also used an equal amount of words and utterances in their conversations for all profiles. There were, however, significant differences in word categories used and other linguistic aspects of the text input. Among others, compared with the positive profile, the negative profile elicited, as expected, less assent (e.g., ok, yes, agree) from users, fewer positive emotion words, more anger-related words, and utterances assessed as significantly less positive by the sentiment and ANEW classifiers (see Fig. 14.3). The positive profile, on the other hand, elicited accordingly more positive emoticons, more positive emotion words (e.g., love, nice, sweet), more user statements, and

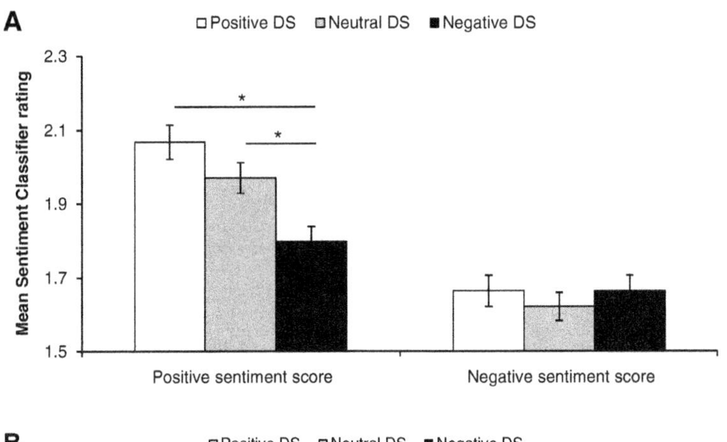

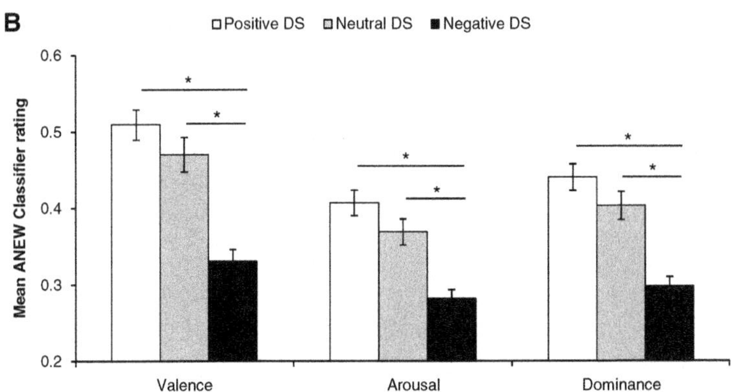

Fig. 14.3 Valence, arousal, and dominance ratings in user exchanges with the dialog system (DS). Panel (**a**) shows the mean positive and negative Sentiment Classifier score per condition. Panel (**b**) shows the mean valence, arousal, and dominance scores based on the ANEW lexicon (Skowron et al. 2011a)

less closed questions to the system. The two latter findings might indicate more information disclosure and less questioning from users towards the positive profile, compared to the negative profile.

14.5.3 *Study 3*

Interactions were completed by 75 participants (38 female, 37 male), aged between 18 and 32, in all three experimental settings resulting in 225 interaction logs.

14.5.3.1 Effect of Communication Scenario on System's Evaluation

A multivariate repeated measures analysis of variance showed the expected absence of effect of communication scenario on users' evaluations of the system (Wilk's $\lambda = 0.90$, $F(14, 276) = 1.07$, $p = 0.39$). In other terms, participants judged the three communication scenarios (neutral, getting acquainted, and social sharing of emotion) as equally enjoyable, coherent, realistic, emotionally connecting, etc. (univariate tests: all $Fs(2, 144) < 2.47$, $ps > 0.09$).

14.5.3.2 Effects of Communication Scenario on Users' Interaction Style

Words and Timing Participants conversed significantly more in the "social sharing of emotion" scenario, despite the time limitation. There was a main multivariate effect of communication scenario on indicators of conversation length (Wilk's $\lambda = 0.84$, $F(6, 284) = 4.48$, $p < 0.001$). Univariate analysis showed that each length indicator, i.e., character, word, and utterance count, was affected by the communication scenario ($Fs(2, 144)>3.03$, $ps \leq 0.05$). Post-hoc comparisons, shown in Table 1 (Skowron et al. 2013), revealed that participants wrote significantly more when sharing an emotional episode.

Confirming these results, we found a main effect of communication scenario on response time ($F(2, 144) = 8.29$, $p < 0.001$). Specifically, participants wrote significantly faster in the social sharing of emotion scenario, compared to the two other conditions.

Dialog Act Classes Participants wrote significantly less statements overall in the social sharing scenario, and significantly more statements about ordering food and drinks in the neutral scenario ($Fs(2, 144)>5.63$, $ps<0.01$).

14.5.3.3 Effect of Communication Scenario on User's Expression of Affective States

Sentiment Classifier, and ANEW Lexicon There was a main effect of communication scenario on both positive sentiment score ($F(2, 144) = 5.59$, $p < 0.001$) and ANEW ratings (Wilk's $\lambda = 0.60$, $F(6, 284) = 13.77$, $p < 0.001$). As depicted in Fig. 14.4, participants wrote significantly less positive, less arousing, and less dominant utterances in the social sharing of emotion scenario; note the negative context of the sharing introduced by the system's "personal experience". The largest effect size is found for ANEW valence ratings ($\eta_p^2 = 0.22$) first indication of the

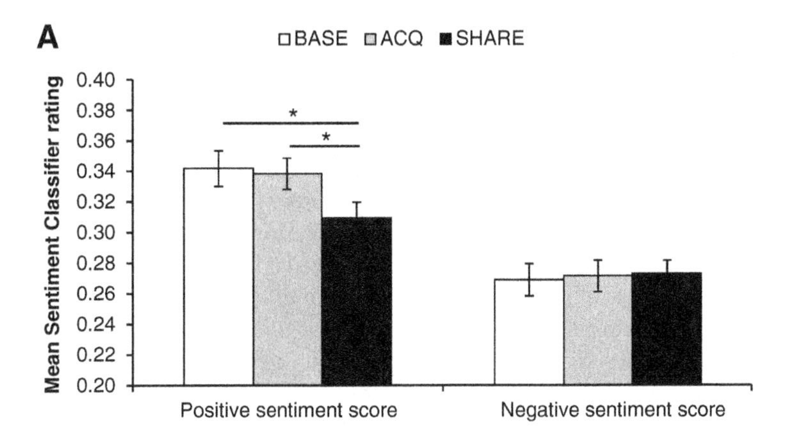

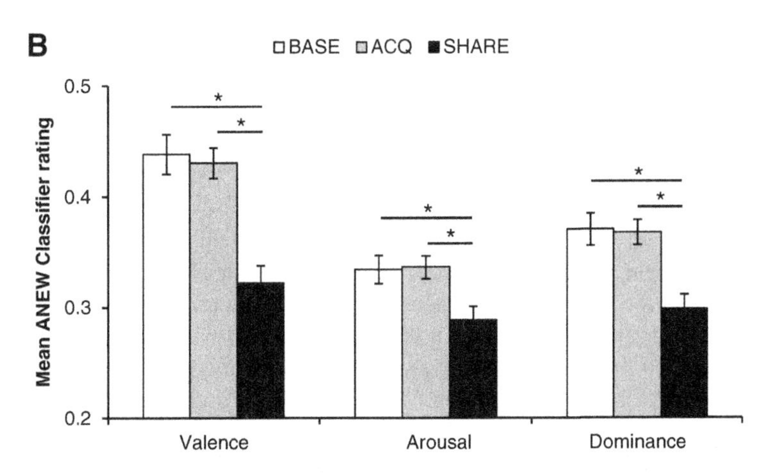

Fig. 14.4 Study 3: Effect on expressions of affective states and sentiment. Communication scenarios: BASE–neutral communication scenario, ACQ–getting acquainted scenario, SHARE–social sharing of emotion scenario. Panel A shows the mean positive and negative sentiment classifier score; panel B shows the mean valence, arousal, and dominance scores based on the ANEW lexicon (Skowron et al. 2013)

successful creation of a *social sharing of emotion* situation, where participants shared a negative experience, which did not happen in the two other conditions.

14.5.4 Summary of Experimental Results

In the presented experiments, a variant of IAS, the Affective Dialog System, was applied to study communication processes and the impact of affect and social processes on users in online interactions. The experiment specifically focused on challenges of structuring emotional interactions in open-domain dialogs limited to the textual modality, and on investigating the impact of such interactions on users. The first study, conducted in a WOZ setting, validated the system's ability, on par with a human operator regarding realistic and enjoyable dialog as well as for establishing an emotional connection with users in a short interaction. The second study confirmed a significant influence of the system's affective profile on users' perception of a conversational partner, on the changes of emotional states reported by the users, and on the textual expressions of affective states. The third study proved a successful application of the dialog system for eliciting social sharing of emotion and for realizing a communication scenario of 'getting acquainted'. The experimental results show the impact of the realized communication scenario on the communication style of users, their expressions of affective states and self-reported emotional changes experienced when interacting with the system.

The reported experiments applied realizations of IASs in dyadic and triadic settings. The next step in our research, the application of IASs to group interactions creates a new set of challenges related with, e.g., simultaneous communication with multiple users (beyond triadic settings), capacities to interact in a way which intentionally follows or violates the typical communication patterns of members of a particular online community.

14.6 Scaling up to Multiple Users Environments

To date, different realizations of IASs have been applied in experimental setups where the system interacted in dyadic and triadic settings. The purpose of these experiments was to evaluate the system's ability to participate in a coherent dialog, and to establish and maintain an emotional connection with users. These experiments shed light on how affective system profiles and fine-grained communication scenarios impact the self-reported emotional changes of users, their communication style, and their textual expressions of affect. The next step, the application of IASs to group interactions created a new set of challenges related with, e.g., simultaneous communication with multiple users (beyond triadic setting), capacities to interact in a way which intentionally follows or violates the typical communication patterns of members of a particular online community. This included factors such as the affective dimension of interactions, and the ability to observe such a behaviour in other participants. These functionalities impact the system's capacity to generate consistent or intentionally inconsistent interactive behaviour, the required affective coherence and the event-dependent adaptation of its interaction patterns to other members in a group. Simultaneously, the system needs to perceive, represent and model discussions and emotional exchanges, at the individual and group levels. This prerequisites abilities to: (i) predict the possible outcomes of the observed group dynamics, (ii) simulate the effects of system's interactions with individuals or a group, and (iii) to assess the real effect of its interventions and to correspondingly update the used models.

To address the requirements related with transferring IASs from dyadic and triadic to multi-user interaction settings, the proposed approach integrated experience gathered in conducted experiments with insights from a wide range of studies on the role of emotions in online communication: psychological studies and experiments on perception and generation of emotionally charged online content (Kappas et al. 2010; Küster et al. 2011; Küster and Kappas 2014), event-based network discourse analysis (Hillmann and Trier 2012), agent based models of emotions (Schweitzer and Garcia 2010; Garas et al. 2012), valence trends (Chmiel et al. 2011a,b) and agent based model on bipartite networks (Mitrović et al. 2011; Gligorijević et al. 2013), presented in the earlier chapters. For more extensive presentation of the models, their transfer and integration in IASs refer to Skowron and Rank (2012) and Rank et al. (2013).

14.6.1 Role of Simulations in IAS

The role of agent-based modelling and simulation in IASs is twofold: (i) to provide cues on information provided to participants, and (ii) to serve as decision support. For example, based on a request from a particular e-community the tools can support it with analysis on affective dimension of their interactions and suggest ways for counter-acting negative tendencies observed in a group, e.g., growing hostility between members, or a decrease of cooperation. Simulation results can also indicate targets for interventions. This entails several requirements that relate to the results of simulation runs, runtime characteristics and the adaptability of the simulation based on data collected during previous interactions. Here, several questions relevant as a potential input for the systems' decision making mechanisms were identified (Rank 2010):

- Which individual in a group will be most likely to provide an accurate response to probing about the group's emotional state, and which one will be most reliable?
- What influence can individuals have on the evolution of the collective emotions in an e-community, and which of the specific participants is likely to have the biggest influence?
- Can potential escalations, both in the negative and in the positive direction, be detected early on?
- What influence will a specific intervention of the system have at the current moment, and which style of intervention is most effective?

Running a simulation on demand to query about the above questions adds the requirement of timely, or possibly anytime, responses but also the need to parameterise the simulation to promptly respond to the current state of an online community, ideally using the recorded history as input. An important part of the decision-making structures of IASs is the modelling of conversation participants. This component of the agent control structure is analogous to adaptive user modelling in standard HCI: the system initially has a default model of the interaction partner, adapts it over time, and complements missing information based on the knowledge derived from interaction events. In the case of multi-user environments, this includes modelling several participants, simplifying the employed models and abstracting from specific individuals. The modelling eventually serves the purpose of deciding on utterance selection, utterance modification, timing of utterances, and the selection of conversational partners in multi-user environments. As such, the main questions that modelling efforts helps to answer for the purposes of affective interactive systems are, from general to specific:

- What potential influence will certain interventions have on the collective state of an online community?
- What is the influence of particular interventions on the future development of a specific group?
- What type of intervention (affective charge, topic, timing) will have which effect?
- What relation does a particular individual have to the state of a specific group?
- Which intervention is most appropriate when addressing a particular individual of a specific group?

14.6.2 Input from Theoretical Modelling and Analysis

The questions presented above relate the decisions of individual agents to the collective emotions of the group, and require a dedicated approach that can deal with the relation between individual and collective levels. We extended our simulation of emotions in IAS with a model for the real-time evolution of emotional states, based on the modeling framework presented in Chap. 10 (Schweitzer and Garcia 2010). Within this framework, we can define the dynamics of individual emotions based on the concept of *Brownian agents* (Schweitzer 2003), integrating empirical data with analytical results that show the emergence of collective emotions in multi-user environments.

Applying this framework, we can integrate previous findings from very different online communities, including product reviews communities (Garcia and Schweitzer 2011), social networking sites (Šuvakov et al. 2012), and blogs (Mitrović and Tadić 2012). A particularly relevant application for IASs is the model for emotional persistence in online chatting communities presented in Chap. 10 (Garas et al. 2012), which provides a description of individual emotion dynamics for the users of a real-time group discussion.

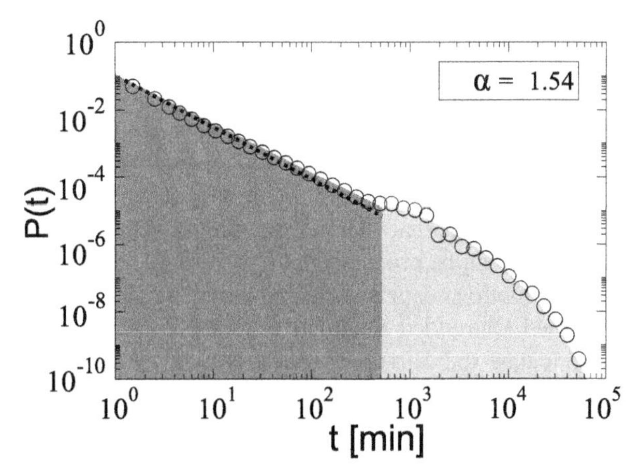

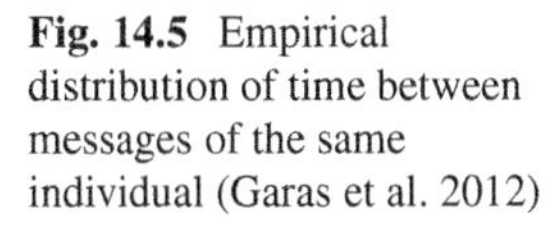
Fig. 14.5 Empirical distribution of time between messages of the same individual (Garas et al. 2012)

14.6.2.1 Activity Patterns in Time

The individual dynamics of emotions in our model are based on the empirical analysis of user interaction in real general discussions in Internet Relay Chat (IRC).[3] We analyzed the time component of the interaction in the chatroom through the distribution of time intervals between messages of the same user, shown in Fig. 14.5.

Similarly to the case of short message service communication (Wu et al. 2010), this distribution has two modes, one corresponding to the time intervals between bursts of interaction (light gray), and another one containing the time intervals within an interaction burst (dark gray). The latter is characterized by a power-law distribution of the form $P(\Delta t) \sim \Delta t^{-\alpha}$ for $\alpha = 1.54$, where the former is better explained by a lognormal distribution. To provide the most realistic behavior possible, the agent actions in our model are chosen to emulate this empirical distribution. For the case of IASs, every reaction has an additional delay Δt, sampled from the power-law regime of the distribution, i.e. the inter-activity times associated with group discussions.

The addition of this stochastic delay changes the properties of the IAS in a substantial manner that is commonly ignored in this kind of system. A usual design decision driven by intuition is assuming that the time between messages of an agent can be sampled from a distribution with a given mean (e.g. normal). On the other hand, the empirical distribution of time intervals between messages in a group IRC discussion is very different, following a power-law of exponent 1.54, which does not have a finite first moment. Therefore, empirical evidence contradicts the assumption mentioned before, and consequently the model used in IASs has a time behavior closer to the real patterns in online interaction.

[3]The analysed data-set included 2.5 million posts acquired from EFNET IRC chats: http://www.efnet.org, covering a range of topics including music, casual chats, business, sports, politics, computers, operating systems and specific computer programs.

14.6.2.2 Emotional Persistence

Sentiment analysis techniques like the ones described in Chap. 6 can extract the emotional content of short chat-room messages, giving a value of polarity per message. Using this method a sequence of messages would be represented as an ordered set of ones for positive messages, zeroes for neutral messages, and minus ones for negative messages. Assuming a sufficiently large data set, these kinds of sequences can be processed with tools from statistical physics, revealing properties of the emotions expressed at individual and group levels. One of these techniques is called Detrended Fluctuation Analysis (Peng et al. 1994), which can be used to calculate the Hurst exponent (Hurst 1951; Rybski et al. 2009), a measure of persistence in a time series. This persistence is reflected in the time series as consistent fluctuations around its mean, revealing states of consistent biases towards positive or negative emotions.

Persistence can be calculated for an individual user, when processing the sequence of emotional expressions of that user, or at the discussion level, when computing it over all the messages in a group discussion. If a user behaves in a fully random way, with fixed probabilities of each sentiment polarity but no memory, it would have a persistence of 0.5 and a mean emotional expression according to the given probabilities. A user with some emotional momentum, who tends to express emotional states similar to previous ones, would have a persistence above 0.5. This case of persistent users is the most common one, which can be appreciated in the kernel density plot of Fig. 14.6 in combination with the distribution of mean emotional expression. Most of the users show a bias towards positive emotional expression, with persistence in the way these emotions are expressed though the discussion.

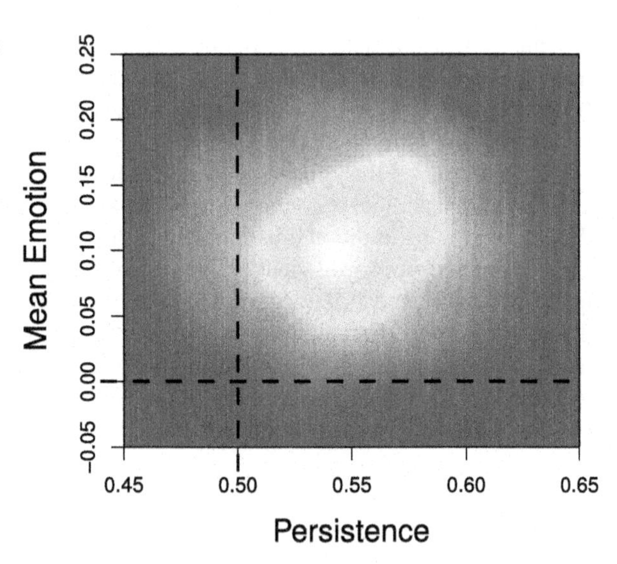

Fig. 14.6 Kernel density plot of the distribution of individual persistences and mean emotional expression, for all users with more than 100 messages in the observed period (Garas et al. 2012)

We measure the collective emotions in the whole discussion by computing the persistence over all the messages in the group chat. The empirical analysis of 20 IRC channels (Garas et al. 2012) showed persistence values above 0.5 for all channels, revealing the existence of collective emotional states at the aggregated level. This appears in the discussion as fluctuations around the mean emotional content, which is also in general biased towards positive emotions. This collective positivity and persistence shows that users influence each other through their expression of emotions, and that certain rules of emotional expression are present even in the anonymous and ephemeral discussions of IRC chat-rooms.

14.6.2.3 Brownian Agents Model

Although a persistence pattern for an individual user can be easily modeled with a biased random walk, the emergence of such collective persistence does not trivially follow from individual emotional expression. If we design a model which only ensures persistence in the behavior of an individual, the combination of expressions of many agents would show no persistence at the group level. Our aim is to provide realistic emotion dynamics, useful for IASs, in which collective persistence appears in a discussion with many agents, which we reproduce with the Brownian agents model.

The agent dynamics of this model are outlined in Chap. 10, where the arousal dynamics is given by the time interval distribution shown in Fig. 14.5. Agents in this model have an externally observable variable s, which corresponds to the messages posted by the IAS in the collective discussion. This variable of expression is activated according to the arousal, and its value is given by the internal valence of the agent. The communication in the group discussion is modeled through an information field, which aggregates the emotional expression of all the agents. This field influences the dynamics of individual valences, creating a herding effect that reproduces collective persistence from the interaction of many individuals. Simulations of this model show how mean emotional expression and persistence depend on the parameter values (Garas et al. 2012), allowing us to simulate discussions that reproduce the observed group behavior. In terms of individual dynamics, the emotional profile of an agent can be sampled from the distribution shown in Fig. 14.6, choosing values of emotional tendency and persistence for the agent. Such Data-driven simulations with the corresponding parameter values lead to the desired individual and collective persistence effects.

14.6.2.4 Model Extensions

The Brownian agents model outlined here is a first approximation to the dynamics of emotions in e-communities that can be improved and extended with additional features within the same modelling framework. The assumptions followed by models like the one explained above are currently being tested in experimental setups like the ones introduced in Chap. 5, which already revealed the patterns of influence of online interaction in internal emotions. In this model, interaction takes place in a publicly accessible discussion in which messages are not directed to individual users. Some of the questions listed in Sect. 14.6.1 aim at driving IASs in a way such that they can interact with individual conversation participants, by previous analysis of their position and influence in the overall discussion. Some models within this framework include a network component that precisely deals with this kind of user heterogeneity, as explained in Chap. 11 (Mitrović and Tadić 2012; Šuvakov et al. 2012).

In real-time interaction, the network mapping approach (Gligorijević et al. 2012a) analysis[4] accounts for the properties of activity patterns and underlying network topologies characteristic for various types of users, including those identified as important/influential in a given online interaction environment. In IASs, the spanning trees analysis can be specifically applied to analyze: (i) the user's activity, i.e.: by the creation and analysis of evolution of the network links, e.g., positive, negative, and (ii) users collective behaviour patterns. As demonstrated in high-resolution analyses of user-to-user communication in IRC channels, only certain links survive over 1 day period and support a particular type of network structure. Based on this observation, the presented system realizations and application scenarios presented in Sect. 14.2.1 are primarily targeted at serving on-demand information requests of e-communities or individuals, i.e., establishing a relatively short-time direct communication links. As experimental evidence demonstrates, such direct, limited in time interactions with users, also contribute to an overall higher level of the perceived realism, dialog coherence, the feeling of an emotional connection and chatting enjoyment (Skowron and Rank 2014). In practical application and deployment scenarios, this often translates to a higher acceptance rate of interactive systems in online communities.

14.7 Ethical Considerations

Endowing interactive systems with the capability to model complex affective processes and to account for affective dynamics in networks opens up a range of applications for assisting e-communities; however, it also allows for malicious misuse. "Short-sighted" commercial, political, or ideological causes of individuals or groups could lead to the use of the technology for generating, amplifying, or reducing naturally occurring collective emotions. Such misuse could result in an "arms race" between individuals or groups with competing interests resulting in a decrease of trustworthiness of online communication channels.

This leads to a question on how these technologies should be developed, tested, and applied in real-world e-communities. The aim of such a debate is early

[4]Analysis conducted on extensive data-sets from Ubuntu IRC channels: http://irclogs.ubuntu.com/.

anticipation of risk and identification of likely consequences, both beneficial and harmful related to field of research and its real-world application in the line of Novotny et al. (2003) who postulate: "against a background of inherent uncertainty about the future state of knowledge (and almost everything else) from which, of course, scientific potential is derived, it is necessary to reach beyond the knowable context of application to the unknowable context of implication. Here knowledge producers have to reach out and anticipate reflexively the implications of research processes". The discussion further requires public dialog, broader than user involvement, while the role of researchers in this aspects shall relate to analysis and explanation of dystopian and utopian scenarios, identification and development of strategies to avoid likely harmful scenarios, and promote likely beneficial ones (Prescott 2013).

In a scientific setting, the acquisition of data and the engagement of experimental subjects should always be accompanied with ethical considerations and the adherence to a code-of-conduct that is intended to protect the rights of the individuals involved. In our opinion, the usage of such technologies in non-laboratory settings requires the consideration of rules of engagement similar to a code-of-conduct that specifies the scope of interventions and ascertains an equivalent of informed consent both from individuals and from e-communities as a whole as far as possible. The new opportunities offered need to be balanced with the potential threats of mis- and abuse. As a minimal rule of engagement, we assume *open identification of such systems* when interacting with users so that they are informed about the presence of this influence.

In a range of application scenarios, a prerequisite for a successful application of such interactive systems could be also the ability to adjust (or at least foresee the outcomes of an intentional violation of) one's communication behavior or affective stance according to: the overall mood detected in a group; individuals' preferences to various entities or fellow participants; the established or evolving "social norms"; or dynamic changes in a hierarchy of interaction patterns of users. The ability to follow, i.e., perceive, model and adhere to *social norms*[5] used in a given e-community could be treated as an additional prerequisite for the deployment of IASs to the real-world online groups.

Such a convention can be advantageous for application development too, since the acceptance of artificial systems in online communities depends on the perceived *benefits* and *costs* associated with the systems' activities. For the purpose of IASs, this dependency is translated in the proposed systems directly into their communications policy: the system communicates directly with users in situations where the relevance of a potential contribution, i.e. added informational or affective value—*contribution value*, is estimated to be sufficient with high confidence. These estimates are based on user and group models for the impact of specific content. Additional *action costs* are associated with interaction scenarios involving larger numbers of users or, progressively more costly actions, which include the

[5]In an absence of other, higher level design principles.

community as a whole, e.g. where posts need to be emitted to all users at the same time. This is also reflected in an IAS's ability to perceive and model the affective stance of individuals and groups towards itself and to adapt in order to improve its perceived usefulness.

Further specific areas for discussion also include question on how e-communities could control the extent to which artificial systems influence their communication patterns including both the exchange of information, an area that has been studied for quite some time, as well as the affective dimension, a relatively new area with a potentially stronger impact.

A positive example for an emerging application scenario is the application of IASs for analysing and acquiring information on online dynamics in interaction environments where large number of young user are involved, e.g., detecting the formations of affective dynamics specific for cyber-bulling or monitoring for other anomalies which can be observed by analysing and modelling of affective communication patterns, or changes in the underlying network and communication structures.

14.8 Conclusions and Outlook

The rapid expansion and evolution of ICT-mediated communication channels requires methodologies tailored for studying collective emotions in e-communities and the development of tools that can support the evolving needs of these communities. Agents' online interactions, mediated over a range of communication channels, have the potential to influence the processes of formation and evolution of collective emotions. Emotional effects can rapidly spread across the virtual realm regardless of the physical location of peers or their offline-world group identification. Consequently, even a single agent's actions and interactions with other members of e-communities can have a strong impact on collective emotions there. This applies both for the human and artificial agents and the presented experiments demonstrate that artificial systems can be useful for studying this impact experimentally and in an interactive way that complements other forms of study.

The system applications presented in this chapter focus on text-based, real-time system-user communication aimed at the detection, acquisition and modeling of users' affective states. The scope of communication with the system was not limited to a specific domain or one particular e-community. Clearly, interactive systems like these fall short in their conversational abilities in comparison with humans, in particular in longer lasting interaction scenarios, potentially integrating open- and closed-domain dialog and discourse processing. However, as the experimental evidence demonstrates for relatively short communication scenarios, the system could match the WOZ results in terms of dialog realism, chatting enjoyment and the ability to establish an emotional connection with subjects. The analysis of activity patterns and affective dimensions of users' communication in multi-user environments revealed that the majority of links are established only temporarily and are used primarily to exchange relevant information, to share or respond to a sentiment expressed. These findings support the application scenarios where an IAS establishes communication links similarly to human counterparts: briefly and on demand. Further, the system should be able to predict likely outcomes of its own behaviour, either following or intentionally violating established or evolving social norms, communication conventions, or the communicative style in a group, in order to adjust its behavior accordingly. The modelling of self, including the affective reception of content provided, both in 1-on-1 and multi-users interactions, allows for adjustments of thresholds used for calculating action benefits and costs. Correspondingly, as a starting point, we proposed application scenarios where systems communicate directly with users in situations where high confidence scores for a potential contribution's relevance, i.e. added informational or affective value—*contribution value*, can be foreseen. Estimates are based on outcomes of simulating the reception of a specific content by a particular individual or a group. Further, additional *action costs* should be accounted for in interaction scenarios where posts need to be emitted to a large number of participants or to the whole e-community. In this chapter, we also presented design choices and specific models for applying agent-based simulation as part of the decision mechanism in the presented systems.

The development and application of Interactive Affective Systems relates to a range of scientific challenges from affect detection, modelling, and generation, to the modelling of interactive, affective and social dimensions in individuals and groups over time. In parallel, models need to account for the dynamics and complexity of underlying social networks as well as of aspects of human-computer interaction. The type of system introduced here can act, e.g., as e-community advisers aiming at increased cooperation in a group, counteracting growing hostilities, and yielding measurable social benefits. However, there are also application scenarios where such systems can potentially be abused for achieving adverse effects. As the study and application of artificial affective systems is expanding, there is also an urgent need to define suitable ethical rule-sets, i.e. code of conduct for real-world use in online communities.

References

Altman, I., Taylor, D.: Social Penetration: The Development of Interpersonal Relationships. Holt, Rinehart and Winston, New York (1973)

Bates, J.: The role of emotion is believable agents. Commun. ACM **37**(7), 122–125 (1994). doi:10.1145/176789.176803

Chmiel, A., Sienkiewicz, J., Thelwall, M., Paltoglou, G., Buckley, K., Kappas, A., Hołyst, J.A.: Collective emotions online and their influence on community life. PLoS ONE **6**(7), e22207 (2011a). doi:10.1371/journal.pone.0022207

Chmiel, A., Sobkowicz, P., Sienkiewicz, J., Paltoglou, G., Buckley, K., Thelwall, M., Hołyst, J.A.: Negative emotions boost user activity at BBC forum. Physica A **390**(16), 2936–2944 (2011b) doi:10.1016/j.physa.2011.03.040

Ellsworth, P.C., Scherer, K.R.: Appraisal processes in emotion. In: Davidson, R.J., Scherer, K.R., Goldsmith, H.H. (eds.) Handbook Of Affective Sciences, pp. 572–595. Oxford University Press, Oxford (2003)

Forgas, J.P.: Affective influences on self-disclosure: mood effects on the intimacy and reciprocity of disclosing personal information. J. Pers. Soc. Psychol. **100**(3), 449–461 (2011) doi:10.1037/a0021129

Frijda, N.H.: The Laws of Emotion. Lawrence Erlbaum Associates Publishers, Mahwah (2007)

Garas, A., Garcia, D., Skowron, M., Schweitzer, F.: Emotional persistence in online chatting communities. Sci. Rep. **2**, 402 (2012). doi:10.1038/srep00402

Garcia, D., Schweitzer, F.: Emotions in product reviews – empirics and models. In: Proceedings of 2011 IEEE International Conference on Privacy, Security, Risk, and Trust, and IEEE International Conference on Social Computing, PASSAT/SocialCom, pp. 483–488 (2011). doi:10.1109/PASSAT/SocialCom.2011.219

Gligorijević, V., Skowron, M., Tadić, B.: Evolving topology on the network of online chats. Technical Report TR-2012-14, Österreichisches Forschungsinstitut für Artificial Intelligence, Wien (2012a)

Gligorijević, V., Skowron, M., Tadić, B.: Structure and stability of online chat networks built on emotion-carrying links. Physica A **392**(3), 538–543 (2013). doi:10.1016/j.physa.2012.10.003

Gobron, S., Ahn, J., Silvestre, Q., Thalmann, D., Rank, S., Skowron, M., Paltoglou, G., Thelwall, M.: An interdisciplinary VR-architecture for 3D chatting with non-verbal communication. In: Coquillart, S., Steed, A., Welch, G. (eds.) JVRC11: Joint Virtual Reality Conference of EGVE - EuroVR, pp. 87–94. Eurographics Association (2011). doi:10.2312/EGVE/JVRC11/087-094

Hamilton, W.: The genetical evolution of social behaviour. J. Theor. Biol. **7**(1), 1–16 (1964). doi:0.1016/0022-5193(64)90038-4

Hatfield, M., Cacioppo, J., Rapson, R.: Emotional Contagion. Cambridge University Press, Cambridge (1994)

Hillmann, R., Trier, M.: Sentiment polarization and balance among users in online social networks. In: Joshi, K.D., Yoo, Y. (eds.) Proceedings of the 18th Americas Conference on Information Systems AMCIS 2012, Seattle, WA, p. 10 (2012)

Holm, S.: A simple sequentially rejective multiple test procedure. Scand. J. Stat. **6**(2), 65–70 (1979)

Hurst, H.: Long-term storage capacity of reservoirs. Trans. Am. Soc. Civ. Eng. **116**, 770–799 (1951)

Kappas, A., Küster, D., Theunis, M., Tsankova, E.: Cyberemotions: subjective and physiological responses to reading online discussion forums. Poster presented at the 50th Annual Meeting of the Society for Psychophysiological Research, Portland, Oregon, September 2010 (2010)

Kemper, T.: Predicting emotions from social relations. Soc. Psychol. Q. **54**(4), 330–342 (1991)

Kuppens, P., Oravecz, Z., Tuerlinckx, F.: Feelings change: accounting for individual differences in the temporal dynamics of affect. J. Pers. Soc. Psychol. **99**(6), 1042–1060 (2010). doi:10.1037/a0020962

Küster, D., Kappas, A.: Measuring emotions in individuals and Internet Communities. In: Benski, T., Fisher, E. (eds.) Internet and Emotions, pp. 48–64. Routledge, New York (2014)

Küster, D., Kappas, A., Theunis, M., Tsankova, E.: Cyberemotions: subjective and physiological responses elicited by contributing to online discussion forums. Poster presented at the 51st Annual Meeting of the Society for Psychophysiological Research, Boston, MA, September 2011 (2011)

Lester, J.C., Converse, S.A., Kahler, S.E., Barlow, S.T., Stone, B.A., Bhogal, R.S.: The persona effect: affective impact of animated pedagogical agents. In: Pemberton, S. (ed.) CHI '97 Proceedings of the ACM SIGCHI Conference on Human Factors in Computing Systems, pp. 359–366. ACM, New York (1997). doi:10.1145/258549.258797

Levenson, R., Ruef, A.: Empathy: a psychological substrate. J. Pers. Soc. Psychol. **63**(2), 234–246 (1992) doi:10.1037/0022-3514.63.2.234

Locke, K., Horowitz, L.: Satisfaction and interpersonal interactions as a function of similarity in level of dysphoria. J. Pers. Soc. Psychol. **58**(5), 823–831 (1990). doi:10.1037/0022-3514.58.5.823

Marsella, S., Gratch, J., Petta, P.: Computational models of emotion. In: A Blueprint for Affective Computing - A Sourcebook and Manual, pp. 21–46. Oxford University Press, Oxford (2010)
Mitrović, M., Tadić, B.: Patterns of emotional blogging and emergence of communities: agent-based model on bipartite networks (2011). http://arxiv.org/pdf/1110.5057
Mitrović, M., Tadić, B.: Dynamics of bloggers' communities: bipartite networks from empirical data and agent-based modeling. Physica A **391**(21), 5264–5278 (2012). doi:10.1016/j.physa.2012.06.004
Novotny, H., Scott, P., Gibbons, M.: Mode 2' revisited: the new production of knowledge - introduction. Minerva **41**(3), 179–194 (2003) doi:10.1023/A:1025505528250
Paltoglou, G., Gobron, S., Skowron, M., Thelwall, M., Thalmann, D.: Sentiment analysis of informal textual communication in cyberspace. In: Proceedings of ENGAGE 2010. LNCS State-of-the-Art Survey, pp. 13–25. Springer, Heidelberg (2010)
Peng, C.K., Buldyrev, S.V., Havlin, S., Simons, M., Stanley, H.E., Goldberger, A.L.: Mosaic organization of DNA nucleotides. Phys. Rev. E **49**(2), 1685–1689 (1994). doi:10.1103/PhysRevE.49.1685
Petta, P.: The role of emotions in a tractable architecture for situated cognisers. In: Trappl, R., Petta, P., Payr, S. (eds.) Emotions In Humans And Artifacts. pp. 251–288. MIT Press, Cambridge (2003)
Prescott, T.: Sunny uplands or slippery slopes? the risks and benefits of using robots in care. In: Proceedings of UKRE Workshop on Robot Ethics (2013)
Preston, S., Waal, F.: Empathy: its ultimate and proximal bases. Behav. Brain Sci. **25**(1), 1–20 (2002). doi:10.1017/S0140525X02000018
Rank, S.: Docking agent-based simulation of collective emotion to equation-based models and interactive agents. In: McGraw, R., Imsand, E., Chinni, M.J. (eds.) Proceedings of the 2010 Spring Simulation Multiconference on - SpringSim'10, p. 6. Society for Computer Simulation International, San Diego (2010). doi:10.1145/1878537.1878544
Rank, S., Skowron, M., Garcia, D.: Dyads to groups : modeling interactions with affective dialog systems. Int. J. Comput. Linguistics Res. **4**(1), 22–37 (2013)
Reeves, B., Nass, C.: The Media Equation: How People Treat Computers, Television, and New Media like Real People and Places. Cambridge University Press, Cambridge (1996)
Rimé, B.: Emotion elicits the social sharing of emotion: theory and empirical review. Emot. Rev. **1**(1), 60–85 (2009). doi:10.1177/1754073908097189
Rosenblatt, A., Greenberg, J.: Examining the world of the depressed: do depressed people prefer others who are depressed. J. Pers. Soc. Psychol. **60**(4), 620–629 (1991). doi:10.1037/0022-3514.60.4.620
Rybski, D., Buldyrev, S.V., Havlin, S., Liljeros, F., Makse, H.A.: Scaling laws of human interaction activity. Proc. Natl. Acad. Sci. USA **106**(31), 12640–12645 (2009). doi:10.1073/pnas.0902667106
Schweitzer, F.: Brownian agents and active particles. Collective Dynamics in the Natural and Social Sciences. Springer Series in Synergetics, 1st edn. Springer, Berlin (2003). doi:10.1007/978-3-540-73845-9
Schweitzer, F., Garcia, D.: An agent-based model of collective emotions in online communities. Eur. Phys. J. B **77**(4), 533–545 (2010). doi:10.1140/epjb/e2010-00292-1
Skowron, M.: Affect listeners: Acquisition of affective states by means of conversational systems. In: Esposito, A., Cambell, N., Vogel, C., Hussain, A., Nijholt, A. (eds.) Development of Multimodal Interfaces—Active Listening and Synchrony: Development of Multimodal Interfaces: Active Listening and Synchrony. Lecture notes in computer science, vol. 5967, pp. 169–181. Springer, Berlin (2010). doi:10.1007/978-3-642-12397-9_14
Skowron, M., Rank, S.: Affect listeners - from dyads to group interactions with affective dialog systems. In: Proceedings of AISB/IACAP World Congress 2012, LaCATODA 2012, pp. 55–60 (2012)
Skowron, M., Rank, S.: Interacting with collective emotions in e-communities. In: Scheve, C.V., Salmela, M. (eds.) Collective Emotions. Oxford University Press, Oxford (2014). doi:10.1093/acprof:oso/9780199659180.003.0027

Skowron, M., Theunis, M., Rank, S., Borowiec, A.: Effect of affective profile on communication patterns and affective expressions in interactions with a dialog system. In: D'Mello, S., Graesser, A., Schuller, B., Martin, J.-C. (eds.) Affective Computing and Intelligent Interaction: 4th International Conference, ACII 2011, Memphis, TN, USA, October 9–12, 2011, Proceedings, Part I. Lecture Notes in Computer Science, vol. 6974, pp. 347–356. Springer, Berlin (2011a). doi:10.1007/978-3-642-24600-5_38

Skowron, M., Pirker, H., Rank, S., Paltoglou, G., Ahn, J., Gobron, S.: No peanuts! affective cues for the virtual bartender. In: Murray, R.C., McCarthy, P.M. (eds.) Proceedings of the 24th International Florida Artificial Intelligence Research Society Conference, pp. 117–122. AAAI Press, Menlo Park (2011b)

Skowron, M., Rank, S., Theunis, M., Sienkiewicz, J.: The good, the bad and the neutral: affective profile in dialog system-user communication. In: D'Mello, S., Graesser, A., Schuller, B., Martin, J.-C. (eds.) Affective Computing and Intelligent Interaction: 4th International Conference, ACII 2011, Memphis, TN, USA, October 9–12, 2011, Proceedings, Part I. Lecture Notes in Computer Science, vol. 6974, pp. 337–346. Springer, Berlin (2011c). doi:10.1007/978-3-642-24600-5_37

Skowron, M., Theunis, M., Rank, S., Kappas, A.: Affect and social processes in online communication - experiments with affective dialog system. IEEE Trans. Affect. Comput. **4**(3), 267–279 (2013). doi:10.1109/T-AFFC.2013.16

Skowron, M., Rank, S., Świderska, A., Küster, D., Kappas, A.: Applying a text-based affective dialogue system in psychological research: case studies on the effects of system behaviour, interaction context and social exclusion. Cogn. Comput. (2014). doi:10.1007/s12559-014-9271-2

Thelwall, M., Buckley, K., Paltoglou, G.: Sentiment in twitter events. J. Am. Soc. Inf. Sci. Technol. **62**(2), 406–418 (2011). doi:10.1002/asi.21462

Thelwall, M., Buckley, K., Paltoglou, G., Skowron, M., Garcia, D., Gobron, S., Ahn, J., Kappas, A., Küster, D., Hołyst, J.A.: Damping sentiment analysis in online communication: discussions, monologs and dialogs. In: Gelbukh, A. (ed.) Computational Linguistics and Intelligent Text Processing, 14th International Conference, CICLing 2013, Samos, Greece, March 24–30, 2013, Proceedings, Part II. Lecture Notes in Computer Science, vol. 7817, pp. 1–12. Springer, Berlin (2013). doi:10.1007/978-3-642-37256-8_1

Šuvakov, M., Garcia, D., Schweitzer, F., Tadić, B.: Agent-based simulations of emotion spreading in online social networks (2012). http://arxiv.org/abs/1205.6278

Watson, D., Tellegen, A.: Toward a consensual structure of mood. Psychol. Bull. **98**(2), 219–235 (1985) doi:10.1037/0033-2909.98.2.219

Wu, Y., Zhou, C., Xiao, J., Kurths, J., Schellnhuber, H.J.: Evidence for a bimodal distribution in human communication. Proc. Natl. Acad. Sci. USA **107**(44), 18803–18808 (2010). doi:10.1073/pnas.1013140107

Glossary

Adjacency matrix A $N \times N$ square matrix representation of binary network (see **Network**, **Link**). If there is a link between nodes i and j then $\mathbf{A}_{ij} = 1$, otherwise $\mathbf{A}_{ij} = 0$.

Affect Listeners Conversational systems aiming to detect and adapt to affective states of users, and meaningfully respond to users' utterances both at the content- and affect-related levels

Agent-based model Following a complex systems approach, systems are comprised of a large number of strongly interacting subsystems or entities which are commonly denoted as agents. Depending on the system under consideration, agents can represent very different entities with specific properties, ranging from software processes to humans. Agents are usually seen as autonomous and not entirely controlled by other elements. They are characterized by internal variables determining their actions which may vary across agents—denoted as agent's heterogeneity. Depending on their internal complexity, agents can behave reactively (response to a given situation) or proactively (generating a situation), they may also have the ability to adapt (learning). Multi Agent Systems (MAS) allow to study the emergence of systemic properties which result from the interaction of a large number of agents, rather than from single agents.

Agent-based modelling and simulation A computational model for simulating the actions and interactions of autonomous individuals in a network in order to re-create and/or predict the actions of complex phenomena. One important element is the use of Multi-Agent Systems.

Appraisal (1) Central term of the cognitive appraisal theories of emotion. The process(es) that turn(s) stimuli into events-as-appraised, i.e. subjective meaning. Direct, immediate, and intuitive evaluations, as well as more complex ones, to account for qualitative distinctions among emotions. Many appraisal theories describe appraisal as an evaluation of stimuli along a set of criteria, such as

J.A. Hołyst (ed.), *Cyberemotions*, Understanding Complex Systems,
DOI 10.1007/978-3-319-43639-5

novelty, intrinsic pleasantness, relevance, responsibility, coping potential, standards compliance. (2) How do emotions come about? Currently, it is believed that internal or external events are being evaluated with regard to several crucial dimensions, such as—"is it relevant for me?" or "is it good for me?", or "can I deal/cope with the event". The key researcher introducing the term appraisal for this process, Magda Arnold, postulated 50 years ago that much of appraisal is automatic, intuitive, and outside of awareness. Recent research is strongly in support of this notion—while we also think consciously about whether something is good or bad for us. There appear to be one or more parallel processes that occur very rapidly in the brain that process the same stimulus or event. Sometimes, automatic and explicit (i.e., conscious) appraisals come to different conclusions—for example when you eat a grape fruit, you might say that you enjoy it, but your face shows that part of you is repulsed by the sour fruit. This highlights the difficulty of measuring things that are partly available to our conscious experience and partly not. Theorists differ with regard to how many dimensions of appraisal there are, but overall, there is considerable agreement. Appraisal theory is playing an increasing role in computer science to model emotions.

Arousal Emotions are metaphorically not cold, but hot. They are linked to actions. To facilitate actions various physiological systems interact—for example, the heart might beat faster, muscles prepare to contract… this is often referred to as "physiological arousal". Some emotional states are characterized by *high* (for example FEAR), others by *low* arousal (SADNESS). There had been a debate in the literature whether emotional states would be distinguishable by specific patterns of arousal (example: heart rate up, systolic blood pressure down, temperature down, etc.) or whether arousal is "general" in the sense either you are activated or not, but that there are not different ways of being activated. The truth appears to be somewhere in-between: People can report about their arousal using questionnaires, but the relationship between objective measures (e.g., measuring heart-rate, as we do in our laboratory) and asking them (did your heart rate increase?) is surprisingly low.

Author That is, on Blogs etc.—a person who writes a post or a comment on some other post or comment (see **User**, **Blogger**).

Autonomous Virtual Agent A Virtual Agent able to play a role by itself; it has the property of self-governing. An autonomous Agent is sometimes called a Virtual Actor.

Avatars An avatar is a representation of a real human being evolving usually in a virtual environment. This person has full control of his or her avatar. Avatars can have various forms from the simplest to the most complex. For instance, an avatar can be: (1) A box with the name of the user or the representation of the user's head associated to the current user's emotion (see **Emoticon**) (2) A 2D cartoon image (3) A 3D Virtual Character (4) A 3D Virtual Character with very realistic appearance and motion.

Bipartite network A network containing two types of nodes (see Network, Node), with links connecting nodes of different type. By definition, in bipartite network linking between nodes of the same type is forbidden. In an example of bipartite networks of Blogs (see **Blog**): network of users is one partition, and posts & comments—the other one.

Blog, Blogsite Is a web page (website), usually maintained by an individual, with regular entries of commentary, descriptions of events, or other material such as graphics or video. Entries are commonly displayed in reverse chronological order. "Blog" can also be used as a verb, meaning to maintain or add content to a blog. Relevant for the data analysis apart form different languages is that the structure of recorded data, varies depending on the policy of the Blogsite. Many blogs provide commentary or news on a particular subject; others function as more personal online diaries. A typical blog combines text, images, and links to other blogs, web pages, and other media related to its topic. The ability for readers to leave comments in an interactive format is an important part of many blogs. Most blogs are primarily textual, although some focus on art (artlog), photographs (photoblog), sketchblog, videos (vlog), music, audio (podcasting) are part of a wider network of social media. Micro-blogging is another type of blogging which consists of blogs with very short posts. Apart from the text other types of material (video, music,...) are taking parts in Blogs.

Blogger An active user and/or person who keeps and updates a blog (see **Author**, **Blog**, **User**).

Chatbot (chatterbot) Is a type of conversational agent, a computer program designed to simulate an intelligent conversation with one or more human users via auditory or textual methods. Although many appear to be intelligently interpreting the human input prior to providing a response, most chatterbots simply scan for keywords within the input and pull a reply with the most matching keywords or the most similar wording pattern from a local database. Chatterbots may also be referred to as talk bots, chat bots, or chatterboxes. A good understanding of a conversation is required to carry on a meaningful dialog but most chatterbots do not attempt this. Instead they "converse" by recognizing cue words or phrases from the human user, which allows them to use pre-prepared or pre-calculated responses which can move the conversation on in an apparently meaningful way without requiring them to know what they are talking about.

Comment That is, on Blogs etc, is generally a verbal or written remark, with an added piece of information, or an observation or statement. A comment on a post (or another comment) is written text in which a blogger (see **Blogger**) is expressing his opinion about the subject of the post or other comment(s) on that post.

Computer-mediated communication Human communication that is enabled via computer systems, often involving the Internet. This form of communication can include two or more devices.

Control parameter In non-linear dynamics the parameter which can tune the course of the dynamics.

Data Model For each event several properties are captured. For example, the time stamp of each message event is recorded as a message property. Hence, the sequence of messages and the change in relationship structure or strength is represented as a series of relational events in the data model. Examples of further event properties are keywords, contents, coded communication types (e.g. socialization vs. task organization), or evaluations. We further can capture actor attributes as variables, e.g. job position, location, evaluations, or other variables. We represent these in the dynamic graph using size, color, saturation, rings, and labels. All attributes can then be used for contentoriented analysis, searching for subnetworks, or for similarity based grouping of actors.

Degree of node in a network k_i A number of edges incident with the node (see **Link**, **Node**). If the graph is directed, the degree of node can be defined with respect to the number of outgoing links $k_{out}(i)$, and number of incoming links $k_{in}(i)$.

Degree distribution $P(k)$ Is defined as the probability that a node chosen uniformly at random has degree k (see **Node**, **Degree**). For many real networks the degree distribution has power law (see Power law) shape, $P(k) \sim k^{-\gamma}$, with exponent γ in range between 2 and 3. These networks are referred as scale free networks, contrary to random networks which have a Poisson degree distribution.

Dimensional emotion theories Laypersons often intuitively think of terms such as "happiness" and "anger" when asked to describe how they feel, and in fact these are prominent terms in emotion theories that aim to distinguish a certain number of primary or "basic" discrete emotional states. Other emotion researchers, however, assume that the core of emotional states or "single simple feelings" is even more basic than a collection of conceptually distinct emotion constructs. The idea here is that emotional experiences can be mapped onto a small number of underlying "core" dimensions rather than an N-dimensional space in which each discrete emotional state would have to be treated separately from all others. This general idea is, in fact, not at all new. Already in 1897, Wilhelm Wundt proposed three dimensions, a notion that was picked up again in the 1950s by Harold Schlosberg and, later, by James Russell and others who suggested a "Circumplex Model" structure in which core emotional states can be distributed in a low-dimensional circular space. In recent versions of this type of model, it is proposed that all core emotional states can be represented in a simple space comprising valence, arousal, and sometimes dominance. Much of the work presented in the present volume is based on such a dimensional conceptualization of emotional states because it offers several theoretical as well as empirical advantages. For example, present-day work on sentiment mining has roots in much earlier work by Osgood and others on the semantic differential that has been used by psychologists since the late 1950s. In general, dimensional views of emotion facilitate comparability of results even across disciplinary and thematic boundaries, in particular when only a limited scope of data is available in a certain field, or when a reliable assessment of emotion dynamics is required. For example, while clear and distinct expressions of a discrete emotional state such as "disgust", may only be identified rather infrequently in a random data stream, assessments of valence (pleasantness vs. unpleasantness) can be made more continuously and on more diverse types of data.

Dialog System Is a computer system intended to converse with a human, with a coherent structure. Dialog systems have employed text, speech, graphics, haptics, gestures and other modes for communication on both the input and output channel. There are many different architectures for dialog systems. What sets of components are included in a dialog system, and how those components divide up responsibilities differs from system to system. Principal to any dialog system is the dialog manager, which is a component that manages the state of the dialog, and dialog strategy. Based on modality the dialog systems can be divided in the following groups: text-based, spoken dialog system, graphical user interface, multimodal.

Distribution of a variable A description of the relative occurrence of each possible outcomes of a stochastic variable in a number of trials. The function describing the probability that a given value will occur is called the probability distribution (or probability density function, abbreviated PDF), and the function describing the cumulative probability that a given value or any value smaller than it will occur is called the cumulative distribution function (abbreviated CDF).

Dominance From the perspective of dimensional models of emotions, much of the variance observed in the laboratory can be explained by the perceived valence and arousal of a stimulus or feeling state. In other words, the questions answered by appraisals of how good vs. bad (valence) something is for us combined with an appraisal of how "hot" and activating it is for us (arousal) already say a lot about how we feel, and how we are likely to respond. In some cases, however, valence and arousal may not be enough to characterize an emotional experience. This is where dominance/power becomes relevant as a third dimension. For example, anger and fear can sometimes be very similar in so far as they can both be perceived as unpleasant and highly activating. In the animal kingdom as well as in the human case, it may sometimes be a fine balancing point that makes the difference on whether an organism tries to run from a threat or, instead, goes onto the offensive. And yet, the extent to which we perceive a situation as within our power to cope can, sometimes, make all the difference. Depending on the specific researcher and theoretical framework, this "dominance" dimension can sometimes alternatively be referred to as "power" or "coping potential". Like valence and arousal, dominance can be measured by questionnaires as well as physiological measures (e.g., blood pressure reactivity). Further like valence and arousal, self-reported dominance and physiological measures do not always agree. In fact, most participants in the laboratory find it more difficult to report about this dimension, and further explanations are sometimes required. In many of the available standardized stimuli used in the laboratory this dimension may further not be directly applicable from the perspective of the subject.

E-community (on a network) Community of users related to posts on Blogsite, etc. Also users+posts (and their comments) can be considered as e-community on bipartite networks.

Edge See **Link**.

Emoticon We define an emoticon as the textual expression representing the face of a writer's mood or facial expression. These character codes can change depending of the continent. For instance, to be confused and sad/unhappy represented as `:s` and `:-(` in western societies corresponds to `(o.O)` and `(;_;)` in eastern societies. Emoticons are often automatically replaced by tiny images.

Emotion Subjectively, from the point of view of a lay person, we appear to focus on "discrete emotions"—this means a small number of well identifiable and clearly distinguishable (=discrete) states, such as happiness, sadness, anger, or fear. There is no agreement how many such emotions exist—for example, some would count surprise or interest as emotions. What could be a criterion? Based on the research of Charles Darwin, who actually published a book in 1872 called "The Expression of the Emotions in Man in Animals", many researchers believe that for a "real" emotion, in the sense of a general thing that is shared by all humans, there has to be a universal sign/expression. We do know, for example, that everywhere on earth people smile. This has been taken as a sign that happiness is universal and a "basic emotion". Of course, this poses some problems. These are that (a) sometimes people smile when they are not happy, and (b) there are emotions (e.g., guilt), that do not appear to have a universal expression, and yet, they seem to exist in most if not all cultures. This has lead some researchers to focus on more general aspects of emotions, such as to what degree an emotional state, at a given moment, is positive or negative, regardless what label we might give it. This approach is referred to as the dimensional view, versus the categorical view of "basic emotions" theorists described above (the most prominent being Paul Ekman). We believe that categories such as anger are useful, but to think of these as dimensions (e.g., a negative state that is accompanied by activation and might involve confrontation) can be useful as well—they are simply different levels of description.

Emotional content (in data set) For a textual file, net probability of objectivity (0) or emotions negative/positive ($pm1$) determined by the emotional classifier routine or by direct measurements of human emotional reactions related to the subject.

Emotional classifier routine A text classifier which able to classify texts of certain according to their emotional content. Classifier assigned to the text binary value, depending on how much and what emotions are manifested in it. The text can be objective (value 0) or subjective, which than can be classified either positive (value $+1$) or negative (value -1).

Equilibrium This phrase can have several meanings. (1) static equilibrium: all system parameters are constant; examples: (i) a ball staying in a hole, (ii) a person has fixed his/her opinion on a certain issue, (2) dynamical equilibrium: a system has approached its asymptotic time-dependent trajectory. Such a trajectory can be periodic or chaotic (deterministic chaos). Examples (i) a rigid body falling down in the air/liquid approaches its asymptotic velocity after some transient time, (ii) a person has become accustomed to his/her daily and weekly activities, (iii) a person is oscillating periodically or unperiodically between different emotional states (3) statistical equilibrium is considered for a system (ensemble) of many similar objects that can possess some different features (e.g. velocity of atoms). The equilibrium in such system means that the probability distribution describing distribution of these features has approached its asymptotic (usually stationary) form. Since the system is described in a probabilistic way thus we are observing fluctuations around mean values. Examples: (i) equilibrium distribution of gas particle velocities (Maxwell distribution) (ii) stationary distribution of price fluctuations at the market (log-normal or Levy distributions) (iii) stationary distribution of opinions in society (iv) stationary distribution of various emotions expressed at internet discussion groups (v) stationary distribution of transitions between different emotional states.

Event-based dynamic network analysis Communication structures and online communities can be studied with Social Network Analysis (SNA). A social network is defined as a set of actors and the relationships between them. Its explicit focus on quantitatively analyzing interdependent patterns of social relationships differentiates SNA from traditional statistics and data analysis. Typical measures to quantitatively describe a network include centrality of a node, density, clustering coefficients, degree of a node, diameter, etc. The approach uses network graph visualization ("Sociograms") extensively to represent, describe, and analyze communication matrices of interrelated actors. However, currently is widely acknowledged that almost all SNA research can be regarded rather as static and cross-sectional than dynamic. To overcome current model restrictions, we use a richer model that is disaggregating relationships into individual relational events. In communication network analysis such relational events are created by exchanging messages with others. From these events, relationships can be aggregated. This approach of event-based dynamic network analysis consists of (1) a data model that contains information about the network including the timing of network events, and (2) a dynamic graph visualization based on a 2D/3D spring embedder.

Experiment In psychology, experiments are a widely used scientific method of empirical knowledge acquisition. They typically involve testing hypotheses and are often characterized by different experimental conditions systematically varying ("isolating") the factor of interest. They may for example be used to test the effect of pictures with different content (independent variable) on people's feelings when being confronted with a specific picture (dependent variable).

Facial Expressions Going back to Darwin, facial activity appears to be a behavior that is closely related to emotions. While we can move our faces on command—thus, control them with our will, there are also movements that are very spontaneous. Our personal research agenda over many years has been to identify how factors other than emotions influence what we show on our faces—yet, we are nonetheless quite convinced that emotions are an important and critical driver of our face and other expressive systems. Under clearly controlled conditions, such as we can create in the laboratory, certain movements, for example the pulling together of the brows, appear to be associated with how positive or negative we perceive an object or an event. To a lesser degree, the muscles of the smile indicate if we like something. One thing is clear: The face is not just a readout of how we feel, and measuring facial activity is not a way to read minds—but it is a good way to estimate certain aspects of emotional processes and we can measure facial behavior without having to ask people questions or interrupt what they are doing, for example while they interact online.

Facial Electromyography (EMG) Using small surface sensors that are glued to the skin, we can measure the activation of specific muscles at a temporal and spatial resolution that goes far beyond the naked eye or even video-based analysis. For example, if we were to ask you about your last vacation (assuming that your vacation was pleasant), we would be able to show a relaxation of the muscles that pull the brows together and down, even if we could see nothing with the naked eye. The EMG produces a continuous measure of the averaged activation of a group of muscle fibers over time.

Feeling Emotions typically are characterized for an individual as a particular feeling state—"I feel sad". However, emotion scientists see this feeling only as one component of several that make up an emotion. Others are: physiological changes (this refers to brain activity and the body at large; expressive changes), changes in behavior; and changes in motivational states. Among physiological changes, facial and vocal expression, in particular, appear to have acquired important functions over the course of evolution. Motivational states are difficult to observe and we have to think of them as changes in tendencies to behave in some way. For example, we might be almost ready to run away, but we are not yet running. These things can actually be measured objectively! Measuring feelings is hard: We can simply ask you, "how happy are you on a scale from 0 to 5?"—but we actually know that your answer will be affected by many different processes that have little to do with how you feel, for example, which question you answered just before saying how happy you are!

Graph Mathematical object which consists of links and nodes which may have additional properties (see Network).

Histogram A bar graph of a frequency distribution of a variable in which the widths of the bars are proportional to the classes into which the variable has been divided and the heights of the bars are proportional to the class frequencies.

Information Extraction (IE) Is a type of information retrieval whose goal is to automatically extract structured information, i.e. categorized and contextually and semantically well-defined data from a certain domain, from unstructured machine-readable documents. A broad goal of IE is to allow computation to be done on the previously unstructured data. A more specific goal is to allow logical reasoning to draw inferences based on the logical content of the input data. A typical application of IE is to scan a set of documents written in a natural language and populate a database with the information extracted. Typical subtasks of IE are: (1) Named Entity Recognition: recognition of entity names (for people and organizations), place names, temporal expressions, and certain types of numerical expressions (2) Coreference: identification chains of noun phrases that refer to the same object. For example, anaphora is a type of coreference (3) Terminology extraction: finding the relevant terms for a given corpus (4) Relationship Extraction: identification of relations between entities, such as: PERSON works for ORGANIZATION (extracted from the sentence "Bill works for IBM."), PERSON located in LOCATION (extracted from the sentence "Bill is in France.")

Information Retrieval (IR) Is the science of searching for documents, for information within documents and for metadata about documents, as well as that of searching relational databases and the World Wide Web. Automated information retrieval systems are used to reduce what has been called "information overload". Many universities and public libraries use IR systems to provide access to books, journals and other documents. Web search engines are the most visible IR applications.

Likert Scale Type of empirical assessment, widely used in self-report questionnaires. Responses on a "Likert scale" are given along a range of answer options, for instance, in order to specify the level of agreement with a given statement on a range from 1 (strongly disagree) to 5 (strongly agree). The inventor is psychologist Rensis Likert.

Link A connection between two nodes in the network (see Node, Network), for example hyperlink between two web pages (WWW), interaction between two proteins in protein protein interaction network, interaction between users or blogger in on-line social communities. Link can be binary (1) or weighted ($\mathbf{W}_{ij}$). Link weight is equal to the strength of interaction between nodes, for instance the correlation between genes in gene expression network, the number of common comments (posts, movies, books) that two users have in technological social networks. Depending on the symmetry of interaction, link can be undirected or directed. In a directed graph, the order of the two nodes is important: e_{ij} stands for a link from i to j, and e_{ij} is not equal to e_{ji}. Weighted network is a network in which each link carries a numerical value measuring the strength of the connection. Unweighted networks have a binary nature, meaning the edges between nodes are either present or not.

Linkevent A linkevent will be an exchanged message between users or nodes within the network. It can be anything from an email to a forum post or something similar. Linkevents can be aggregated to links between nodes (users).

Longitudinal research Rather than collecting data at one point in time only, one may be interested in seeing how e.g. people' opinions or behavior change over time.

Therefore, longitudinal studies involve different waves of data collection in order or to enable comparison over time.

Message- and thread-based communication networks The examined networks or datasets consists of exchanged messages or linkevents. However there are generally two different types of text-communication in the internet. The communication can be message- or thread-based. Message based communication is a network, where one or more users (nodes) are sending messages to concrete group of recipients. The second type of relationship, thread-based relationships, occur whenever a linkevent is originated by one or more nodes but not directly addressed to one or more other nodes, e.g. creating a new thread in a discussion group by posting a message. In this case, nodes interlink whenever one linkevent refers to another linkevent, e.g. whenever a post in a discussion group refers to another post. Combinations of both types of linkevent relationships are possible as well, especially if different types of linkevents are mixed, e.g. a discussion group allowing direct messages being send between authors.

Mood Many researchers use the term mood to refer to emotion related changes that are not linked to particular emotion-eliciting material/stimuli, and that last for a long time. Emotions, in turn, are typically linked to an external or internal stimulus and last only seconds or perhaps minutes. For example, you missed the bus. Despite the bus driver having seen you, he simply took off and left you standing there. Now—if you feel angry "all day", it is believed that you get angry again, every time you remember this—in this case, the memory is an internal stimulus that causes a similar cascade of emotional changes. In contrast, you might wake up and your partner asks you after a while, why you are grumpy—but you might not even have noticed that you behaved in a particular way. Many researchers believe that moods are, for example, effects of chemical changes—there is even experimental evidence that the weather systematically affects mood. However, we know very little how emotions might add up to changes in mood. We therefore tend to avoid the term mood in our research!

Motif A pattern of interconnections occurring either in a undirected or in a directed graph G (see **Network**) at a number significantly higher than in randomized versions of the graph, i.e., in graphs with the same number of nodes, links and degree distribution as the original one, but where the links are distributed at random. As a pattern of interconnections, motif is usually meant as a connected (undirected or directed) n-node graph which is a subgraph of G.

Multi-Agent System A system composed of multiple interacting (intelligent/autonomous) agents used to solve (or model) problems that are difficult or impossible for a monolithic system.

Network (graph) $G(V, E)$ A set of nodes (vertices) V (see **Node**) connected by edges (links) E (see **Link**). Two nodes joined by a link are referred to as adjacent or neighboring. Networks can be directed or undirected; weighted or unweighted, depending on the properties of it links (see **Link**).

Node An entity that describes building unit of the system which is represented by network (see **Network**). For example node can represent router or computer in Internet network, gene in gene expression network, a person (user or blogger) in off-line and on-line social networks, post (story, movie, comment) in technologically-mediated social networks.

Opinion mining Opinion mining is the computational study of opinions, sentiments and emotions expressed in texts. The discipline of opinion mining is at the crossroads of information retrieval and computational linguistics.

Online survey Empirical methodology referring to a collection of questionnaires which is implemented on a specific platform on the Internet. Online surveys are often used to reach a broad audience of participants over the Internet.

Pattern That is, users activity patterns in time.

Post An entry on a blog written by blog author (see **Author**, **User**, **Blogger**).

Power-law Broad distribution with one parameter describing the algebraic decay, for example Levy distribution. Often distribution of real exhibit only power law tail.

Power-spectrum For a given time signal (see Time series), the power spectrum gives a plot of the portion of a signal's power (energy per unit time) falling within given frequency bins. The most common way of generating a power spectrum is by using a discrete Fourier transform. In time-series with correlated fluctuations (see Fractal time series) the power-spectrum obeys a power-law behavior as $S(f) \sim f^{-\beta}$, with the exponent β related with the fractality exponents of the time-series.

Probability density function Of a continuous random variable is a function that describes the relative likelihood for this random variable to occur at a given point in the observation space.

Preferential attachment rule Is a widely used concept of complex network growth. The process is described by the following probabilistic approach. Probability that a certain event takes place at time t (we call it a success) increases with the number of previous successes. For example if balls can be randomly placed in certain boxes then a box where at given time moment more balls are placed would be more attractive for next balls. The most famous example of this rule is a network emerging in Barabasi-Albert model. In every time step a new node is attached randomly to the old network and the probability that an old node will be selected by a link to the new node is proportional to a temporary number of links of the old node.

(Relational) Action Tendencies (In emotion theories, e.g. Nico Frijda's:) A state of readiness to achieve or maintain a given kind of relationship with the environment. Emotions are typically characterized by a change in action tendencies. In modeling practice, an action tendency could be an internal incitement to perform a certain kind of behavior without determining the specific behavior.

Relaxation Is a process of system changes towards its equilibrium position due its time irreversible interactions with environment. Relaxation is usually connected with a time decay of a distance of system state from a system equilibrium. Examples: (i) relaxation (decay) of oscillation amplitude for a damped oscillator (ii) relaxation of electrical potential in an excited neuron (iii) relaxation of abnormal bloggers activity (regarding a special subject in blogosphere) (iv) relaxation of emotional state understood as a physiological, psychological (individual), sociological (collective) phenomenon

Self-esteem is a psychological construct which is defined as the evaluation of one's own self and oneself as a person. Humans are known to strive for a positive self-esteem and employ a variety of methods to enhance their self-esteem.

Self-report is a common way in e.g. psychological experiments and surveys to collect answers from respondents. Self-report can appear in a variety of forms; participants may for example either choose from (a list of) answers provided by the researcher or use a **Likert scale** to indicate their personal agreement with given statements or questions regarding their beliefs, attitudes, feelings or typical behavior in certain moments. For example, feelings using the Positive and Negative Affect scale are assessed via self-report measures. Self-report can be contrasted with "objective" methods of data collection, e.g. observation, physiological data or fmri data.

Social capital is a construct which originated in sociology and has been employed in research on social networking sites to describe the network people acquire on platforms such as `Facebook`. It is typically distinguished in bridging social capital (weak ties) and bonding social capital (strong ties).

Spanning tree A loop-less (see Tree, Cycle) subgraph with same number of nodes as graph. One network can have many spanning trees. Weight of spanning tree is equal to sum of the weights of the edges in the tree. The spanning tree or trees with minimal weight are called minimal spanning tree.

Studying Subnetworks With the attributes we are able to analyze subnetworks in the overall network, which relate to a set of filtering conditions, e.g. a subnetwork on a certain topic (a bag of words), or a subnetwork only with actors of a certain location, or a subnetwork with relationships that are coded to have negative sentiment. For this objective, we can further employ structural clustering and content based clustering (based on text mining). We can thus study the development of the focused subnetwork in a larger network over time in an efficient way by means of network analysis. As shown in the first figure, this elicitation of subnetworks is capable of emphasizing selected parts of a larger network and its evolution over time, e.g. a topical discussion or an area with negative sentiment and its embeddedness in the faded overall network. The visual patterns can further be quantified via dynamic measures like networking activity or structural or topic changes, stabilization of subgroups, changes in central positions etc. The quantified patterns can then be subject to further statistical analysis (e.g. typical transitions, activity peaks, etc.).

Time-series A sequence of data points of dynamical variable, measured typically at successive times spaced at uniform time intervals.

Tree A connected network without cycles.

User A person registered on a Blogsite who writes posts or comments-on-posts (see **Post**, **Comment**) on that site (see **Blog**, **Author**, **Blogger**).

World Wide Web Abbreviated as WWW and W3 and commonly known as The Web, is a system of interlinked hypertext documents contained on the Internet. With a web browser, one can view web pages that may contain text, images, videos, and other multimedia and navigate between them using hyperlinks. Hyperlinking between different Blogs is another aspect of Blog research.

Valence The key dimension of appraisal is whether something is good or bad for you. Basically we like things that are good for us and we do not like what is bad for us. There are exceptions such as the chocolate cake when you are on a diet, or the pain when you do that strenuous exercise regime you are fond of—but overall ("hedonic") valence is both a key element of how we perceive the world and determining what we do. It is believed that even very simple organisms can evaluate and react to different valences of stimuli. In some cases valence appears to be biologically determined (e.g., sweet taste elicits already in neonates a smile, as opposed to lemon juice), in others it is learned or acquired (e.g., Jazz music). For some theorists (e.g., economic models focusing on maximizing utility), emotions can be reduced to hedonic valence: we enjoy and search pleasure, and we avoid pain. The valence dimension can be positive or negative to different degrees. This is not a binary distinction or ternary if we take "neutral" into account, but graded dimension. Furthermore, there is reason to believe that our brain processes positive and negative things differently and in different places. In fact, some things can be described as positive AND as negative at the same time. Thus, some researchers like to measure positive and negative separately instead of a scale that has a negative pole on one end and a positive one on the other! Note that valence is only one dimension of appraisal, but it is the one that explains most of the variance in multidimensional analyses.

Virtual Agents A Virtual Agent is handled by the computer; it is a piece of code that drives the Character; it is not driven by the **User**.

Virtual Characters Virtual characters are articulated bodies animated using a computers

Virtual Humans Virtual Characters (VH) with a Human shape; they may be avatars or virtual agents

Visualization Visualization represents the network data model graphically. In the conventional form, called sociogram (social network graph), actors are represented by nodes and edges represent the relationships. The sociogram extended with additional means for information visualization and the capability to adapt to longitudinal network change yields a dynamic graph termed "communigraph" that can be shown with the software Commetrix. This software also provides a comprehensive filtering approach that lets us visually search and explore the model and thereby obtain insights into complex social corpora (sample output in the first figure and many more at www.commetrix.de). With the approach outlined, we are able to visualize, measure, and analyze the collective self-organization of interactive subgroups and the diffusion of topics through networks of online communication, as found in newsgroups, discussion boards, etc. and the population dynamics around certain topics have been described and traced visually. Such a visual and analytic description of dissemination processes in complex networks of hundreds and thousands of actors will yield significant new insights about patterns of collective emotions and sentiment.

Web 2.0 Term refers to web-based technologies which allow users to generate content, connect, interact and collaborate with several other users without having special technical skills. Examples for such participatory technologies are social networking sites (e.g., `Facebook`) or video-sharing websites (e.g., `YouTube`).

Well-being is typically used as an outcome measure in psychological research to assess temporary life-satisfaction. Several questionnaires can be used, often one-item measures are employed.

CPSIA information can be obtained
at www.ICGtesting.com
Printed in the USA
BVOW07*0749031116
466828BV00009B/3/P

9 783319 436371